FROM BREEDING & FEEDING TO MEDICALIZATION

RURAL HISTORY IN EUROPE

VOLUME 17

From Breeding & Feeding to Medicalization

Animal Farming, Veterinarization and Consumers in the Twentieth-Century Western Europe

Edited by

DANIEL LANERO TABOAS, CARIN MARTIIN, LOURENZO FERNÁNDEZ PRIETO AND LAURENT HERMENT

BREPOLS

THE SERIES RURAL HISTORY IN EUROPE

The countryside forms a complex and evolving universe. Clearly, it exhibits both inertia and ruptures, but it would be an illusion to contrast a contemporary rural world in rapid transformation with a traditional rural world of frozen landscapes, petrified societies, immobile economies and lethargic political contexts. Even if rural societies have been overthrown, rural landscapes have been profoundly transformed and the intervention of the State has considerably strengthened the regulation of production and trade, history must of necessity be considered as the main explanatory factor for our time.

The rural world can also only be fully understood if it is viewed in a manner which transcends national boundaries and if the discrepancies which can best be observed by adopting a broad view are taken into account. The aim must be to create a dialogue between researchers which goes beyond national frontiers, crosses chronological barriers and breaks disciplinary boundaries.

The main objective of the 'Rural History in Europe' collection is thus to provide keys to unlock the changes experienced by present-day European rural societies in the light of their historical experience. It will produce the necessary historical knowledge to make it possible for all to conceptualise the future of European country-dwellers as they face problems of the kind historians have always grappled with in examining societies of the past. How can the changes taking place in present-day Europe be understood without taking into account a past which is still very present, and which determines both structures and behaviour?

The volumes will be published after a peer-review process for each paper, supported by the editorial board, the authors and the editors of each book. In order to ask the relevant questions about the future of the peasantries and rural spaces in transformation, the volumes of this collection will deal with the longue durée and will present either research in progress or a synthesis on a regional or national scale. The outcome of this procedure will be a series giving a detailed overview of historical developments in Europe as a whole.

Cover illustration: Vaccination campaign (1910). Surroundings of A Coruña

D/2023/0095/142
ISBN 978-2-503-58915-2
eISBN 978-2-503-58916-9
DOI 10.1484/M.RURHE-EB.5.119900
ISSN 2032-6084
eISSN 2566-0063

Printed in the EU on acid-free paper.

Contents

List of Illustrations

Carin Martiin

Luca Andreoni and Laurent Herment

Telmo Otero Rodríguez and Daniel Lanero Táboas

Marco Marigliano

Ronald Plantinga, Merijn Knibbe and Marijn Molema

Fernando Collantes

Karl Bruno

Diego Conde Gomez and Lourenzo Fernandez Prieto

Kristina Nordeus

Annick Jacq

Nicolas Fortané

LAURENT HERMENT, LOURENZO FERNÁNDEZ PRIETO, CARIN MARTIIN AND DANIEL LANERO TÁBOAS

1. From breeding & feeding to medicalization, an introduction*

I. The reason why: aims and purposes of this book

Contemporary developments in animal husbandry are worrying on many counts. They call into question our way of life, our health, our ability to control our development, and our relationships with other species. Among the most obvious and critical important issues that breeding raises are human diet imbalances, global warming, deforestation, eutrophication of streams, lakes, and seashores, antibiotic resistance, animal welfare, and zoonosis.

These issues are not new. As early as the 1950s, concern about the potential dangers of antibiotic resistance was being voiced in various countries (Kirchelle, 2018 and 2020). Since then, antibiotics in relation to breeding have been considered as contributors to what Jarrige and Le Roux (2020) call 'the contamination of the Earth'. In the 1960s, Ruth Harrison (1964) also began to pose questions regarding animal well-being.[1]

More recent questions about the climate effects of breeding remain outside the historiographical scope of this book. However, as soon as the question of global warming emerged in the 1970s and 1980s, two major concerns arose regarding the role of livestock in this process: that of gas emissions from ruminants and that of deforestation linked to the expansion of pasture lands in certain tropical countries (Henderson-Sellers, 1981; Fearnside 1983, 1985).[2] The latter

* This book is the result of a workshop organised by the Division of Agrarian History, Swedish University of Agricultural Sciences, Uppsala (2019), as part of the research program of CNRS – GRDI – AAA Network (2016-2019). In 2018 the Network held a first workshop in Santiago de Compostela, organised by HISTAGRA Research Group, University of Santiago de Compostela (USC), that also resulted in a collective publication (Fernández Prieto and Lanero, 2020).

1 Though several decades earlier, Upton Sinclair had denounced the hellscape of the meat-packing industry for men and for pigs in *The Jungle* (1906).

2 The FAO estimated that breeding is responsible for 14.5% of greenhouse gases (FAO, 2016). For estimates of the greenhouse gas emission from cattle breeding in Brazil, which has one of the most

From Breeding & Feeding to Medicalization, Rural History in Europe, 17 (Turnhout: Brepols, 2023), pp. 11–28
BREPOLS PUBLISHERS 10.1484/M.RURHE-EB.5.135019

issue also entails clearing land for feed crops intended to supply intensive livestock farming operations[3].

There are endless examples, and they all tell us something about our present situation. The possibility of disease transmission from animals to humans has been well-identified for over a century, particularly with regard to tuberculosis. The term zoonosis has also become familiar to many of us since the mad cow crisis of the mid-1990s,[4] or the bird flu and the famous H5N1 virus in the first decade of the 2000s. Current events have demonstrated the extent to which our societies can be debilitated by the outbreak of a virus that may have come from a wild animal, though its origin is still debated. Though the recent pandemic is not addressed in this volume, the animal origin of most of our diseases and the role that human contact with wild animals and coexistence with livestock has played in the emergence of new human diseases since the Neolithic era is no longer debated (WHO, 1959; Fiennes, 1978; Hardy 2003).[5] What is especially worrying now, since the industrialization of livestock farming, is how the number and the dimensions of potential infection sites have multiplied.[6] Another issue that remains outside the scope of this book is how meat and dairy production directly or indirectly account for an increasingly large share of the virtual water flow (Duarte, Pinilla and Serrano 2019).

The main aim of this book is to historically analyse European livestock farming and the ongoing transformations in its forms of rearing and feeding. The authors look at questions related to animal husbandry based on several dominant research lines that cross a wide range of national situations. The two-fold ambition is to shed light on the evolution of animal husbandry and to provide multiple readings that will reflect diverse methods and questions. The book demonstrates the existence of historical change and the intensification (in scientific, technological and industrial terms) of animal husbandry in general and dairy farming in particular. After the Second World War, a model of industrial animal farming was established

---

important herds in the world, see Bustamante et al. 2012 and Cardoso et al. 2016. In this case, not only does deforestation release carbon dioxide, but grazing (including biomass burning after deforestation) favours the release of methane (enteric fermentation) and nitrous oxide (from animal excreta in pastures). Nevertheless, as we will see, the shift from grazing to more intensive animal husbandry also raises many issues.

3 On the multiple significances of the tremendous expansion of the area devoted to soybean whose cakes are used to feed livestock see Langthaler 2018 and 2020.

4 On the importance of the mad cow crisis in the implementation of the precaution principle in Europe and related bibliography, see the chapter by Nordéus in this volume.

5 See also Morse (1995); Pierce-Duvet (2006), and the short but useful text of Hart, Bennett and Begon (1999). It is also valuable to devise anthropological approaches to zoonoses that allow us to re-interrogate relations between human and non-human worlds. See for example Levi-Strauss (2001) and Lynteris (2017; 2020, 42-76).

6 Nevertheless, it is important to note the existence of many sites and human activities that could favour the emergence of new zoonosis. Industrial breeding sites are one of them, though not necessarily the most important. Domestic animals are particularly efficient as intermediate hosts, as is intensive animal farming, which acts as a reservoir and device for the spread of new threats.

that is still prevalent today. However, this industrial model coexists (increasingly) with organic proposals that demonstrate the resilience and relevance of historical livestock feeding and rearing practices.

While there is no shortage of historical work on livestock farming and medicalization (Woods, 2011 and 2013), we believe that rural historians (and historians on the whole) should consider this question in a multidisciplinary light. That is the aim of this work: it is a history book, and a little more.

The book chapters reflect the range of the authors' disciplinary and multidisciplinary insights, which cover agrarian and livestock history; economic history; the history of science; microbiological science; sociology and veterinary medicine. Hopefully, the multidisciplinary ambitions of this book will contribute to the merging of perspectives and reduce some of the gaps in the broader history of animal farming. Accordingly, during the workshop and subsequent work on this book, rural historians set questions with the collaboration of expertise from outside their discipline; while sociologists, economists, microbiologists and veterinarians adapted their fields of expertise to a rural history publication.

The book covers a period of nearly 150 years from the late 19th century to the early 21st century. The chapters deal with several Western European countries: Spain, Italy, France, United Kingdom, The Netherlands, and Sweden. Though not exhaustive, it allows us to contemplate an array of contrasting situations and development paths for livestock breeding within that time span, which can be separated into two periods. The first, from 1880 to 1939, was framed by the second wave of industrialization and characterized by innovation and technological advances, the strength of organically-based livestock farming, professionalization of the livestock sector (breeders, public institutions, and technicians) and the consolidation of farmers' and stockbreeders' associations. The second period covers the latter half of the 20th century to the present: a period of rapid intensification dominated by the industrial animal farming model, though it has been challenged increasingly since the end of the last century.

The chapters give attention to the roles of some professions, mainly veterinarians but also agronomists, geneticists, technicians, and other experts. While focused on their specific field, they often lacked awareness of the outcomes, the big picture or other pieces of the puzzle such as animal well-being, stress on farmers, or consumer criticism. Gradually, farmers (and aquafarmers) constructed networks based on their practical understanding of many disciplines. They (intuitively) understood the links between veterinary aspects and indoor environments, for example, or between feeding and sexual reproduction in dairy farming. The roles and work of veterinarians are examined in these pages, both in their capacity as scientists and as a corporative profession with its own interests and relationships (market, enterprises, official public bodies, breeders' associations) in the sphere of cattle improvement. Veterinarians were also involved with the consumer culture, animal health, environmental concerns, and food safety.

The book invites readings through a variety of grids, such as chronology, type of knowledge, or type of breeding. A chronological approach makes it possible to account for the implementation of an increasingly techno-scientific approach to animal husbandry. It shows how gradually, or in leaps, the different stages of activity were infused with scientific discourse. However, a strictly chronological approach would ignore the fact that regions (or countries), sectors, and livestock rearing practices do not advance at the same pace.

An approach based on the type of economic activity is unfeasible for at least two reasons: a lack of space and the fact that issues may be common to several sectors, even if they did not initially follow the same path or advance at the same pace. A typical example of this would be the use of antibiotics, which appears in several chapters. The increasing use of antibiotics eventually has affected every sector, but the picture is much more complicated than a simple shift from a pre-antibiotic world to one where they are ubiquitous. Perennial resistances in some countries make antibiotics a structuring element for the entire sector: not only Western Europe and the USA, but the Soviet block and eventually lower-income countries (Kirchelle, 2018).[7] It affects countries that use them heavily as well as those that try to resist them or implement regulations, such as Great Britain or Sweden.

Adopting an approach based on the type of knowledge implies relying on the expertise of a multitude of scientists who can claim pre-eminence or sovereignty over a particular field of knowledge. We then risk superimposing bodies of knowledge that would remain more or less foreign to each other. More importantly still, a knowledge-based approach ignores that knowledge is contingent, or at least that it is changing. Delphine Berdah (2012 and 2021) has described how much French veterinary knowledge and the place of French veterinarians have evolved. 'Veterinarian knowledge has porous and protean borders' (Berdah, 2012) that induce an almost permanent renegotiation of knowledge and status. The systematization of knowledge is accompanied by a work of 'removing borders intended to establish a double institutional but also cognitive demarcation' (Berdah, 2012). From the end of the eighteenth century through long nineteenth century, French vets had to contend with blacksmiths and define their fields of expertise in relation to medicine, agronomy, biology, etc. This enabled them to collaborate with other fields of expertise, but they also had to preserve their own social and economic position.

In light of these considerations, the editors have adopted a rather methodological approach to organizing this book, as it seems more useful to ask what is normal in rural history, what is less so, and what is not.

Twenty years ago, it was 'normal' for sociologists and historians of economics and science to dialogue with rural historians.[8] But it was not common for scientists to contribute to a volume on rural history. Hopefully, the

---

7 Forecasts for the use of antibiotics in livestock farming are mixed, but never optimistic (see for example Van Boeckel et al. 2015; Tiseo et al. 2020). In the aquaculture sector, the absence of accurate data sometimes renders forecasting difficult (Schar et al. 2020).

8 On the importance of the history of science and technology, see Fitzgerald et al. 2018.

multidisciplinary ambitions of this book will help merge perspectives and reduce some of the gaps in the extensive history of animal farming, while also providing a way to organize the book. The aim is not to separate chapters into those written by historians (mostly rural/agrarian historians) and those written by non-historians.[9]

II. Structure and chapters

II.1. Part I

The first seven chapters are devoted to 'Animal husbandry and dairy production and consumption', in the form of case studies from the United Kingdom, Sweden, The Netherlands, Spain and Italy, along with a comparative look at France and Italy. The topics covered range from changes in the functionality of livestock statistics (sheep) in the UK during the 19th and 20th centuries to satisfying milk protein consumption needs in Spain and how the dairy industry and supermarkets were compelled to diversify the supply of dairy products. They reflect more global and long-term perspectives on the evolution of the dairy sector in Sweden, Friesland (The Netherlands), Galicia (Northwest Spain) and Northern Italy, and the need to adapt cattle feed across Europe to the developing agricultural products market. This is exemplified in the chapter describing the use of oilseed cakes for cattle feed in the late 19th century, in the context of the fin-de-siècle agricultural crisis in France and Italy. These early chapters contain a series of cross-cutting issues or common denominators that will draw the readers' attention. Together, they highlight the historical changes and intensification (in scientific, technological and industrial terms) of livestock farming in general, and the dairy sector in particular, throughout Western Europe from approximately 1870 to 1990. However, the decisive decades for large-scale production specialization were those immediately following the Second World War.[10] In the long historical process of dairy specialization, specific breeds – first the Friesian, later the Friesian-Holstein – became predominant in the intensification process thanks to their high production capacity (in litres per cow per day). This and other forms of intensification and specialization took place at varying rates and times in European countries. There were early-comers (such as Denmark and The Netherlands), late-comers (Italy, Spain, or Portugal) and countries like Sweden, where intensification and

9 The aim is not to record the history of a field of expertise.

10 See Martiin, Pan-Montojo, Brassley (2016) about increased food production in Western Europe during the early post-war years. Many countries, including those under fascist regimes were attempting to modernize as early as the 1930s, as Tiago Saraiva (2016) has emphasized. Indeed, there was a constant effort to industrialize agriculture from the mid-19th century on. Regarding the mechanization of American agriculture, see Fitzgerald (2003), on the emergence of chemical fertilizers in France, see Herment (2021).

scaling up occurred more gradually and small-scale family farming was still clearly dominant until the 1990s.

Livestock improvement (breed morphology, feeding, animal health and hygiene, bovine milk yields) was already underway throughout Europe in the late 19th century, where this book picks up historiographically. The results should not hide the fact that the knowledge and practices related to livestock improvement often originated in the local sphere, with local breeders. They organized themselves into local and regional associations (cooperatives, unions), which in some places became powerful industrial production cooperatives. As specialization and technological innovation continued, these networks became increasingly dense and complex. Local and regional institutions became more involved in promoting livestock farming, along with scientists and technicians, the institutions for which they worked, and eventually state administrations; thus completing the trajectory from the local to the state level. However, we must never lose sight of the importance of this first step, at the local level, and of this first stage, prior to the Second World War. It demonstrates the potential of peasant forms innovation, which are relatable to the challenges of sustainable livestock farming that we face today. Another fundamental vector in these chapters concerns dietary changes in the European population, especially since the Second World War, but equally so from around the 1880s to the 1930s. The growing demand for animal protein – for meat, but mostly for milk and its derivatives (butter, cheese) – decisively conditioned livestock breeding, and in more ways than animal selection and the importation of breeds with higher productive capacities.

In Chapter 2, Karen Sayer critically analyses a century of British livestock statistics, from1866 to 1966, focusing on data reflecting the evolution of the total number of sheep. The apparently stable numbers of sheep hide very significant qualitative changes in production specialization (from wool to meat) and the territorial distribution of the flock. For the author, official statistics, initially designed to count the number of live sheep (and the impact of deaths due to disease) must be understood as cultural artefacts and political products of their time. Among British farmers, statistical data created two antagonistic identities: one was the progressive farmer who opted for scientific management and formal accounting of his farm in the interests of the national economy; and the other was the traditional, individualistic, routine farmer who wanted nothing to do with official statistics and control.

Chapter 3 contains Carin Martiin's ambitious, long-term research (1870-1995), which identifies three models of Swedish dairy farming based on local-level analysis. The first (1870 to the mid-1920s) was characterized by the challenges of cattle feed shortages, low milk yields and the threat of animal malnutrition; the second (mid-1920s to the early 1960s) involved a moderate increase in milk yields despite cattle feed limitations. The third, (1960 to 1995), was strongly influenced by the demands of Swedish economic policy, which indirectly incentivized the disappearance of smaller dairy farms and the survival of larger-scale intensive dairy operations. The author has designed an interesting calculation of the feed required (in terms of energy and protein) to obtain the milk yields of each model.

In Chapter 4, Luca Andreoni and Laurent Herment start with the agrarian crisis of the late 19th century and the challenges for European agriculture stemming from the global integration of agricultural markets. They study the shift towards livestock farming and increasing dairy specialization. In the context of livestock improvement, the authors examine the centrality of concentrated feeds (oilseed cakes) in Italy and France. They look at the role of the agricultural press, farmers' associations, scientists and official institutions in informing farmers about the wide range of oilseed cakes available on the market, their uses (as fertilizers or cattle feedstuff), their nutritional qualities and the fight against adulteration (fraud). In the end, it appears that if scientific expertise was essential, negotiations between industrialists, officials, scientists and farmers' unions were critical in determining what was fraud and what was not.

Chapter 5, by Telmo Otero and Daniel Lanero, explores the historical process of dairy specialization in Galician agriculture (Northwest Spain) from the first decades of the 20th century until the end of the Franco dictatorship (1975). The authors argue that dairy specialization intensified from 1960 on as part of a state-led process of agrarian industrialization. However, significant innovations in livestock improvement also took place prior to the dictatorship (during the 1920s and 1930s), coinciding with dietary changes that led to a greater demand for milk. The Franco regime imposed an agricultural industrialization model (which included milk production) that supplanted peasant innovation and its associative networks of livestock farmers who also worked closely with technicians.

Marco Marigliano's text in Chapter 6, explains how the Friesian dairy cow became the predominant breed in northern Italy from 1960 on. The author looks at two main factors: breeders' associations and technological innovations. He explains how agency in livestock innovation shifted in Italy, from being locally led by private individuals and technicians acting relatively freely in each province, to the Itinerant Teachers, to the State. An important factor in this transfer was the growing demand for milk and cheese after the Second World War, which made it necessary to increase productivity by improving livestock farming. New technological requirements related to genetics also forced the government to take a more active role, which partly explains state promotion of breeders' associations.

Chapter 7 contains the work of Ronald Plantinga, Merijn Knibbe and Marijn Molema. It begins by explaining the early and successful dairy specialization in the province of Friesland (The Netherlands) at the end of the 19th century. However, the authors focus mainly on the 1950-1990 period and the concern of all actors (farmers, cooperative industries, institutions and scientists) to improve the protein content (percentage) of milk. This was fully achieved with the importation of selected Holsteins from North America in the 1980s. Interestingly, the authors demonstrate that livestock improvement in Friesland (or anywhere) must be understood as a multi-causal process that reaches far beyond the achievement of a specific objective. In Friesland, the cultural preference for a traditional type of cow (small and robust) had to be considered, though it might not have been the most productive. Other factors, such as increased farm size and more efficient use of water resources, also helped improve yields.

The first part of the book ends with a contribution by Fernando Collantes in Chapter 8. The author explains the significant overall increase in milk and dairy consumption in the Western world (the 'Global North') after the Second World War. He then compares this overall development with that of Spain, which reached the same point later and through a short and accelerated nutritional transition from 1960 to 1980. The author describes the problems created for the dairy industry and the distribution sector (supermarkets) when consumers took their dairy calorie consumption needs for granted in the 1970s-1980s. This led the industry to diversify its offer (dairy desserts, liquid yoghurt, different types of milk, etc.). Collantes ends his text with a very interesting observation about the impact of inequality and income differences between population sectors on the type of demand (and access) to dairy products, which he identifies as basic or sophisticated.

II.2. Part II

The second part of this book highlights the roles of several professions involved in livestock farming: mainly veterinarians but also agronomists, geneticists, technicians, and other experts. Substantial attention is given to veterinarization in a body of work that provides a broad national scope and diverse but substantially relevant themes. Despite their different approaches, some of the authors draw similar conclusions that point to the impact of veterinarians as a corporative profession with their own interests and relations to markets, enterprises, breeders' associations, and public entities with regard to the cattle improvement process. This included preserving the professional identity of veterinarians as distinct to other professions, such as agronomists. Veterinarians also operated in relation to consumers, the culture of consumption, animal health, the environment, and food safety. In this section, the genesis and historical evolution of legislation concerning hygiene and sanitation for milk and livestock products is also analysed, as well as its implementation through specific policy measures for livestock care or shelter conditions. Accordingly, the veterinary profession is considered in a wider context than that of mending sick animals. Interestingly, parts of this history developed in relatively small-scale farming contexts, before the scaling-up of recent decades. It creates a sort of hen-or-egg dilemma in determining the extent to which antibiotics paved way for scaling-up, or vice versa.

In Chapter 9, Karl Bruno focuses on Sweden, the new role of veterinarians in reproduction and artificial insemination, and reproductive health in dairy cattle in the early and mid-20th century. His paper analyses how 'sexual health control' of dairy cattle developed in Sweden, from *c.* 1910 to 1960. The process involved a set of veterinary practices for monitoring reproductive health in dairy cows. Bruno discusses the controversy surrounding such control and argues that it reflects broader concerns over the distribution of labour and expertise in a changing system of cattle breeding and husbandry. In Sweden, artificial insemination was carried out by veterinarians, while in Denmark, Britain and the United States, independent non-veterinary technicians played a much bigger role.

Chapter 10 deals with the decades prior to the Spanish Civil War (1936-39). Diego Conde and Lourenzo Fernandez Prieto highlight the role of the veterinary profession in livestock improvement during the early 20th century in Spain. The authors analyse state initiatives and those of the veterinary profession, along with its corporative struggles with other professions, such as agronomists, and the search for greater social recognition of veterinary science and professionals in Spain. Veterinarians were integrated into a sphere of public administration that included new corporate and professional roles. According to the authors, 'throughout the period of study, most practitioners were united by an interest in gaining prestige, respect and social dignification for the profession'. Recognition of Spanish livestock farming as a differentiated productive sector in the context of the great transformations of the second wave of industrialization was clearly linked to the consolidation of the veterinary profession and its public and professional configuration in Europe and America.

Chapter 11 focuses on antibiotic growth promoters. Veterinarian Kristina Nordéus examines the processes that led the European Union to ban these substances in 2006. The author argues that the food crises of the 1990s contributed to the remodelling of the EU food safety system, which paved the way for phasing out antibiotic growth promoters. She looks at the role of science in shaping ideas, the actions of the various stakeholders, and the participation of Sweden as a new member state in 1995. She describes the polarization of the scientific community, and how those opposed to the bans accused them of being 'unscientific'. The scenario illustrates the dilemmas that accompany application of the precautionary principle.

Antibiotics occupy centre stage in Chapter 12 also, where Annick Jacq deals with the use of antibiotics in French aquaculture. This area has been overshadowed by earlier concerns about antibiotics in agriculture. The author emphasizes difficulties in bringing this to public attention and analyses the use of antibiotics in French aquaculture (since the 1970s) as a potential cause for the development of antimicrobial resistant bacteria. Various actors, including professional and consumer associations, scientists, veterinarians, regulatory agencies and environmental NGOs have adopted this frame. Jacq also argues that while some scientists question the use of antibiotics in aquaculture, especially since 1990, the French debate has mainly taken place in the scientific and expert arena.

In Chapter 13, Alexandra Waluszewski offers a different perspective on antibiotics by asking: Who pays for the use of antibiotics as an element of the innovation system in pork production? And how was the cost of change distributed, when the main costs were incurred in production spheres and the main benefits were perceived in health care systems, environment and society at large? The author contends that it is possible to radically decrease antibiotics inputs, but raises questions about how to distribute the cost and benefits of system innovation. Along the same lines as the previous chapter, here Waluszewski looks at how the insiders – farmers and the Swedish Farmers' Association – took on a role more commonly represented by 'outsiders', such as interest groups, media debaters and activists. She argues that the strong ties between cooperatives and private companies in Sweden became the pivotal mobilising and coordinating force behind several measures that have contributed to precautionary health and biosecurity.

Finally, in Chapter 14, Nicolas Fortané turns our attention to 'the rise and fall of ecopathology' in France. The critical perspectives of some veterinarians affirm that intensified animal production has deteriorated animal health along with the socio-economic situation of many farmers, while other veterinarians were criticized for being out of touch with livestock farmers. Against this backdrop, ecopathology was developed as an alternative form of veterinary medicine in the 1970s and continued through the 1990s. Fortané argues that ecopathology failed with regard to its initial ambitions, partially because people actually began to support rather than challenge industrial forms of livestock farming. Ecopathological knowledge had also warned of changes in veterinary medicine.

III. Animals as abstract machines? Some reflections on the industrialization of animal farming and its consequences

The practices that began to unfold with the industrialization of agriculture and livestock in the mid-20th century were criticized very early on (González de Molina et al., 2020).[11] One has only to remember the impacting of the works of Rachel Carson and Ruth Harrison. However, breeding practices continues their trajectory almost as if nothing had happened. What can we attribute this to? It may have something to do with how animals are designed.

Thinking of animals as machines is not new.[12] The idea was quite common well before the 19th century and applied particularly well to draft and plough animals. What is undoubtedly new with the industrialized breeding of poultry

---

11 In the authors' words: 'The process of agricultural industrialization has consisted of substituting the biochemical circuits of an agroecosystem with working capital that depends on resources outside the agrarian sector, usually via markets [...]. The reproduction of fund elements was possible through biomass flows in organic metabolic regimes; but under the industrial metabolic regime, external fossil energy flows are widely reproduced by social funds and can cause environmental deterioration when attempting to reproduce biophysical funds, especially agroecosystem services'. We think the idea can be extended to all types of external inputs acquired by farmers and breeders. For breeders, these include antibiotics, feedstuff (including supplemental feedstuff), energy to warm or refresh buildings, electricity to provide light, and insemination material, to name a few.

12 Beldo (2017) reminds us that farmers were encouraged to consider animals as machines very early in the twentieth century.

and livestock, and with aquaculture, is the fact that all these animals are bred to be machines. But what kind of machine are we talking about? And how can talking about a 'machine' help us better understand the contemporary trajectory of animal husbandry, as well as the significance of resistance to these practices among breeders and consumers?

Leaving aside the distinguishing features of each species and the goals associated with each type of breeding, we find breeders asking the same questions since the end of the nineteenth century. On one hand, they want to increase animal growth rates, get the maximum product from each animal (fat, meat, protein, milk, eggs, etc.), feed it economically and so on. On the other hand, they want to prevent diseases and mortality, counteract aggressiveness and spleen issues, and compensate for the lack of sun or fresh air. Of course, there are specific answers for each species, but the logics are often common to all types of breeding, though dependent on chronology and (the sometimes chaotic) scientific advances (antibiotics, use of animal meal, artificial insemination, genetic engineering, etc.). In the end, the individuality of a species seems to reside in highlighting a trait (or a set of traits) that scientists, officials, companies, breeders and their cooperatives seek to promote or combat (% of flesh % of fat in pig, % fat or % milk protein, number of eggs per unit time, growth rate, etc.).[13]

The life of the animal is shortened to maximize its ability to meet the expectations of the breeder at a given moment, 't'. The animal becomes an abstract animal or perhaps a succession of abstract animals. It could be considered as an abstract machine, or more precisely an 'abstract technical object', as described by Gilbert Simondon. In the first chapter of his book, *Du mode d'existence des objets techniques,* Simondon analyses the genesis of internal combustion engines.[14] He concludes that the first engines were abstract in the sense that

> the different parts of the engine are like individuals who could be thought of as working each in his turn without their ever knowing each other [...] the early engine is a logical assembly of elements defined by their total and single function [...] In this case, the integration of the particular unit into the ensemble involves a series of problems to be resolved, problems that are called technical but which, in fact, are problems concerning the compatibility within a given ensemble. (Simondon, 2012 [1958], 24)

Conversely, in modern engines, 'each critical piece is so connected to the rest by reciprocal exchange that it cannot be other than it is' (Simondon, 2012 [1958],

13 The (sometimes) ancient process for selecting specific animal traits should not be confused with the process of industrialising and transforming animals into machines (or 'abstract technical objects', as we shall see). For pigs, see White (2011).

14 It seems more relevant to quote Simondon than to make a paraphrase of his book. The translation was largely inspired by that of Lilian Mellanphy (1980), which was consulted via internet on 22 May 2022. The references are from the French reedition of 2012. We want to thank Clémentine Lessard for reading this part of the introduction and we assume responsibility for any faults or errors.

23). In that sense, he described modern engines as 'concrete'. This may seem rather far removed from agriculture and rearing practices, but Simondon indicates that the abstract technical object, or the 'primitive' technical object bears no resemblance to a natural object, because

> it is a materialization of a set of scientific notions and principles that at the most basic level are unrelated to each other and are only connected by their results that converge to produce a desired outcome. (Simondon, 2012 [1958], 56)

The 'concrete object', the 'advanced object', is close to the 'natural object' because it 'approximates the mode of existence of the natural object. It tends to internal coherence and closure of the system of causes and effects that operate in circular fashion within its boundaries' (Simondon, 2012 [1958], 56). Later, Simondon briefly examines the case of a greenhouse plant with double petals that cannot reproduce itself to emphasize how

> Making a natural object artificial gives results that differ from those effected by technical concretization. [...] The initially coherent system of biological functions has been opened to functions that are independent of each other and that are related to one another only by the gardener's care.

It could then be concluded that 'artificialization is a process of abstraction in the object that is rendered artificial' (Simondon, 2012 [1958], 57).[15]

The distinction between abstract and concrete technical objects introduced by Simondon in this chapter allows us to redefine our category of 'artificial'. It no longer designates the provenance or the essence of objects but their mode of existence. In other words, it is how we treat them and possibly disturb their relations with their environment. The concretization of a technical object implies that it becomes more flexible and available to a greater set of uses.[16] It is then possible to understand the importance of the fact that industrially raised animals are not only subject to an industrial-type chain of operations – an assembly line that slices up their existence – but at every stage they undergo a process of abstraction in which they lose their capacity for self-regulation as a species or as individuals.[17]

---

15 On the relation between Latour and Simondon, see Latour 2010.

16 *Two Lessons on Animal and Man* was a course taught by Simondon in 1963-64 and is not central to our interrogation. See Simondon (2020 [2004]) and the introduction by René Chateau. See also the useful article by Ramos (2019). For a first appraisal of Simondons' philosophy, we recommend readings in the special issue of *Doispontos* (2019), the special issue of *SubStance* (2012) and Schmidgen (2012).

17 In *Aspect psychologique du machinisme agricole* (2014 [1959]), Simondon made this point very clear for tractors as tools for multiple tasks, as opposed to more specialized (and abstract and less flexible) technical objects such as combine harvesters. It is very interesting to highlight Simondons' observation (based on a sociological survey) that Poitvine farmers chose their tools based on

If we adopt this analytical grid, we can draw two conclusions that allow us to better understand the trajectory of breeding practices and the emphasis on the need to adopt a holistic approach to breeding, which appears in several texts of the second part of the book.[18] Firstly, it would be the only way for the industrial animal to become a natural object again rather than an abstract machine (or an abstract technical object). Secondly, if an animal is not a natural object, but an abstract one, then the more artificial it is, the more abstract it is. Thus, whether individually or collectively, artificial animals no longer have power to self-regulate, and humans must intervene at every point of the breeding process to compensate this with increasing amounts of complemented feedstuff, antibiotics, vitamins, light, air, etc. A holistic approach gives more space, more liberty, and more autonomy to the animal. It respects basic animal needs without assessing performance based on a single indicator. The authors promote this as the only way that would envisage or enable the abandonment of antibiotics, for example. It could also be a way to reverse the process that transforms animals into abstract objects.

The idea that industrial agriculture is a process that must constantly repair the damage it causes was recently developed by Romero (2022) in a book about the use of pesticides in US agriculture prior to the Second World War. It could be interpreted as a means of permanently counteracting the artificialization of the natural object. In breeding, the transformation of animals into abstract objects has reached dizzying extremes, and so the need to repair is ongoing.

While the work of Simondon can help us to better understand the trajectory of breeding, it does not tell us the whole story. Fernando Collantes, reminds us in Chapter 8 of the industrial logic behind seeing farm animals as machines. It was a means of lowering the price of protein and conformed to a 'Fordist' model.

Perspectives have also changed regarding hunger and other threats, or the opposite issues of overproduction and problematic surpluses. However, the links between animal care, sustainable breeding practices and production levels are not at all direct. In contexts of food scarcity, attention to animal welfare and production methods may be subordinated to volume growth for supplying meat, eggs, and milk. Inversely, where there is overproduction, or industrial production, farmers and meatpackers seem inattentive to animal welfare and environmental disasters caused by intensive animal farming.

The issue of medicalization is also complex. Medicalization, and especially the use of antibiotics, could be perceived as a means of replacing labour. But medicalization may also have been seen as an advance over the poor hygiene of earlier times.

---

economic considerations rather than any allegedly backward peasant mentality or the desire to distinguish themselves from it.

18 See for example the chapter by Kristina Nordéus, and that of Nicolas Fortané for a methodological perspective.

Finally, despite the dominant industrial rearing and aquaculture practices in Europe, the USA, Brazil, China, South Africa, etc., there are other practices that do not (or at least try not to) transform animals into abstract objects. Not all farmers see the scale and production methods applied today as ideal. Many breeders are seeking a balance between external pressure (techno-scientific or economic), the preservation of animal well-being on the farm and sustainability in farming. The contributors to this book have shown how these aspects vary greatly over time and by regions. What was once considered large-scale dairy farming in Galicia, for example, may have been thought rather small in the British countryside. Indeed, animal farming and animal-based food production are not easy endeavours. The economic context is often problematic, and the work hours are long. Still, many farmers keep on, unwilling to give up, which reflects their interest in the animals and in working with them. Swedish breeders demonstrated this with their involvement in the fight against antibiotics. In this sense, sustainable agriculture and husbandry that pays attention to animal well-being could be thought of as a culture, a way of life, and even a way to resist to the agro-industrial complex. It could also be a way of improving farm profitability. AOP cheese in eastern France, which specialized in cheese production (Comté, Morbier, Bleu de Gex)[19] illustrates this point. Dairy farmers there decided against the use of automated milking machines to protect the need for grazing and preserve the link between farmers and their animals, while also protecting the brand image of their cheeses. But it is striking to realize that to preserve 'traditional' milking and breeding methods, they were obliged to argue that automated milking machines could not preserve the intrinsic characteristics of milk. They diminished the solidity of the fat globules in the milk, thereby altering its quality and that of the cheese.[20]

Though distant from a scientific approach to rearing, these dairy farmers are nonetheless very comfortable with such an approach, which often leads to the industrialization of breeding and the artificialization of animals. Since the nineteenth century, cheese specialization in this region has led to the decline and near demise of cereal production. In some cases, natural meadows have even been replaced by leguminous monocultures. Meanwhile, breeder's associations do not allow milk production to exceed a given quantity per hectare. This illustrates the very complex relations between dairy farmers and industrialization efforts as well as the need to avoid sweeping judgments.

The historical perspectives contained in this anthology were developed under idyllic conditions in the summer of 2019, with researchers from six countries enjoying some intense days of fruitful discussions. They focused on animal farming from different angles, giving special attention to the time and place of various actions as well as differences in regulations, networks, and knowledge. Since then, dramatic international changes have impacted our perspectives on production

19 The AOP Comté is the most important AOP in France and far from anecdotal.
20 Salvi 2022.

methods in animal farming, dependency on feed imports, the global food trade and energy needs. This new panorama sheds fresh light on the development of animal farming that is discussed in these pages. The assumed directions of its development, albeit criticized, may appear differently to future researchers. Will future interpretations shed new light on the intensification, specialization, and dependency on input flows from all over the world that has occurred in our time?

Bibliography

Beldo, Les (2017), 'Metabolic Labor. Broiler chickens and the exploitation of vitality', *Environmental humanities*, vol. 9, no. 1, p. 108-28

Berdah, Delphine (2021), 'Serum therapy against FMD and the development of the French veterinary profession in the 1930s', Review of agriculture, food and environmental studies, vol. 102, 151-69

——— (2012), « Entre scientifisation et travail de frontières: les transformations des savoirs vétérinaires en France, XVIIIe-XIXe siècles », *Revue d'histoire monderne et contemporaine*, vol. 59, no. 4, p. 51-96

Bustamante, M. (2012), 'Estimating greenhouse gas emissions from cattle raising in Brazil', *Climatic Change*, 115: 559-77 DOI 10.1007/s10584-012-0443-3

Cardoso, Abamel S. (2016), 'Impact of the intensification of beef production in Brazil on greenhouse gas emissions and land use', *Agricultural systems*, vol. 143, p. 86-96

Doispontos, Revista dos Departemnetos de Filosofia da Universidade Federal do Paraná e da Universidade de São Carlos volume 16, no. 1

Duarte, Rosa, Pinilla, Vicente and Serrano, Ana (2019), 'Long Term Drivers of Global Virtual Water Trade: A Trade Gravity Approach for 1965-2010', *Ecological Economics*, 156, 318-26

González de Molina (2020) et al., *The social metabolism of Spanish agriculture, 1900-2008. The Mediterranean Way Towards Industrialization*, Environmental History, 10, SpringerOpen

Fearnside, Philip M. (1983), 'Land use trends in the Brazilian Amazon region as factors in accelerating deforestation', *Environmental conservation*, vol. 10, no. 2, 42-52

——— (1985), 'Brazil's Amazon Forest and the Global Carbon Problem', *Interciencia*, vol. 10, no. 4, p. 179-86

Fernández Prieto, Lourenzo and Lanero, Daniel (2020), *Leche y Lecheras en el Siglo XX. De la fusión innovadora orgánica a la Revolución Verde*, Zaragoza, Prensas Universitarias de Zaragoza & SEHA

Fiennes, R. N. T. W. (1978), *Zoonoses and the origins and ecology of human diseases*, London, Academic Press

Fitzgerald, Deborah (2003), *Every farm a factory: The Industrial Ideal in American Agriculture*, New Haven, Yale University Press

——— (2018), 'Roundtable: Agricultural History and the History of Science', *Agricultural History*, Vol. 92, no. 4, p. 569-604

FAO (2016), *Elevage et changement climatique* (available in English)

Hardy, Anne (2003), 'Animals, Disease and man making connection', *Perspectives in biology and medicine*, vol. 46, no. 2, p. 200-15

Harrison, Ruth (1964), *Animal Machine*, London, Vincent Stuart

Hart, C. A., Bennett, M. and Begon, M. E. (1999), 'Zoonoses', *Journal of epidemiology and community health*, vol. 53, p. 514-15

Henderson-Sellers, A. (1981), 'The effect of land clearance and agriculture practices upon climate', *Studies in the Third World Societies*, no. 14, p. 443-85

Herment, Laurent (2021), *La plus belle firme du monde ne peut offrir que ce qu'elle a*, Mémoire d'habilitation pour diriger des recherches, Université Paris 1 Panthéon-Sorbonne avialable on HAL

Jarrige, François and Le Roux, Thomas (2020), *The contamination of the world. A history of pollutions in the industrial age*, MIT Press, Cambridge

Kirchelle, Claas (2020), *Phyrric progress, The history of antibiotics in Anglo-American food production*, New Brunswick, Rutgers University Press

——— (2018), 'Pharming animals: a global history of antibiotics in food production (1935-2017)', *Humanities & Social Sciences Communications*, 96

Langthaler, Ernst (2018), 'The soy paradox: The Western nutrition transition revisited, 1950-2010', *Global Environment*, vol. 11, no. 1, p. 79-104

——— (2020), 'Broadening and Deepening: Soy Expansions in a World Historical Perspective', Historia Ambiental, Latinoamericana y Caribeña, vol. 10, no. 1, p. 244-77

Latour, Bruno (2010), 'Prendre le pli des techniques', *Réseaux*, 2010/5, n° 163, p. 11-31

Levi-Strauss, Claude (2001), 'La leçon de sagesse des vaches folles', vol. 157-58, p. 9-14

Lynteris, Christos (2020), *Human extinction and the pandemic imaginary*, New-York, Routledge

——— (2017), 'Zoonotic diagrams: mastering and unsettling human-animal relations', *Journal of the royal anthropological institute*, vol. 23, p. 463-85

Martiin, Carin; Pan-Montojo, Juan, and Brassley, Paul (2016), *Agriculture in Capitalist Europe, 1945-1960. From food shortages to food surpluses.* London and New York, Routledge

Mercedes M. C. (2012), 'Estimating greenhouse gas emissions from cattle raising in Brazil', *Climatic change*, vol. 115, no. 3-4, p. 559-77

Morse, Stephen S. (1995), 'Factors in the Emergence of Infectious Diseases', *Emerging Infectious Diseases*, vol. 1, no. 1, p. 7-15

Moser, Peter and Varley, Tony (eds.) (2013), *Integration through Subordination: The Politics of Agricultural Modernization in Industrial Europe*, Rural History in Europe vol. 8, Turnhout, Brepols

Pierce-Duvet, Jessica M. C. (2006), 'The origins of human pathogens: evaluating the role of agriculture and domestic animals in the evolution of human diseases', *Biological reviews*, vol. 81, no. 3, p. 369-82

Ramos Souza (de), Silvana (2019), 'Merleau-Ponty e Simondon: sobre o animal e o humano', *Doispontos*, vol. 16, no. 3, p. 137-46

Romero, Adam M. (2022), *Economic Poisoning Industrial Waste and the Chemicalization of American Agriculture*, University of California Press

Salvi, Raphaël (2022), 'Les AOP face aux innovations, améliorer sans altérer. Le cas du robot de traite en AOP Comté et dans les AOP fromagères du massif Jurassien', communication, Paris, 16 June 2022 (unpublished)

Saraiva, Tiago (2016), *Fascist Pig. Technoscientific organisms and the history of fascism*, Cambridge (Massachusetts), MIT Press

Schar, D.; Klein, E.Y.; Laxminarayan, R.; Gilbert, M.; Van Boeckel, T.P. (2020), 'Global Trends in Antimicrobial Use in Aquaculture', *Science Reports*, 2020, 10, 21878

Simondon, Gilbert (2020 [2004]), *Deux leçons sur l'animal et l'homme*, Paris, Ellipses

——— (2014 [1959]), 'Aspect psychologique du machinisme agricole', in *Sur la technique* (1953-1959), Paris, Presses Universitaires de France, p. 255-66

——— (2012 [1958]), *Du mode d'existence des objets techniques*, Paris, Aubier-Montaigne

Schmidgen, Henning (2012), 'Inside the Black Box: Simondon's Politics of Technology', *SubStance*, vol. 41, no. 3, p. 16-31

SubStance (2012), Gilbert Simon, vol. 41, no. 3, p. 3-92

Tello, Enric (2016) 'Opening the black box of energy throughputs in farm systems: A decomposition analysis between the energy returns to external inputs, internal biomass reuses and total inputs consumed (the Vallès County, Catalonia) *c.* 1860 and 1999', *Ecological Economics*, vol. 121, p. 160-74

——— (2014), 'Looking Backwards into a Mediterranean Edge Environment: Landscape Changes in El Congost Valley (Catalonia), 1850-2005', *Environment and History*, vol. 20, no. 3, p. 347-84

Tiseo, Katie (2020), 'Global Trends in Antimicrobial Use in Food Animals from 2017 to 2030', *Antibiotics*, 9, 918; doi:10.3390/antibiotics9120918

Van Boeckel et al. (2015), 'Global trends in antimicrobial use in food animals', *Proc. Natl. Acad. Sci.* USA, 112, 5649

White, Sam (2011), 'From Globalized Pig Breeds to Capitalist Pigs: A Study in Animal Cultures and Evolutionary History', *Environmental history*, vol. 16, no. 1, p. 94-120

Woods, Abigail (2011), 'A historical synopsis of farm animal disease and public policy in twentieth century Britain', *Philosophical Transactions of The Royal Society B.*, Vol. 366, 1573

——— (2013), 'Is Prevention Better than Cure? The Rise and Fall of Veterinary Preventive Medicine, *c.* 1950-1980', *Social History of Medicine*, Vol. 26, 1, pp. 113-31

WHO (1959), *Zoonoses*. Technical Report, Series No 169. Geneva: World Health Organization

Part 1

Animal husbandry and dairy production and consumption

KAREN SAYER

2. Figuring the past

Statistics as cultural artefacts in British agri-culture from the mid-nineteenth to the mid-twentieth century

I. Introduction

Statistical and numerical data, local and national, has been a key source for agricultural historians, its use by the discipline of history being entirely consistent with the period of Modern British agricultural improvement. As Christine Hallas observed in 1999, traditionally agricultural historians have used livestock as a measure for 'farmer responsiveness' to 'market demand', 'where output has been examined in terms of numbers, stock type and animal products'. She went on to provide a cogent critique of the difficulties of this, around 'calculating raw numbers, identification of the type, age and weight of animals', resulting in significant problems in determining 'the size of the national herd' and changes to that herd and production/productivity and offered up local sources as a corrective to the large-scale national datasets (Hallas, 1999: 83-84). In the nineteenth century, underpinning agricultural improvement, as it was known at the time, and 'Practice Through Science' (the slogan of the Royal Agricultural Society of England, founded 1838), British policymakers, statisticians, and agricultural commentators had a widening interest in the collection, analysis and interpretation of numerical data linked to animal husbandry. This data, modelled, interpreted, and reinterpreted was seen at the time as being of national significance, of utility to agriculturalists and economists. In this sense, the interests of agriculturalists and agricultural historians have often converged. But this data was at the time also taken to be representative of the nation and its national values. In other words, it will be argued, livestock data generated by exercises such as the Agricultural Returns became part of the scientific and technical/'sociotechnical' imaginary (Jasanoff and Kim, 2015)[1] of the British nation, that fed into decision-making, and legislative change. On this basis it should be approached as a qualitative, as

1 This approach has been adopted widely in histories of technology, especially on energy, for example, Chatterjee (2020); Yang (2018). And, Jasanoff (2005) has written elsewhere about civic approaches

From Breeding & Feeding to Medicalization, Rural History in Europe, 17 (Turnhout: Brepols, 2023), pp. 31–52
BREPOLS PUBLISHERS 10.1484/M.RURHE-EB.5.135020

This work was supported by the Wellcome Trust [209818/E/17/Z].

well as a quantitative, source i.e., as holding evidence of the culture and ideology of the period.

Utilising journal articles from the periodical press, particularly *The Journal of the Statistical Society* (1838-1886), *The Journal of Agricultural Science* (established 1905), *The Journal of the Royal Agricultural Society of England* (established 1840), agricultural advisory texts, and historiographical material drawn from the *Agricultural History Review* and *Economic History Review*, this chapter will address statistical and numerical data as part of a culture of scientific improvement. I am not seeking to reject the statistical material here. Rather, I am taking a step back to address the emergence of the interest in data considered as essential to the evolution of a fully commercial, competitive, agricultural sector. Data was essential during the long twentieth century to the changes in approach to animal health and disease that we see in husbandry and veterinary practice, discussed elsewhere in this volume. At the same time, it will be argued, the animals enumerated within that data were captured and re-presented within a numerical landscape utilised rhetorically as part of the cultural formation of British national identity linked to the idea of the Land.

In the 1980s Raymond Williams posited that High culture (art and literature) was part of a shifting formation of an idealised, dominant, and powerful understanding of the British countryside, that (in each generation) reflected the societal issues of its period (Williams, 1985). This itself reflected a wider cultural materialist understanding that emerged at the time across a range of disciplines, from geography to literary studies, each focused on constructions of the rural and rural life (Daniels, 1994; Sales, 1983; Maclean, Landry, and Ward, 1999). Art historians such as John Barrell, Ann Bermingham and Christiana Payne suggested (Barrell, 1983; Bermingham, 1987; Payne, 1994),[2] dominant cultural perceptions of the countryside existed in a dynamic relationship with British national identity and politics. This point was subsequently accepted by many rural and social historians including Alun Howkins, Mark Freeman, Jeremy Burchardt (Howkins, 1991; Freeman, 2003; Burchardt, 2002), also rural sociologists, cultural geographers, and ethnographers such as J. Agyeman and R. Spooner, Doreen Massey, and Michael Mayerfield Bell (Massey, 1994; Agyeman and Spooner, 1997; Mayerfield Bell, 1994). And though the impacts were perhaps particularly noticeable in relation to concepts of Englishness, they were never observed solely in relation to Britain (Brettell and Brettell, 1983; Lehning, 1995; Kolodny, 1984; Ford, 2001). As Stephen Daniels argued of the UK, the association of national identity with certain landscapes 'negotiates many other forms of identity: local, regional and international; social, religious and familial' (Daniels, 1994: 8), a point equally applicable to other countries.

---

to agriculture, decision-making and health. But these threads are only now beginning to enter agricultural historical work Zetterstrom-Sharp (2020: 293-313).

2 More recently, see also Pearson and Nasby (2008).

Nevertheless, none of these classic cultural historical accounts addressed numerical datasets, nor discussed their or statistically derived re-presentation as cultural artefacts. Agricultural historians have of course used large-scale series of numerical data to address historical change, offering highly nuanced re-readings of these collections to establish comparable, authoritative sets of figures. They have managed the complex and changing nature of the available data sets skilfully through established well-founded statistical formulae to consider a vast range of subjects related to farms and farming, from livestock husbandry, labour, and farm management through to energy, in the UK are around the world (for example, Dodgshon, 1976; Howkins and Verdon, 2008; Henriksen et al., 2012; Kander and Warde, 2011). But, they did not assess the data that they were using as cultural objects, even when (as in the case of Howkins, for example) they wrote historical accounts of the re-representation of rural life.

However, situated between these bodies of work, in a separate sub-discipline, we see historians of science such as Giuditta Parolini writing of the links between the formation of the discipline of statistics and the development and history of the growth of agricultural experimentation and agricultural science (Parolini, 2014; Paroloini, 2015). Through these accounts we see some of the ideological components associated with the development of statistical methodologies, such as the formation of subfields like eugenics, (Haraway, 1992; Berry, 2015). Drawing on these approaches in this chapter, to place numerical into dialogue with non-numerical cultural artefacts, I will focus on sheep, (beloved by statistician Sir Henry Rew). As will be seen, sheep and sheep production were widely appreciated in Britain by non-specialists as well as agriculturalists throughout the nineteenth and twentieth centuries and were often linked intrinsically with British national identity and the idea of 'the Land'. As such, throughout this period, British sheep were situated as figures in an international political landscape (Woods, 2017: 2, 5, 52-77).

Though widely used in economic and agricultural history, the livestock Returns have not been interpreted other than as statistical data sets to date. Yet these data sets, like other discursive objects in which we see sheep re-presented (advisory literature, press reports, parliamentary papers, etc.), were co-produced by livestock, farmers, merchants, agricultural commentators, and politicians. Approaching these sources in this way i.e., as co-curated cultural artefacts, as would be done by environmental historians, animal historians and animal geographers, and those working in animal studies, (Cragoe and Mcdonagh, 2013; Fudge, 2013: 13; Haraway, 2015) enables us to gain a better understanding of the drivers for data that dominated the commercial and specialist rhetorics of progress that surrounded livestock farming and its systems in Europe.

The chapter will begin by addressing the historiographical relationship between the Annual Returns (a key large-scale data set in Britain from 1866) and British agricultural history writing. It will then relate them, as artefacts that re-presented the national flock, to the place of sheep within the wider landscape of British agricultural production and the idea of 'the Land'.

II. Historiography

In 1969 a review (one that is a salutary lesson in the necessity of addressing historiographical – as opposed to historical – change), T. W. Fletcher observed that 'the contents of *Agricultural Statistics* should have changed over the years is inevitable, but that an official publication celebrating the centenary of a remarkable enterprise should give no inkling of what has been lost is to be regretted'. Though an authority, as a reviewer rather than as an author, Fletcher's reading runs the risk today of being considered somewhat idiosyncratic. But in his view, *A Century of Agricultural Statistics: Great Britain, 1866-1966* (H.M.S.O. 1968) 'provide[d] the historian with little more than a statistical skeleton', because the 'emphasis on the economic interests of today obscures the changing nature of the agricultural statistics themselves'. As a result, Fletcher opted for a wholesale critical rejection of *A Century of Agricultural Statistics,* removing it from the oeuvre of history to 'the shelves of all teachers, advisers, and administrators concerned with the economics of British agriculture' (Fletcher, 1969: 574-76). But if the *Century of Agricultural Statistics* can be said to have been driven to oversimplification in the face of, what was seen in 1968, as the very significant progress the UK had made to increase its food production nationally (especially following the Agriculture Act 1947), the Returns themselves have proved themselves to be essential sources, while other statistical summaries were steadier, more detailed and more measured.[3]

The Agriculture Act (1947) shifted British agriculture after the 1939-1945 war onto a footing aligned to international policymaking on food production and security, a process that had begun pre-war in the UK, in Europe and USA with the formation of the International Institute of Agriculture, then later continued with the 1943 Hot Springs Conference.[4] It was a focus that arguably cemented a shared interest in data between agricultural historians and policy makers reaching into the mid-twentieth century, as both sought signs of progress against the objectives of the period and continued shortages. But a timeline mapping out the historiographical interest in the origins, development and value of British agricultural statistics would begin decades before Fletcher's review of *A Century of Agricultural Statistics.* Hence, even in his 1945, 'written at the height of the flying-bomb period' (Fussell, 1945: 79 fn1) survey of the field of agricultural history in general, and of the history of agricultural statistics in particular, G. E. Fussell referred to the work of 'J. A. Venn [who] discussed the subject at some length in his *Foundations of Agricultural Economics* (1923)', the then key text on the subject. Fussell, drawing on Venn, argued that it was because a 'more precise knowledge of the productive capacity of the industry, as well as education for the aspirant to farming, had long been desired' that 'the propaganda for the collection of reliable agricultural

3 For example, Marks and Britton (1989), Burrell et al. (1990) and for metadata Peters (1988).

4 The impacts of this on UK policy were reflected in debates within Parliament such as: Debate, 'Post-War Agricultural Policy', Commons Sitting of Tuesday, 3rd August 1943, *Hansard*, 1942-42, Fifth Series, Vol. 391, page column 2057-2266, p. 2080.

statistics grew steadily, culminating in the publication of the first official volume in 1867'. (Fussell, 1945: 79 fn1; Venn, 1923: 429).[5] In other words, during World War II the (teleologically-oriented and empirical, if not economic) foundational agricultural historians who wrote about this data saw it as an artefact contingent on a supposed need to drive policy for economic gain. This perceived value also led to the development of large-scale surveys at around this time, such as the Farm Management Survey (now called the Farm Business Survey) in 1936 that shared the Annual Returns' intellectual roots (Brassley et al., 2013). The agricultural historians who wrote about the Returns in the Second World War also saw them as reflecting a particular set of technical demands connected with what they considered the necessary and vital emergence of formal agricultural education and professionalization. Agricultural statistics were therefore intertwined with the case for and the processes of agricultural education and examination, tied both to the economic interests of the day and the morality of political economy, centred on and grounded in the progressivist, utilitarian, positivist empirical values of the societies and institutions that came to dominate British Agriculture in the nineteenth century: the Highland and Agricultural Society of Scotland, the Royal Agricultural Society of England, the Agricultural Club, the Farmers' Club and their county counterparts.[6]

The interest in the collection of quantitative data by agriculturalists, including record keeping by farmers, similarly, predated the 1866 Returns. In fact, though the detail of its meaning has shifted, the word 'statistics' originated in political economy, specifically the political economy of agricultural production, in the eighteenth century (OED Online, June 2021).[7] Moreover, French statisticians were cited in publications such as the British periodical *The Farmers' Magazine'* in its first (1834) volume, (Anon, May 1834, Vol1: 420).[8] Indeed, the breadth and the nature of the available data in the period predating the Returns is

5 Biographical information about Venn available at Pickles, John D. 'Venn, John Archibald (1883-1958)', John D. Pickles *Oxford Dictionary of National Biography*. Online ed. Ed. David Cannadine. Oxford: OUP, 2004. 25 April 2023 <https://www.oxforddnb.com/view/article/40972>.

6 For accounts of the origins of these societies, see Goddard (1981, 1983 and 1988); Hudson (1975); MacDonald (1979); Miskell (2012).

7 This included a Scottish agricultural improver -- specialist in sheep farming and member of the Highland Agricultural Society – Sir john Sinclair. He helped establish the 'Society for the Improvement of British Wool' at the turn of the nineteenth century, and is cited in the OED as follows: '1798 J. Sinclair Statist. Acct. Scotl. XX. App. p. xiii. In 1786, I found, that in Germany they were engaged in a species of political inquiry, to which they had given the name of Statistics; and, as I thought that a new word might attract more public attention, I resolved on adopting it'. See 'statistics, n.'. OED Online. June 2021. Oxford University Press. https://www.oed.com/view/Entry/189322?rskey=v85Ukd&result=2 (accessed 25 April 2023). I am indebted to Nicole Gosling for bringing this to my attention; Mitchison, Rosalind. 'Sinclair, Sir John, first baronet (1754-1835), agricultural improver, politician, and codifier of "useful knowledge"', *Oxford Dictionary of National Biography*. 23 Sep. 2004; accessed 25 April 2023. https://www.oxforddnb.com/view/10.1093/ref:odnb/9780198614128.001.0001/odnb-9780198614128-e-25627.

8 E.g. 'A French Statistical Writer has lately…' Anon, *The Farmer's Magazine*, Vol. 1, May 1834, p. 420.

indicated in the extensive discussions that have taken place among agricultural historians about accessing and utilising data sets fit for the purposes of historical analysis (Turner, et al., 1996). As the subdiscipline of agricultural history began to develop, we see John Venn, who worked from 1917 as a statistician and was later an advisor to the Ministry of Agriculture, writing a chapter entitled 'Agricultural Statistics' on the history of the arguments surrounding the origins of the legislative machinery of national data collection, covering the period from the Norman Doomsday survey to his present (Venn, 1923: 424). Ultimately, Venn argued, it was an 1864 Parliamentary Resolution in the House of Commons led by James Caird (1816-1892)[9] – from the 1840s the leading proponent of so-called 'High Farming' (Caird, 1880) – that made a case for the collection of agricultural data annually, which carried the day for the instigation of the Agricultural Returns. This is quite accurate, if we look at that Resolution as reported in *Hansard* 7th June 1864 (Caird, 7th June 1864). And, reflecting on this in his 1880 Inaugural Address as President of the Royal Statistical Society, Caird was proud to highlight the value and increasing significance of the Agricultural Returns over the intervening period. 'Twelve years ago, in addressing this Society', he concluded:

> I described the contents of a little blue book which I held in my hand, the first number of the Agricultural Returns. Under Mr. Giffen's care, that annual book has doubled in size. The prefatory remarks, introduced by Mr. Valpy, have been continued and extended, many comparative tables have been introduced, and much new matter has been added. It has become not only a useful guide to the legislator as well as the landowner and farmer, but to the importer and purveyor of food, and to all persons interested in its distribution and consumption. Those who took an interest in obtaining from parliament the means of collecting these returns, will join with me in expressing our warm appreciation of the increasingly useful and instructive manner in which they are now placed before the public. (Caird, 1880: 571-72)

But, according to Venn, those who argued for agricultural statistics to be collected had long believed that they had significance beyond an increased market advantage, or increased home production (both key elements in Caird's original rhetorical armoury in the 1860s[10]). They also had belief in their longevity as authoritative and useful knowledge. In the future, John Hannam (Secretary to the Yorkshire Agricultural Society) wrote in the mid-nineteenth century, historians would need them:

---

9 Mingay, G. E. 'Caird, Sir James (1816-1892), agriculturist and writer' *Oxford Dictionary of National Biography*. 23 Sep. 2004; accessed 25 April 2023. https://www.oxforddnb.com/view/10.1093/ref:odnb/9780198614128.001.0001/odnb-9780198614128-e-4339.

10 Caird celebrated the success of getting the Agricultural Statistics into production in this address: James Caird. 1868. 'On the Agricultural Statistics of the United Kingdom', *Journal of the Statistical Society of London* 31 (2): 127-45. doi:10.2307/2338858.

> How great will be the interest that will attach to such statistics, when experience shall have pointed out to us the proper questions to ask and the best modes of obtaining the answers. How useful for all the purposes of enquiry. And when in fifty years hence it shall fall to the lot of some person to address himself to an audience on the statistics of agriculture, the volume will be open to him from which he may draw at once the history of the past, the moral of the present and the prophecy of the future as regards agriculture. (Venn, 1923: 429)

It was therefore assumed that there would be a shared empirical, but also moral, political economic, interest between agriculturalists and historians, that statistics would serve the interests of both, and that historians would help farmers in due course by analysing the data.

British agricultural historians (and economists) have, as Hannam predicted, subsequently used the *Agricultural Returns*, along with other data sets such as the *Report on Wholesale and Retail Prices in the U.K.*, and analyses published in the *Journal of the Statistical Society* and *Journal of the Royal Agricultural Society of England*, to assess the extent and nature of change in agricultural production. Some economists, such as W. W. Rostow, not only used historical data to build influential models of economic growth, but also wrote agricultural history (Rostow, 1941; Sayer, 2019). In 1961 T. W. Fletcher himself published a nuanced (and foundational) article on the question of 'The Great Depression' in Britain 1873-1896 that drew on this material to assess its extent and impact, which explains his familiarity with the sources and was one reason why the *Economic History Review* believed he had the authority to review *A Century of Agricultural Statistics* (Fletcher, 1961).[11]

Like many state-sponsored data sets, including the UK's decennial population census, the framing of the Agricultural Returns altered depending on the political demands of the period. In 1867 the *Agricultural Statistics* began with introductory observations about the USA, France and Russia. Under the auspices of the Board of Trade and then from the 1880s the Board of Agriculture (but, 1866-1918, always gathering data enumerated by Customs and Excise officials), the number of sections then increased in each iteration until, just before the First World War, the report had come to include 'Foreign and Colonial Statistics' as well as those by county. Then after that war, they were handed over to the sole control of the Ministry of Agriculture, *Agricultural Statistics* and started decreasing in volume and scope. Because they constantly changed, the difficulties of handling these

---

11 Nb in setting out the established historiographical case for a Depression, as context for his own work, Fletcher cites J. A. Venn, before arguing against him: 'Venn, discussing the supply of gold and the price level, believed agriculture to have suffered more than most industries, "being plunged indeed into the abyss"'. (p. 417, fn 2). He does not refer to Fussell. Fletcher argued that the Agricultural Depression dethroned the large-scale corn producers of the Eastern counties, who had dominated the industry up to that point, in favour of a new type of farmer interested in turning a profit via fruit, dairy and sheep production. Fletcher (1961: 417-32).

sorts of sources are manifold, and this of necessity formed a key line of enquiry within the subfield. As noted, the history of the Agricultural Returns had already been established by Venn thirty years before the time that Fletcher was writing. Aside from Venn in the 1920s, and Fussell in 1945, J. T. Coppock also wrote about the subject in the immediate post WWII period in two very detailed articles for the *Agricultural History Review*. Despite them, however, the consensus for Coppock and Fletcher was that it could be taken for granted, as Hannam had suggested it might be that, in Fletcher's words (writing for the *Economic History Review* in 1969), that the 'historian's continuing need for the specific and the concrete is to a degree catered for in the pre-1914 volumes'. In other words, by the post-Second World War period, for the historian as for the farmer, the statistical data had become a necessary evil: nineteenth-century producers and twentieth century-users of statistics did indeed share a professional interest in the empirical data. Despite their failings, Coppock argued, the Returns were a unique and (at the time he was writing) (Coppock, 1956a: 4; Coppock, 1956b: 79), still focused on change as the lifeblood of history, they believed that the numerical data delineated and offered a window onto a very specific historic period.[12]

III. Agricultural Statistics and Animal (Production) Histories

The difficulties in using the Agricultural Returns (and other data sets) as sources of numerical data are therefore well-known, and there is no need to rehearse them here. What I want to do instead is to suggest that we can use the Returns, their flaws, and the responses to them as cultural artefacts that had an impact beyond markets, generating for example the formation of different farming cultures in nineteenth-century Britain, and also revealing that variations in interpretation, resistance and changes in expectations, as indicated by the use of differing categories and questions over time – i.e. the problems with the 'proper questions' and how best to get the information – actually constitute the evidence. The creation of the Agricultural Returns reflected for example, an interest in the perfection of agricultural production at a national and thence competition at an international level, as Venn identified, with the purpose of harnessing the knowledge of farm production for the competitive advantage of the nation. As one 1867 *Times* leader stated, reflecting on the difficulty of getting English farmers to provide data for the Returns:

> if the British public could have foreseen this time last year the prices corn was to reach before next harvest, it would have secured betimes, and at a lower

12 See Desrosières (2013: 8-23) on a related point on the entanglement of producers and users of statistics re nineteenth and twentieth century statisticians, and twentieth century economic historians sharing an empirical footing.

> price, a larger proportion of last year's produce. France, it is notorious, beats us in this respect, and by that means of a more complete system of Agricultural Returns. In fact, France enters the market with her eyes open, we with our eyes wilfully shut; and the result is what might be expected. (*Times*, Tues 18th Jun 1867: 11)

This was a failure from the point of view of the market, certainly, but the inability to collect accurate data at farm level was addressed in instances such as this as a reflection of the English farmer's character, as against the French (or, elsewhere in the same piece, the Irish or Scottish) farmer who was said to be far better at filling in forms and therefore engaging with the domestic or overseas market through the gathering of consistent data. Utility was deemed to rest upon successful standardisation, 'as the Returns will be worthless if they are not made on a perfectly uniform system' in the *Times'* view. But, neither the *Times* nor *Punch* held out much hope of that taking place because of the innate (national) character of the English farmer, as they saw it. Both, for example, noticed that part of the problem may have been due to the English farmer's hostility to 'The Exciseman' caste as 'a disagreeable visitor' in his eyes by the *Times* (*Times*, 18th Jun, 1867: 11), and in *Punch* represented through a poem entitled 'Agricultural Returns: "Put that in your Pipe and Smoke it"' (in so-called Mummerset doggerel) as designed to 'git a new tax', and a distraction from farming at a key point in the agricultural calendar (haysel):

> 'Commiss'ners of Inland Revenny remarks
> As how we insults and opposes their clerks,
> A gwaiun their rounds 'mongst he yeomen and squires,
> Agricultural statistics as comes and requires.
> ...
> Was we up in London, myself, mates, and you.
> And had need to Zummerzet House for to goo,
> And question them just when they'd got most to write,
> For all they was busy they'd still be perlite.
> They're called civil servunts, the reason for why
> Is their always gieun a civil reply;
> Which shows us a pattern what answers to gie
> At hay-harvest when they comes questionun we
> (*Punch*, Sat 17th April 1869: 156)

Punch was a Metropolitan publication, but evidently considered Caird's success in Parliament worth reporting and disingenuously assumed its readers would understand that the subject of 'collecting and publishing Agricultural Statistics ... [was] extremely important and perfectly uninteresting, except to the producers of the nation's food, and to the smart London men, who' had bought country property and pretended extensive knowledge of farming (Anon, *Punch* Saturday, June 18th, 1864: 249). This would suggest that there was lively interest in the subject at national level, at least among the literate elite, as the *Times* was even able to mention the prospect of Agricultural Statistics in a knowing aside during a (non-agricultural) book review that mentioned the *Doomsday Book*:

> to which, with its marvellous record of every single hide, every single virgate of land, and all that grew or fed upon it, the Saxon chronicler refers as conclusive proof that the Conqueror was a petti fogging tax-gatherer as well as a tyrant: "Not even", he cries (as indignantly as a modern farmer called upon for agricultural statistics), "it is shame to tell, though it seemed to him no shame to do, an ox, or a cow, or a swine was left that was not set down". (Anon, *Times*, 3rd Nov, 1865: 8)

Together the two sources therefore tell us why some of the final Returns may have been less than reliable, at least in the early days (Coppock, 1956a; Coppock, 1956b). And, elsewhere, it is also clear that concerns about the Returns among agriculturalists, despite Caird's arguments, were manifold. However, reports like this in the national press also created a perception that there was a hierarchical, two-tier culture among farmers.

As The Hon. Percy Wyndham, M.P. was reported to have suggested in the *Times* via the published record of the West Cumberland Agricultural Society annual meeting, farmers distrusted the Returns because they assumed (as in the *Punch* poem) that the government were setting out to improve their information for the purposes of increasing taxation, but he argued they were also worried that their neighbours would steal a march on them if they shared this information. He went on to an attempt to allay their fears and in so doing emphasised the important of the Returns as the complete versions 'would give a great advantage to the Government and to those who farm upon scientific principles, because they could calculate upon what amount of any particular kind of crop was likely to come into the market. (Cheers)'. (Anon *Times*, Tues, 8th Oct 1867: 5)[13] The fact that his argument was met with 'Cheers' suggests that the audience of farmers at the West Cumberland Agricultural Society, like those of other county societies and the national societies were those who sought to farm on 'scientific principles', as opposed to 'Mummerset' colloquially-cod-voiced yeoman too busy to fill in forms, distracted by getting the hay in and their Anglo-Saxon suspicions.

IV. Sheep in British Culture

As agricultural historians have demonstrated, the development and sale of breeding stock, in response to specific market demands achieved through local specialisation and management, was increasingly profitable at this time (Bowie, 1987; Bowie, 1990; Copus, 1989). In Britain the preferred breeds, grazing, methods,

13 Hon. Percy Wyndham, (1835-1911), Conservative M.P. for Cumberland West 1860-1885.

and management regionally were often highly specific, if interconnected, through commerce and breeding from the mid-nineteenth to the mid-twentieth century (Clark, 1878; Cooper, Thomas, 1965; Carlyle, 1979; Dodgshon, 1976). And a heightened scientific interest in sheep and their products persisted up to the First World War at least (Bailey and Engledow, 1914). For those who had the resources to invest within a context of continued agricultural experimentation and innovation, as outlined by Rebecca Woods, during the nineteenth century British breeders gradually formed societies and established preferred visual markers, based on each animal's lineage, expected use, via prize-giving, portraiture and herd books that listed sires, dams, and progeny. These herd books standardized the animals, and that standardisation gave them a market value. But each breed was also supposed to have developed inherent qualities and traits that suited the weather, available on-farm feed, and other features of their locality that could be put to commercial use for wool or food production. Surplus animals and ewes who had lambed multiple times were sold off farm, or even moved to be fattened for meat and other products of slaughter[14] at lower levels on the same farm, thereby linking upland and lowland production at incredibly nuanced scales. As Woods has discussed, amongst British agriculturalists in the nineteenth century the idea that each regional breed had developed within and belonged to specific landscapes and climatic conditions was taken as a given. It was as much part of each breed's story as its lineage as shaped by its human so-called improvers. As Harriet Ritvo has outlined, this idea subsequently persisted in debates about the significance of breeds that came to be designated as rare in the mid twentieth century. This supposed close association between environment and beast shaped discussions about sheep farming in relation to the economy, nation and even nationhood and 'race'. While there was experimentation with crossing in the Victorian period, as Ritvo has shown, and a mature export trade for breeding stock and attempts to develop breeds suited to all parts of the British Empire, as outlined by Woods, there was also debate about the potential impacts of other climates on breeds exported overseas, and whether (given those impacts) those bred outside Britain could still be included in the breed's flock book (Woods 2017: 2, 5, 52-77; H. Ritvo, 1997: 100, 114, 115; Ritvo, 2010; Rodriguez-Ledesma, et al., 2011: 79).

The national landscape of British livestock husbandry was thus interconnected culturally and in practice across what was dubbed more broadly 'the land' (traditionally capitalised 'the Land'[15]). Taken for granted by the Victorians, we can better see and understand this key component of the thinking behind much of their agricultural commentary, and determination to represent this within large-scale datasets such as the Annual Returns, via the exploration of this concept

14 The British government kept detailed records via the Board of Trade on its tallow imports and exports which have been discussed by Clifford (2019) re soap making.

15 For typical varied uses of 'land' and 'the land' see also H. Rider Haggard (1902), *Rural England*, https://archive.org/details/ruralenglandbeino2hagguoft/mode/2up?ref=ol&view=theater (last accessed 25 April 2023).

as it came up for discussion at the turn of the nineteenth and twentieth centuries. The idea of the Land had a lasting influence, even among agricultural scientists such as R. G. Stapledon, Professor of Agricultural Botany and Director of the Welsh Plant Breeding Station, who took up its defence repeatedly from an environmental point of view whenever British agricultural policy came up for scrutiny (Stapledon, 1936; Stapledon, 1942). The idea of 'the Land' for British agricultural commentators included the geographical bounds/acreage of the UK, the available resources (woodland, cultivated land, grazing etc.), the soil and underlying geology, but also something inherently of the nation and its history that lay beyond the 'human' world of individual landowners and farmers, though connected to the legacy of their work with it. As Henry Rew put it in 1913:

> The Land remains, [the cultivation of the soil goes on]...beyond the interests of individuals, above even the interests of the present generation, is the interest of the Land itself. ... the land must be fairly dealt by, and the maintenance of its fertility should, in the national interest, be the paramount consideration. (Rew, 1913: v-vi)

Though the place of sheep in relation to 'the Land' was discussed in terms of their appearance, breed and region, they were also seen as key to supporting its fertility (the land – in lowercase) through their behaviour and their work with it in a very direct way. The conception of the productive value of the national flock was therefore also, at least in part, based on that flock's concrete, physiological relationship with soil and plants. Sheep were in other words part of the *system* of agricultural production and could be assessed beyond the quantification of their bodies as simple outputs (for breeding stock, or wool, meat, horn or tallow etc.). Many lowland sheep, for instance, were fattened in temporary folds on arable fields after the main crop had been cut and carted, or on root crops such as swedes or turnips that were grown to be 'sheeped'. In an 1839 article by Phillip Pusey 'On the Present State of the Science of Agriculture in England' for the *Journal of the Royal Agricultural Society of England* sheep were credited as having improved light, sandy, arable land, bringing it slowly during one lifetime into a 'high condition' (so that it would yield ten-times more wheat) 'by the droppings and the tread of the sheep which are fed on the turnips that now grow in garden-like order' (Pusey, 1839: 12). The sheep worked the fields to be used for arable crops by destroying the weeds by their 'tread', and by manuring the land, and through their work co-produced it, a practice that had been in place since the eighteenth century in many areas. Sheep were especially well suited to this work because their hooves could pierce the ground repeatedly, cutting and crushing weeds' runners, and in wet weather puddled the soil which exposed their roots. Though many farmers fertilised their land using artificials from the 1820s, this practice of using arable flocks to sustain the land was sufficiently effective on certain soils that it could still be observed in operation over a hundred years later, for example in Norfolk in the late 1940s, despite the changes in method and technology of the intervening period. A century after Pusey was writing, agricultural scientists assessed the physical effect of sheep folding on the soil in the 1930s, and G. H Bates, Principal of the Staffordshire Farm Institute, calculated that an average 100lb folded arable sheep working an area of 12sq. inches, acted with 8 1/3lb. pressure per square inch simply standing and twice as much when walking. Something that he believed even a roller could not replicate (Rayns, 1948: 93; Hall, 1909: 274; Keen and Cashen: 1932; Bates, 1948: 54-55).[16]

Though newly assessed (mechanically) in the 1930s-40s, over the course of the previous hundred years this knowledge had been disseminated through the national and local agricultural societies' meetings and publications, and written up in qualitative accounts of livestock husbandry, such as *Sheep: Their Breeds, Management And Diseases*, by William Youatt (1776-1847). Whereas endemic (end even exotic) livestock diseases had been taken for granted or even occasionally exploited[17] by breeders, graziers and farmers in the early nineteenth century, Youatt sought to establish professional schooling and qualifications for livestock veterinarians in the UK and set up a veterinary committee while involved with the Royal Agricultural Society of England (RASE) established in 1838 (Clarke, Ernest, and Hall, 2004). As demand increased for meat, over the next thirty years the risks to livestock and public health of a range of diseases brought in by imported livestock became of increasing concern at a national level (Medical Officer of Privy Council, 1862). But, more significantly, the books people like Youatt wrote, and the articles about sheep husbandry published in the journals of the agricultural societies, became part and parcel of the advice literature of British agriculture, as essential to farmers as the market data published in *The Mark Lane Express*. This literature, which today forms a core primary source for British agricultural historians, consistently valued sheep as an expression, through their relationship with 'the Land', of the nation's past, and productively, through their relationship with 'the land', for their ability to fatten on and fertilise light arable land, grow well on sparse uplands, and grow wool and/or meat (normally quantified as live weight) (William Youtt, 1837: 173, 234; Wrightson, London, 1898: 45). This was the rhetorical territory that the annual Agricultural Returns and other data sets entered when debated, promoted, utilised or criticised.

If we look at the wider context of discussions about livestock *c.* 1863-64, then it also becomes clear that the demand for a national UK census for livestock was partially a product of an immediate need to assess the extent of animal deaths from disease, nationally. More data was available on numbers of imported animals than 'home grown' ones, so that there was considerable difficulty in compiling statistical data to assess the impact of livestock disease at the time (beyond insur-

---

16 A. D. Hall, (1909), *Fertilisers and Manures*, https://archive.org/details/fertilisersmanuroohallrich/mode/2up?ref=ol&view=theater (last accessed 25 April 2023), a pivotal early text on soil science read widely by agriculturalists in Britain, sets out the soil chemistry of folding.

17 The use of sheep-rot to falsely fatten sheep just before sale at market was a practice sufficiently well-known to merit literary attention. See West (2016).

ance data).[18] The increased demand for an annual agricultural census was also shaped (because statistics were an emergent agricultural technology supporting the growth of commercial practice) by the formation of empirical knowledge that was supposed to be of direct practical value in animal husbandry. Yet, in the sorts of political discourse that we see in the House of Commons, when Caird won his Resolution, the value of counting agricultural products lay more clearly in the demonstration at a national level that British agriculture, and 'the Land', was manifestly productive.

The annual census/Agricultural Returns was therefore more useful in the end for comparing head of livestock internationally, as a form of nationally conspicuous production, than as a reliable tool to generate useful data on the ground. That rhetorical utility was part and parcel of the political debate that led to the establishment of the Annual Returns and had a reach into the mid-late twentieth century. But, because data sets are political rhetorically, and the head of livestock they capture as much societal or public as private goods, they are evidence of much more than numerical continuity or change. Sheep producers, like other agriculturalists, were encouraged in the advisory literature and farming periodicals to adopt data-rich methods of agricultural management harnessed to this emergent system via farm records, accounts, maps (of their farms) and herd books. In the advisory literature and periodicals by the late-nineteenth century the imaginary of data had become paramount. Everything implicated in British agricultural production, it would appear, ought to be counted and tabulated. There were extensive data-rich reviews of British farming in periodicals such as *The Agricultural Gazette*[19] graphed findings of the depth to which rain percolated and the temperature of the soil varied by season at different depths at Rothamsted Experimental Station, disseminated in textbooks such as A. D. Hall's *The Soil* (1903) and articles with tabulated production data that investigated agricultural practices overseas in the *Journal of the Royal Agricultural Society of England*.[20] But, this data grounded in empirical observation, then interpreted and re-presented, was a cultural product entangled in politics, ideology and policy, and in the imaginary of national identity.

18 As Caird observed, the insurance data in disease skewed the statistics as farmers were more likely to take insurance out where a disease was likely to have an impact. Index to the report from the Select Committee on Cattle Diseases Prevention, and Cattle, &c. Importation Bills. BPP 431 431-I, VII.1, 235, 2, 24.

19 I have discussed this elsewhere, see Sayer (2019: 911-40).

20 The Royal Agricultural Society of England published reports on production overseas, especially of its nearest competitors, in every edition, e.g. J. Wilson, (1868), 'Danish dairy Farming', *Journal of the Royal Agricultural Society of England*, (2nd series) Vol. 4: 312-20.

V. Conclusions

This chapter has explored the relationship of twentieth and nineteenth-century data, husbandry, and politics, via a focus on British sheep production. British sheep were captured as figures in this landscape, as their bodies worked the land, so they seemed to generate vast quantities of numerical data for the statistical gaze. When dealing with statistical data as historians we are tasked to think about who was involved in its production, the ideological influences, political parties, or regimes implicated, the questions that were asked in its generation, how it was then used to form policy and it impact. What work was being done by the data and for whom? In this respect it is striking that a report by Rew from 1924 shows (in a question raised by Bowley, an equally authoritative statistician), the statistical parameters and tools utilised in the analysis of British agricultural data had gained traction on the international stage, with many statisticians' interests aligning with organisations such as the International Institute of Statistics, with a focus firmly on the data as a tool in the shaping of the biological (Rew, 1926: 83; Porter, 1986: 316-18; Parolini, 2015; Berry, 2015).[21]

Designed in part to count live animals (akin to the decennial human census), to trace the impact of a disease, the Annual Returns that have formed the focus of this chapter were not useful to the economists at the time who wanted to know about dead weight at slaughter. We can clean the data and use it in economic and agricultural histories despite that, but what we can also see is that they can equally be read as an official response to and shaped by imported microorganisms that moved around the country from market to market, farm to farm, as well as sheep, or part of the armoury of political rhetoric and national imaginary. Though we should therefore still trust and use quantitative data, it should also be recognised that it is a cultural artefact – i.e. its ideological and cultural formation may be read: part of what it yields is qualitative evidence.

All farming publications during the nineteenth century offered their readers market reports on prices, records of weather, market information about overseas competitors etc., but it seems that at the point that the Agricultural Returns were being debated and began, not all farmers were engaged in 'practice through science', and not all of them wanted to share their data. This was the rub when it came to establishing the Returns, but the debate also added to the formation of two identities in the national mind: the scientific and the non-scientific farmer, i.e. the professional, prepared to tackle the emerging mechanisms of farm management, including forms and tables, to get a nationally-competitive edge, as opposed to the self-interested local farmer, who merely farmed according to established

21 At this time data generated in bulk by experiments at agricultural research stations was also being used to create new forms of statistical methodology. R. A. Fisher Professor of eugenics at University College London from 1933), Statistician at the Rothamsted Experimental Station from 1919, for example, collaborated in the design of experiments of heritability with agriculturalists there and used the results to tackle variance.

prejudice and vernacular knowledge. If we use the Annual Returns as cultural artefacts, as products of politics, ideology and debate, traces of microbiological pasts and animal labour, then they generate new forms of evidence for us as historians (other than the statistical) and new findings. In this instance, we see that the Victorian non-farming public had a variety of farmers in mind within their political discourse. And they understood not all of them had signed up to the project of 'progress'.

Bibliography

Agyeman, J. and Spooner, R., (1997), 'Ethnicity and the Rural Environment', in Cloke, P., and Little, J., (eds) *Contested Countryside Cultures: otherness, marginalisation and rurality*, (London: Routledge)

Anon, (1834), *The Farmer's Magazine*, Vol. 1, May, p. 420

———, (1855), 'Agricultural Statistics (Scotland), Report of the Highland Society to the Board of Trade on the Agricultural Statistics of Small Holdings in *Scotland*, for 1854, 26 June

———, (1862), Medical Officer of Privy Council: Fifth Report, Public health. Fifth report of the Medical Officer of the Privy Council. With appendix Vol. XXV.1, BPP 161, 277-80 https://parlipapers.proquest.com/parlipapers/docview/t70.d75.1863-039165?accountid=13651

———, (1864), 'Punch's Essence of Parliament', *Punch* Saturday, June 18: 249

———, (1865), 'Mr Hardy's Descriptive Catalogue', *The Times*, 3rd Nov: 8

———, (1867), 'England is Fertile in Paradoxes, and Not the Least' Leader, *The Times*, Tues, 18th Jun: 11

———, (1867), 'West Cumberland Agricultural Society', *The Times*, Tues, 8th Oct: 5

———, (1867), *The Times*, 18th Jun:11

———, (1869), 'Agricultural Returns: "Put that in your Pipe and Smoke it"', *Punch*, Sat. 17th April: 156

———, 'Index to the report from the Select Committee on Cattle Diseases Prevention, and Cattle, &c. Importation Bills'. BPP 431 431-I, VII.1, 235

Anon., 'statistics, n.'. OED Online. June 2021. Oxford University Press. https://www.oed.com/view/Entry/189322?rskey=v85Ukd&result=2 (accessed 25 April 2023)

Bailey, P.G., and Engledow, F.L., (1914), '"Quality" in Wool', *The Journal of Agricultural Science*, 6(3): 349-70. doi:10.1017/S0021859600001891

Barrell, J., (1983), *The Dark Side of the Landscape*, (Cambridge: CUP)

Bates, G. H., (1948), *Weed Control*, (London: Farmer & stockbreeder Ltd., E. & F. N. Spon Ltd)

Bermingham, A., (1987), *Landscape and Ideology: The English Rustic Tradition, 1740-1860*, (London: Thames and Hudson)

Berry, D., (2015), 'Agricultural Modernity as a Product of the Great War: The Founding of the Official Seed Testing Station for England and Wales, 1917-1921', *War & Society*, 34:2: 121-39, DOI: 10.1179/0729247314Z.00000000051

Bowie, G. G. S., (1987), 'New Sheep For Old – Changes In Sheep Farming in Hampshire, 1792-1879', *Agricultural History Review*, 35.1: 15-24

———, (1990), 'Northern Wolds and Wessex Downlands: Contrasts in Sheep Husbandry and Farming Practice, 1770-1850', *Agricultural History Review*, 38.2: 117-26

Brassley, P., Harvey, D., Lobley, M., and Winter, M., (2013), 'Accounting for Agriculture: The Origins of the Farm Management Survey', *Agricultural History Review* 61, no. 1, May: 135-53

Brettell, R. and C. R., (1983), *Painters and Peasants in the nineteenth century* (Geneva: Albert Skira)

Burchardt, J., (2002), *The Allotment Movement in England 1793-1873* (London: Royal Historical Society)

Burrell, A. , (1990), *Agrifacts: A Handbook of U.K. and EEC Agriculture and Food Statistics* (London: Pearson Education)

Caird, J. Mr., MP, (1864) Agricultural Statistics, Resolution, House of Commons *Hansard* third series, Vol. 175, 7th June, Col. 1362-1383 https://parlipapers.proquest.com/parlipapers/docview/t71.d76.cds3v0175p0-0020?accountid=13651 (last accessed 25 April 2023)

———, (1868), 'On the Agricultural Statistics of the United Kingdom', *Journal of the Statistical Society of London* 31 (2): 127-45. doi:10.2307/2338858

———, (1880), 'The Inaugural Address of James Caird, Esq., C. B., F.R.S., President of the Statistical Society, Delivered on Tuesday the 16th of November, 1880', *Journal of the Statistical Society of London* 43 (4): 559-72. doi:10.2307/2339025

Carlyle, W. J., (1979), 'The Changing Distribution of Breeds of Sheep in Scotland, 1795-1965', *Agricultural History Review*, 27.1: 19-29

Chatterjee, A., (2020), *Conflict and Identity in the Social Life of Electricity in Colonial Calcutta, c. 1880-1925*. Unpublished PhD thesis, Leeds Trinity University

Clark, J. A., (1878), 'Practical Agriculture', *Journal of the Royal Agricultural Society of England*, Series 2, Vol. 14: 582-620

Clarke, E., and Hall, S. A., (2004), 'Youatt, William (1776-1847), Veterinary Surgeon', *Oxford Dictionary of National Biography*. September 23, (Oxford University Press)

Cooper, M. Mcg., Thomas, R. J., (1965), *Profitable Sheep Farming*, (Ipswich: Ipswich Farming Press Books)

Coppock, J. T., (1956a), 'The Statistical Assessment of British Agriculture', part I, *Agricultural History Review*, 4.1, 4-21

———, (1956b), 'The Statistical Assessment of British Agriculture', part II, *Agricultural History Review*, 4.2, 66-79

Copus, Andrew K., (1989), 'Changing Markets And The Development of Sheep Breeds in Southern England, 1750-1900', *Agricultural History Review*, 37.1: 36-51

Cragoe, M., and Mcdonagh, B., (2013), 'Parliamentary Enclosure, Vermin and the Cultural Life of English Parishes, 1750-1850', *Continuity and Change*, vol. 28, no. 1: 27-50, doi:10.1017/S0268416013000118

Daniels, S., (1983), *Fields of Vision, Landscape Imagery and National Identity in England and the United States* (Cambridge, 1994); R. Sales, *English Literature in History 1750-1830, Pastoral and Politics*, (London: Hutchinson)

Debate, 'Post-War Agricultural Policy', Commons Sitting of Tuesday, 3rd August 1943, *Hansard*, 1942-42, Fifth Series, Vol. 391, page column 2057-2266, p. 2080

Desrosières, A., (2013), 'The History of Statistics as a Genre: Styles of Writing and Social Uses', *Bulletin de Methodologie Sociologique*, 119: 8-23

Dodgshon, R. A., (1976), 'The Economics of Sheep Farming in the Southern Uplands during the Age of Improvement, 1750-1833', *The Economic History Review*, 29.4: 551-69 https://doi.org/10.2307/2595343

Fletcher, T. W., (1961) 'The Great Depression of English Agriculture 1873-1896', *The Economic History Review* 13 (3): 417-32. doi:10.2307/2599512

———, 'Review', *The Economic History Review* 22, no. 3 (1969): 574-76

Ford, C., (2001), 'Landscape and Environment in French Historical and Geographical Thought: New Directions' Review Essay, *French Historical Studies*, 24/1 (Winter): 125-34

Freeman, M., (2003) *Social Investigation and Rural England 1870-1914*, (Woodbridge: Royal Historical Society)

Fudge, E., (2013), 'Milking Other Men's Beasts', *History and Theory*, no. 4: 13-28

Fussell, G. E., (1945), 'English Agriculture from Cobbett to Caird (1830-1880)', *Economic History Review*, Vol. 15, No. 1-2: 79-85

Goddard, N., (1981), 'Agricultural Societies' in G. E. Mingay (ed) *Victorian Countryside* Vol. I (London: Routledge)

———, (1983), 'The Development and Influence of Agricultural Periodicals and Newspapers, 1780-1880', *The Agricultural History Review* 31: 116-31

———, (1988), *Harvests of Change: The Royal Agricultural Society of England 1838-1988* (London: Quiller Press)

Haggard H. R., (1902), *Rural England*, https://archive.org/stream/ruralenglandbeino2hagguoft?ref=ol (last accessed 25 April 2023)

Hall, A. D., (1909), *Fertilisers and Manures*, https://archive.org/stream/fertilisersmanuroohallrich?ref=ol (last accessed 25 April 2023)

———, (1915), *The Soil: An Introduction to the Scientific Study of the Growth of Crops*, (London: John Murray; First edition, 1903) https://archive.org/stream/soilintroductionoohallrich?ref=ol (last accessed 25 April 2023)

Hallas, C., (ed.) (1999), *Rural Responses to Industrialization: The North Yorkshire Pennines 1790-1914* (Frankfurt, New York, Paris, Wien: Peter Lang, Bern, Berlin)

Haraway, D., (1992), 'Teddy Bear Patriarchy: Taxidermy in the Garden of Eden, New York City, 1908-1936', reproduced in Haraway, *Reader*: 151-97

———, (2015), 'Anthropocene, Capitalocene, Plantationocene, Chthulucene: Making Kin', Environmental Humanities 1 May; 6 (1): 159-65

Henriksen, I., (2012), 'The Strange Birth of Liberal Denmark: Danish Trade Protection and the Growth of the Dairy Industry since the Mid-Nineteenth Century', *The Economic History Review*, vol. 65, no. 2: 770-88, www.jstor.org/stable/41475606. (last accessed 25 April 2023)

Higgs, E., (1996), 'The Statistical Big Bang of 1911: Ideology, Technological Innovation and the Production of Medical Statistics'. *Social History of Medicine* 9: 409-26

Howkins, A., (1991), *Reshaping Rural England, A Social History 1850-1925*, (London: Harper Collins)

———, and Verdon, N., (2008), 'Adaptable and Sustainable? Male Farm Service and the Agricultural Labour Force in Midland and Southern England, C.1850-1925', *The Economic History Review*, vol. 61, no. 2: 467-95

Hudson, K., (1975), *Patriotism with Profit: British Agricultural Societies in the Eighteenth and Nineteenth Centuries*, (New York: the British Book Centre)

Jasanoff, S., (2005), 'Civic Epistemology', in Sheila Jasanoff, *Designs on Nature: Science and Democracy in Europe and the United States* (Princeton: Princeton University Press)

———, and Sang-Hyun K., (eds) (2015), *Dreamscapes of Modernity: Sociotechnical Imaginaries and the Fabrication of Power* (Chicago: University of Chicago Press)

Kander, A., and Warde, P., (2011), 'Energy Availability from Livestock and Agricultural Productivity in Europe, 1815-1913: a New Comparison', *The Economic History Review*, vol. 64, no. 1: 1-29. JSTOR, www.jstor.org/stable/27919479 (last accessed 25 April 2023)

Keen, B., & Cashen, G., (1932), 'Studies in Soil cultivation. VI. The Physical Effect of Sheep Folding on the Soil (With Five Text-Figures.)', *The Journal of Agricultural Science*, 22(1): 126-34. doi:10.1017/S0021859600053144

Kolodny, A., (1984), *The Land Before Her: Fantasy and Experience of the American Frontiers, 1630-1860* (Chapel Hill and London: The University of North Carolina Press)

Lehning, J. R., (1995), *Peasant and French: Cultural Contact in Rural France during the Nineteenth Century* (Cambridge: Cambridge University Press)

MacDonald, S., (1979), 'The Diffusion of Knowledge Among Northumberland Farmers, 1780-1815', *Agricultural History Review*, 27.1: 30-39

Maclean, G., Landry, D., & Ward, J. P., (eds) (1999), *The Country and the City Revisited: England and the Politics of Culture, 1550-1850*, (Cambridge: Cambridge University Press)

Marks H. and Britton D. K., (1989), *A Hundred Years of British Food and Farming: A statistical survey* (London: Taylor and Francis)

Massey, D., (1994), *Space, Place and Gender* (Cambridge: Polity)

Mayerfield Bell, M., (1994), *Childerley: Nature and Morality in a Country Village*, (Chicago & London: University of Chicago Press)

Mingay, G. E., (2004), 'Caird, Sir James (1816-1892), Agriculturist And Writer', *Oxford Dictionary of National Biography. Retrieved 6 Apr. 2023, from* https://www.oxforddnb.com/view/10.1093/ref:odnb/9780198614128.001.0001/odnb-9780198614128-e-4339

Miskell, L., (2012), 'Putting On A Show: The Royal Agricultural Society Of England And The Victorian Town, C. 1840-1876', *Agricultural History Review*, 60.1: 37-59

Mitchison, R., (2004), 'Sinclair, Sir John, First Baronet (1754-1835), Agricultural Improver, Politician, and Codifier of "Useful Knowledge"', *Oxford Dictionary of National Biography*. 23 Sep. (last accessed 25 April 2023) https://www.oxforddnb.com/view/10.1093/ref:odnb/9780198614128.001.0001/odnb-9780198614128-e-25627

Parolini, G., (2014), 'The Emergence of Modern Statistics in Agricultural Science: Analysis of Variance, Experimental Design and the Reshaping of Research at Rothamsted Experimental Station, 1919-1933', Journal of the History of Biology. Printed in 2015 in Vol. 48(2): 301-3510.1007/s10739-014-9394-z

———, (2015), 'Charting the History of Agricultural Experiments', History and Philosophy of the Life Sciences, Vol. 37(3): 231-41

———, (2015), 'In Pursuit of a Science of Agriculture: The Role of Statistics in Field Experiments', History and Philosophy of the Life Sciences, Vol. 37(3): 261-81

Payne, C., (1994), *Toil and Plenty: Images of the Agricultural Landscape in England, 1780-1890* (New Haven and London: Yale Centre for British Art, Yale UP)

Pearson, C., and Nasby, J., (2008), *The Cultivated Landscape: An Exploration of Art and Agriculture*, (Montreal & Kingston, London, Ithaca: McGill-Queen's University Press)

Peters, G. H., (1988), *Agriculture, Reviews of UK Statistical Sources*, (London: Springer)

Pickles, J. D. (2004), 'Venn, John Archibald (1883-1958), Historian and College Head'. *Oxford Dictionary of National Biography* (last accessed 25 April 2023) https://www.oxforddnb.com/view/10.1093/ref:odnb/9780198614128.001.0001/odnb-9780198614128-e-40972

Porter, T. M., (1986), *The Rise of Statistical Thinking, 1820-1900*, (Princeton, New Jersey, USA: Princeton University Press, New edition, 2020)

Pusey, P., (1839), 'On the Present State of the Science of Agriculture in England', *Journal of the Royal Agricultural Society of England*, Vol. 1: 1-21 https://www.biodiversitylibrary.org/item/163263#page/28/mode/1up (last accessed 25 April 2023)

Rayns, F., (1948), 'The Manuring of Arable Crops', in H. C. Long et al., *Profit from Fertilisers* (London: Crosby, Lockwood and Son ltd., 3rd edn)

Rew, H., (1913), *An Agricultural Faggot: A Collection of Papers on Agricultural Subjects*, (Westminster: P. S. King & Son)

———, (1926), 'The International Statistical Institute and Its Sixteenth Session', *Journal of the Royal Statistical Society*, vol. 89, no. 1: 70-95

Rew, R. H., (1902), 'Memorandum on Some Estimates Made by Various Authorities of the Production of Meat and Milk', *Journal of the Royal Statistical Society*, vol. 65, no. 4: 666-79. JSTOR, www.jstor.org/stable/2339699 (last accessed 25 April 2023)

Ritvo, H., (1997), *The Platypus and the Mermaid: And Other Figments of the Classifying Imagination*, (Cambridge, Massachusetts and London: Harvard university Press)

———, (2010), 'Counting Sheep in the English Lake District: Rare Breeds, Local Knowledge and Environmental History', *Nobel Cows and Hybrid Zebras: essays on animals in history*, (Charlottesville and London: University of Virginia Press)

Rodriguez-Ledesma, A., A. Waterhouse, C. Morgan-Davies and F. Bruelisauer, (2011), 'Structural Assessment of the Scottish Stratified Sheep Production System', *Small Ruminant Research* 100, no. 2-3 (October 1): 79-88. doi:10.1016/j.smallrumres.2011.06.003

Rostow, W. W., (1941), 'Business Cycles, Harvests, and Politics: 1790-1850', *The Journal of Economic History*, vol. 1, no. 2: 206-21. JSTOR, www.jstor.org/stable/2113474

Sayer, K., (2019), 'The Changing Landscape of "Labour": Work and Livestock in Post-Second World War British Agriculture' *History* Dec, Vol. 104, Issue 363: 911-40 http://dx.doi.org/10.1111/1468-229X.12920

Stapledon, Sir R. G., (1936), *The Land Now and Tomorrow*, (London: Faber & Faber)

———, (1942), *The Way of the Land* (London: Faber & Faber)

Turner, M. E., J. V. Beckett and B. Afton, (1996) 'Taking Stock: Farmers, Farm Records, and Agricultural Output In England, 1700-1850'. *Agricultural History Review*, 44: 21-34

Venn J. A., (1923), *Foundations of Agricultural Economics* (Cambridge: Cambridge University Press)

West, A., (2016), ''Rot the Genuine': Moral Responsibility and Far from the Madding Crowd's Cancelled Fragment', *Journal of Victorian Culture, Vol. 21, Issue 3, 1* September: 387-404, https://doi.org/10.1080/13555502.2016.1188409

Williams, R., (1985), *The Country and the City*, (London: the Hogarth Press)

Wilson, J., (1868), 'Danish dairy Farming', *Journal of the Royal Agricultural Society of England*, (2nd series) Vol. 4: 312-20

Woods, R. J. H., (2017), *The Herds Shot Round the World: Native Breeds and the British Empire, 1800-1900* (Chapel Hill: University of North Caroline Press)

Wrightson, J. (1898), *Sheep, Breeds and Management* (Vinton, London) https://archive.org/stream/sheepbreedsmanag00wrigrich?ref=ol (last accessed 25 April 2023)

Yang, C. (2018), *Sociotechnical Imaginary and Rationality: Political Factuality and Public Authority in Taiwanese Energy Politics* Unpublished PhD, Lancaster University. https://Doi.Org/10.17635/Lancaster/Thesis/335

Youtt, W., (1837), *Sheep: Their Breeds, Management and Diseases, to Which Is Appended the Mountain Shepherd's Manual* (London: Library of Useful Knowledge, Robert Baldwin)

Zetterstrom-Sharp, J. (2020) 'Imagining Future Agricultural Landscapes in a New Sudan: Entitled Expertise, Cultural Intransience and Fine Warm Rain in the English Wilds', *History & Anthropology* 31 (3): 293-313. doi:10.1080/02757206.2019.1669591

CARIN MARTIIN

3. Shifting needs for feed in Swedish dairy farming from the late 19th to the late 20th century*

I. One hundred years of dairy farming with varying emphasis

The initial idea behind this chapter was to compare needs for feed in different kinds of Swedish dairy farming from the late 19th century and a century ahead. What did the oldest form, with poor feeding, low yields, and long-lived but meagre cattle look like in terms of total feed and milk compared with well-fed high yielding but short-lived cattle in the late 20th century? And how about more modest forms in between? The final version has been given a wider content that also includes the mutual interplay between dairy farming and Swedish society, reflected in consumer demand, and political aims and regulations. The point of departure is taken in average milk yield per dairy cow from 1871 to 1995, Figure 3.1. The solid line shows the yearly average yield of all dairy cows in Sweden, and the dotted line is the yearly average of dairy cows in so called official milk control, described later. The incomplete lines reflect a lack of statistics. The diagram serves as the basis for the identification of the main forms of dairy farming in this study. In this way the chapter is a kind of methodological study, too, beginning with the performance of the dairy cows, and linking this to wider perspectives, instead of vice versa.

* The author is most grateful for financing by the Axel and Margaret Ax:son Johnsons Foundation.

Carin Martiin • Swedish University of Agricultural Sciences, SLU, Uppsala, Sweden
carin.martiin@slu.se; martiincarin@gmail.com

From Breeding & Feeding to Medicalization, Rural History in Europe, 17 (Turnhout: Brepols, 2023), pp. 53–76
BREPOLS PUBLISHERS 10.1484/M.RURHE-EB.5.135021

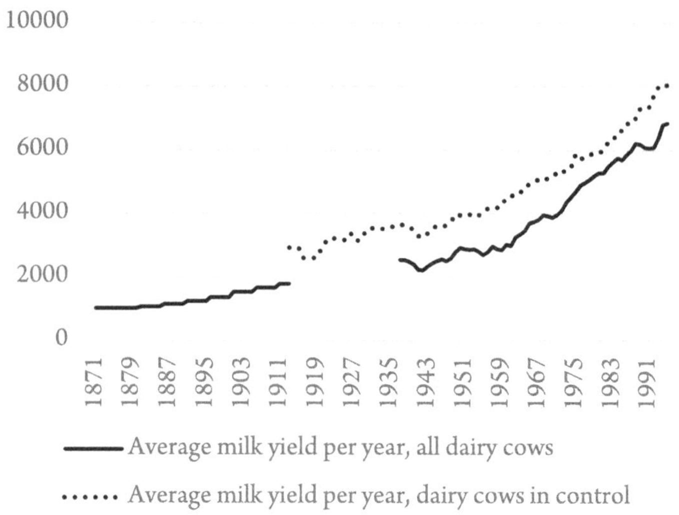

Figure 3.1. Milk per dairy cow and year, 1871-1995, kilos
Sources: 1871-1913: SOU 1930:4: 2; 1914-1995: Statistical Yearbook of Sweden, tables 'Livestock', 'Livestock production', and 'Milk recording' (varying formulations of the tables over time).[1]

The studied time period includes both 'traditional' dairy farming, with far distant roots, and the more recent, with more commercial forms of production. The beginning in 1871 is due to the earliest found officially published figure about total Swedish milk production, while the end in 1995 is motivated by Sweden's membership in the European Union, which brought about too many changes to be included in this chapter.

Three main yield levels can be distinguished in Figure 1 indicating an end of form A around 1913, form B from the 1920s to the late 1950s, and C from 1960 onwards. The direction from low to high yields shall not automatically be understood as going from bad to better, the performance has to be related to the possibilities and expectations of the time. Moreover, it could at least be considered as requiring the management of low producing dairy cattle under poor conditions in the late 19th century versus the comparably high producing dairy herd under good material conditions a century later.

1 'SOU' is short for Swedish Government Official Reports. SOU investigations have been published since 1922, and deal with the most different issues of political interest. The investigations were made on inquiry of the primarily concerned ministry, could take several years, and were often carried out by leading names in the field. The published SOU was then circulated for comments by concerned authorities and organisations, and there after revised into a Government Bill for final decision by the Swedish *Riksdag*. In this study, SOU reports were used for their extensive reasoning, that reflects the current issues and puts them into a broader contemporary context.

I.1. Swedish dairy farming in brief

Generally speaking, the first half of the studied period coincides with large-scale changes in Swedish society such as industrialization, urbanization, and improved infrastructures.

In 1927 as many as 98% of all farm holdings had dairy cows, and 93% in 1949 (SOS *Jordbruk och boskapsskötsel*, 1949:35). Until the mid-twentieth century being a farmer almost always meant being a dairy farmer, in Sweden as well as in neighbouring Finland (Peltonen, 2016: 233-47). This was especially the case during the decades when the B form dominated Swedish dairy farming. Most farms were small. In 1944, about 25 per cent of the dairy cows were kept at holdings with less than 5 hectares of arable land, but only 6 per cent at farms with more than 100 hectares of arable land (Statistical Yearbook of Sweden). The number of dairy cows peaked at two million heads in 1930, that would compare to a total human population of 6.1 million (Statistical Yearbook of Sweden). In comparison, Britain had 2.4 million dairy cows in 1955 (Woods, 2014: 298). Milk was a matter of year-round production in Sweden, and was not subject to the seasonal to year-round production changes that took place in Denmark in the late 19th century (Henriksen & O'Rourke, 2005: 520-54). By the mid-century approximately one and a half million people in the Swedish countryside lived close to dairy cattle and the daily work with milking and feeding that was practiced within the four walls of a cowshed in the wintertime, and almost everywhere in the landscape in the summertime. Dairy farming was an everyday reality for a myriad of people and dairy cattle who turned labour, feed and other inputs into fresh milk. Furthermore, dairy farming played important roles in Sweden's transition from a mainly rural to mainly urban society. The regular sale of milk gave jobs and income to the countryside, and provided food to the growing urban population.

Going back to the 19th century and earlier times, dairy farming was often subordinated to arable farming, and was to a great deal based on meagre feed from meadows and forests. This began to change in the late 19th century driven by successful butter exports, inspired by the even more successful Danish example (SOU 1930:4: 2; Historical Statistics 3: 202, Table 3.2). As a result, the production of milk gained increased political and economic attention. Among other things public dairy consultants were stationed at the county agricultural societies (*Mejerihantering* [Official Dairy Statistics], *1913*:1). The opportunity to make money on milk and butter spurred proprietors at estates and other big farms to expand and upgrade their dairy herds, improve feeding, and invest in new fancy cowsheds. This contributed to the parallel existence of two different kinds of dairy farming in the late 19th and early 20th centuries: a continuation of the previously dominant form here called 'A', and the newly developed 'Early commercial' form.

Butter exports decreased during WWI and the early 1920s, after which some political efforts were made to regain Sweden's earlier position at the international butter market. However, the attention was soon turned to a growing domestic demand for liquid milk and other dairy products, which was met with some relief among the authorities who considered the domestic market comparably easier to deal with (SOU 1930:4: 10-11).

The growing Swedish demand increased farmers' trust in becoming dairy suppliers, instead of being dependent on a more volatile international market, or subsistence producers only. Moreover, a growing contemporary political interest in supporting both production and consumption of milk made the domestic market comparably safe (SOU 1938:6: 39).

From the mid-20th century, political interest in dairy farming took another direction. Like many other parts of society, the agricultural sector was put under gradual pressure to rationalize, to save labour and produce more per unit but, in the case of dairy production, to keep the same total volume as before (SOU 1946:42: e.g. page 110; SOU 1966:30, frequent references). From 1946 to 1995 as many as 94 per cent of the dairy suppliers and 74 per cent of the dairy cows disappeared, while milk yields rose and herd sizes increased (Statistical Yearbook of Sweden). The processes escalated from around 1960, which marks the beginning of the form of dairy farming that is here called C. At the same time, the previously high consumer demand began to decline.

I.2. The impact of increased domestic demand

Official dairy statistics include the sale of liquid milk from dairy plants, but does not show the total consumption, which also included substantial direct consumption, in the farm households, or via direct sale to neighbours and others in the countryside. For example, only 61 per cent of the total milk production was supplied to dairy plants in 1938 (Statistical Yearbook of Sweden 1943, Table 79 and 82).

The following reasoning is thus based on the consumption of dairy processed liquid milk, only, which is assumed to have been primarily a matter of urban demand. This increased substantially as Sweden was urbanised from the late 19th century onwards. In 1870 only 13 per cent of the Swedish population lived in towns and municipalities; 25 per cent by the turn of the century; almost 40 per cent in 1930; and nearly 60 per cent in 1955. In numbers, the urban population increased from 1.3 million in the year 1900 to 2.4 million in 1930, and 4.2 million in 1955 (Historical Statistics I: 66). At the same time the sale of liquid milk was more than tenfold, with an especially high increase during the 1930s. Figure 3.2 demonstrates the average consumption of fresh milk per person living in towns and municipalities, from 1900 to 1960. At the same time, the total demand for fresh milk from the dairies was influenced by the ongoing urbanization, and was slightly higher in 1960 than in 1950, despite the reduced consumption per person (Statistical Yearbook of Sweden).

According to Figure 3.2 the average individual liquid milk consumption in town increased from 56 to 241 kilo per person per year from 1900 to 1945. The peak by 1945 may be explained by the fact that liquid milk was one of few kinds of food that were excluded from the rationing of foods during WWII (Åmark

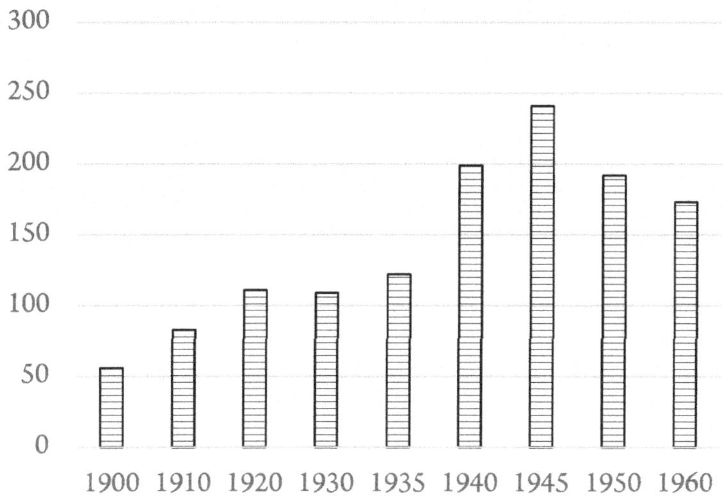

Figure 3.2. Average consumption of dairy processed liquid milk in Sweden.
Sources: Historical Statistics I, Population, p. 66; Statistical Yearbook of Sweden 1940, p. 105, Table 84; SÅ 1946, p. 105, Table 88; SÅ 1949, p. 103, Table 87 and SÅ 1956, p. 86, Table 102.

1952: 338). The figures for Sweden remind one of those for Denmark and the Netherlands regarding the period of 1900-1930, but thereafter the Swedish urban consumption rose to a level more similar to Switzerland (Hernández Adell & Pujol, 2017: 64).

What about driving forces behind the increased urban demand for liquid milk? Did the urban migrants bring rural milk consumption habits to town, or was the urban increase something rather new? The question was highlighted by Hernández Adell & Pujol who pointed at 'the leading role of cities' where consumer preferences were encouraged by new influences such as improved milk hygiene, and new knowledge about vitamins and nutritious food (Hernández Adell & Pujol, 2017: 63).

In the case of Sweden, liquid milk was far from new on the tables. With dairy cows everywhere, people in the countryside were used to almost constant access to milk. Traditionally, milk was used as food (butter and cheese, whole milk, skimmed milk, milk mixed with beer, fermented milk, cottage cheese, cakes of cottage cheese, milk for bread baking, gruel, porridge, and pancakes); as animal feed, gifts, exchange articles; and/or for sale (*Mjölkhushållning*, a questionnaire about milk economizing). Yet, traditional use of milk does not exclude new motives for consumption, like nutrients and vitamins, together with improved availability for urban dwellers, thanks to milk distribution structures and higher standards of living. However, this does not explain the increased demand for liquid milk during the second half of the 1930s. The number of rural migrants was similar throughout the 1920s and 1930s, and the number of (milk-drinking) children was comparably low (Historical Statistics I: 6, Table 14; 78, Table 19). Another, more plausible suggestion is that the increased urban demand for liquid milk was driven by combinations of new findings about vitamins and other nutritious aspects together with colourful propaganda by the producer-driven association 'The Milk Propaganda', with sister organizations in many other countries. In the 1930s the association was supported by public grants (Martiin, 2010: 213-32). This reflected a positive public attitude on milk production and consumption as a win-win concept, for the rural as well as urban populations (SOU 1938:6: 39).

Finally, pasteurization, or lack of it, must have been of great importance, even though Sweden was late to apply this process. Some dairy plants pasteurized the milk voluntarily, but a parliamentary decision on obligatory pasteurization was not taken until 1937, and in force in 1939 (SOU 1936:3). All in all, the substantially increased demand in parallel with the urbanization, meant that almost all farmers in Sweden became dairy suppliers, instead of producing milk only for subsistence use. Alternatively formulated, the increased demand was of greatest importance for the transition from the primarily subsistence-oriented A to the more commercial but still modest form B.

II. Main kinds of dairy farming in the late 19th to late 20th century Sweden

The identification of different kinds of dairy farming on the basis of Figure 1, include interpretation of two kinds of average yields. The solid line is calculated as total production divided by the total number of dairy cows. According to the solid line milk yields increased modestly; 31 per cent, from 1927 to 1959, typically the form B, to compare with period C, when the average yield rose 112 per cent from 1960 to 1991, and as much as 140 per cent to 1995 (Statistical Yearbook of Sweden). In comparison, the Swedish average for all dairy cows was 4,700 kilos in 1975, when Dutch dairy cows yielded 4,900 kilos (Bieleman, 2005: 232), whereas British dairy cows produced 3,414 kilos on average in 1975/76 (Woods, 2014: 298).

The dotted line shows average yields for dairy cows connected to official milk control, which was substantially higher compared to the national average. The first Swedish milk control association was founded in Scania in 1898, a small local association inspired by Danish colleagues (Roos, 2004: 221). Similar associations were gradually founded all over the country. The number changed somewhat over time but was, for example, 726 local associations in 1927, and 1087 in 1950/51 (Statistical Yearbook of Sweden). As an example, in Jönköping County, the first local association was founded around 1903. Farmers paid a fee, but some costs were paid by the country agricultural society, plus 0:25 SEK state support per dairy cow, a grant that was approved by the Swedish *Riksdag* in 1905 (Nyberg,

1917: 296-97). Regarding practical control, each associated dairy herd was visited by a local milk control assistant, once a month or every second week. During his visit, the assistant measured the yield from each dairy cow in the evening and morning, followed up the feeding, noted calving dates etcetera, and supervised the farmer, and or the employees, concerning further improvements (Nyberg, 1917: 554-55).

The importance of these milk control associations should not be over-, or under-estimated. Still in 1921/25 only 10 per cent of all dairy cows in Sweden were associated, primarily herds larger than the average. In 1927 the average herd size in official control was 20 dairy cows, but the Swedish average was around 5 dairy cows per herd (Statistical Yearbook 1930, Table 87 and 89). To attract more farmers, less detailed forms of control were introduced in the 1930s, but the interest was still modest until the emergence of the C form of dairy farming, from 1960 onwards (Brandesten, 2005: 263). In 1960 27 per cent of all Swedish dairy cows were associated, and in 1995/96 it was as much as 82 per cent (Statistical Yearbook of Sweden).

In spite of the limited interest, the official milk control had a substantial impact; directly on the dairy economy at the individual member farm, and indirectly in that the yield figures were official and generated a kind of competition in milk yields. According to the instructions for Jönköping County in 1907 the (male) assistant should bring a book with figures from all members, to show all members (Nyberg, 1917: 556). To join the control was to expose the cowshed and its pros and cons, to neighbours as well as professionals in dairy cattle husbandry. The dotted graph in Figure 3.1 can obviously be seen as a kind of 'frontier'. To limit the study to dairy cows under control, only, would be to exclude substantial numbers of dairy herds, in particular among lower socioeconomic classes. The level in official control in the late 1930s for example, was not reached at average level until the mid-1960s (figure 3.1).

II.1. Scarcity of material resources, typical for the kind of dairy farming called A

The A form of dairy farming appears to have been generally applied in Sweden at least until the mid-19th century, but in the late 19th and early 20th century it was primarily related to the lower rural classes, or to remote holdings in roadless mountainous areas up in the north (Martiin, 2017: 24). In the summers of the late 19h century small herds of the rural poor were kept grazing roadsides, in ditches and similar marginal land resources, and in winter time the cattle were fed with hay, straw, chaff, leaves and other matters from meadows and forest land, but very little from arable fields (Martiin, 2005: 171-95, 276-77 summarised in English). This kind of low feeding – low yield dairy farming can be derived from far back in time, with indications of roots in the Middle Ages (Myrdal, 2013). It was also found in a feeding plan from 1832, analysed by the author (Martiin in Cserhalmi,

2003: 60-68), and in the study of an estate, where the milk yields were half as high in the 1860s as in the 1890s (Martiin, 2005: 135 and 136).

Accordingly, it is not surprising to find strikingly low milk yields to the left in Figure 1, ranging from 1,020 to 1,785 kilo per year, or 3 to 5 kilo per day (SOU 1930:4: 2). Still, the yields may have been a little too high in relation to the low content of energy and protein in the feed. Most probably many cattle were thus permanently or periodically underfed. An undernourished dairy cow could have difficulties to get pregnant, and even eating, e.g. due to ketosis (related to increased amounts of ketone bodies in the blood), caused by problems the animal's energy balance (Plym Forsell 1997:163-64). Keeping dairy cows under such conditions may have been at least as challenging as under better material conditions, such as the at about the same time practiced 'early commercial' form.

II.2. Early commercial dairy farming, in parallel with A

The butter export that began in the 1860s generated an early commercial form of dairy farming, operated under far better conditions such as generous feeding with oil cakes, new spacious cowsheds, and improved breeding of the dairy cattle (Martiin, 2005: 65, 154, 248, and 176). The solid and dotted lines by 1913 in Figure 1 illustrate the different worlds of dairy farming by that time, on average 1,785 kilos per dairy cow and year, but 2,918 kilos under the watchful eye of the milk control assistant, who primarily visited this kind of farms. Sharp & Lampe have presented corresponding figures for Danish estates from 1880 to 1900, ranging from 2 – 2.5 tonnes per dairy cow per year (Lampe & Sharp, 2014: 5, 9).

In many Swedish parishes the two contrasting forms appeared within sight of each other. It was no secret that the feeding table in the estate's cowshed was filled with nutritious grains, and oil cakes of peanut, sesame, sunflower, coconut or cottonseed from around the world, tastier and more nutritious than the food on ordinary peoples' tables. The strong commercial orientation, high investments, and amounts of purchased feeds made the early commercial form have more in common with the later practiced form C than with the more modest B.

II.3. Increased sale to dairy plants, characteristic for the B form of dairy farming

Over time, the socioeconomic inequalities declined in Swedish society, and so did the differences in dairy farming that was gradually transformed into the B form, generally related to the late 1920s to 1960. A most important driver was the increased connection to dairies. The process was late in Sweden compared to Denmark where 'By 1900, most of the milk in Denmark was processed by cooperative creameries' (Sharp & Lampe, 2014: 18). Despite increased sales, the levels of production were comparably modest, per dairy cow and supplier. This made room for many suppliers, and distributed the benefits from the rising urban

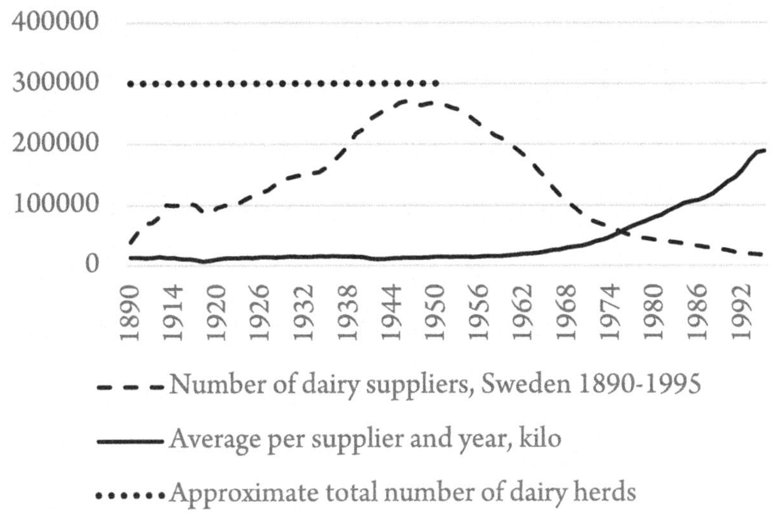

Figure 3.3. Herds, suppliers and average deliveries, Sweden from 1890-1995. Sources: Statistical Yearbook of Sweden, 1914-1997.

demand. A large number of farm households across the country enjoyed a regular income, e.g. every second week, and the accompanying potential to develop the production, and some private consumption.

According to the dashed line in figure 3.3 the number of dairy suppliers grew from 125,000 in 1927 to a peak at 271,000 in 1946 (Statistical Yearbook of Sweden). The increase was most probably due to several coinciding drivers, improved access to dairy plants, and increased share of cooperative dairy plants, from 18% in 1895 to 37% in 1913/15, to 49% in 1931/35 and 83% in 1948 (Statistical Yearbook of Sweden). The cooperatives were small and local, and potential new members appear to have been urgently encouraged to join. Taken together, farmers' confidence in the dairy supply rose with the access to cooperatives, with positive public attitudes, and rising consumer demand, and with the fact that the perishable liquid milk was sheltered from foreign competition by the surrounding sea.

Moreover, the dotted line in figure 3.3 outlines the declining number of farms over time, with dairy cows for subsistence use only. According to the solid line the average supply per farm remained stable and quite modest for a long time. The average, around 13,000-14,000 kilos per year, not more than 37 kilos per day, coincides with photos of wooden tables with milk cans standing by the road side, soon on their way to the dairy. The lack of average increase over time is an unexpected finding. A possible explanation is that those who became suppliers late, may have been comparably smaller, and/or with lower yields, and that this counteracted increases among other suppliers, e.g. at the exemplified farm later on in this chapter. Moreover, the continued importance of subsistence

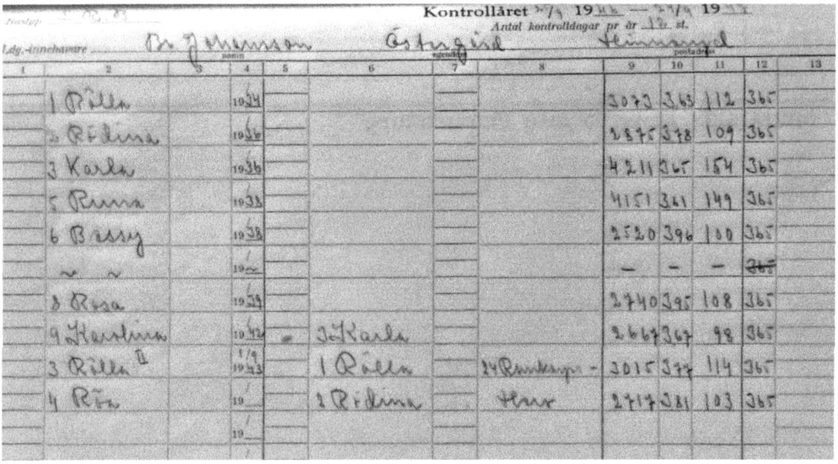

Figure 3.4. Individual results for each dairy cow at *Hinneryd Östergård* in 1928. Names in column 2; milk yields in column 6; per cent fat in column 7, butter fat in column 8 and date of calving in the final column.

use may have played a role. The aforementioned 61 per cent supplied to dairies in 1938 had increased to 73 per cent ten years late, but was still considerable (Statistical Yearbook of Sweden 1943: Table 79 and 82; 1951: Table 87 and 94). The B form of dairy farming shall obviously not be understood as far-reaching commercialization, but as combining commercial and subsistence dairy farming. This was also the case at the family farm in the following example of the form B, based on milk control accounts from 1928 and 1946/47. The example includes similarities as well as deviations compared to Swedish averages.

II.4. Dairy farming at Hinneryd Östergård

Hinneryd Östergård, a family farm of common size in a comparably meagre, forested region in *Kronoberg* county in the south-west of Sweden, was run by August Johansson and his family in 1928, and by two of the sons, Åke and Gustav in 1946/47. The records were handed over to the author by the then aged Gustav in the early 21st century, with the comment 'I have saved this, it may be of some use for your research'. The books are in the possession of the author. The dairy herd was of the common SRB breed (Swedish red) with 7.8 dairy cows in 1928 and 9.0 dairy cows in the mid-1940s. In spite of the comparably modest size, the dairy herd was associated with the local milk control association.

The average milk yield was 1801 kilos per cow in 1928, based on a wide span, from 2,852 kilos for the highest yielding 'Lilja', to 1,090 kilos for the lowest yielding that year, 'Rölla'. This was below the Swedish average of 2,176 kilos for all dairy cows, and substantially below 3362 kilos for controlled herds (Statistical Yearbook

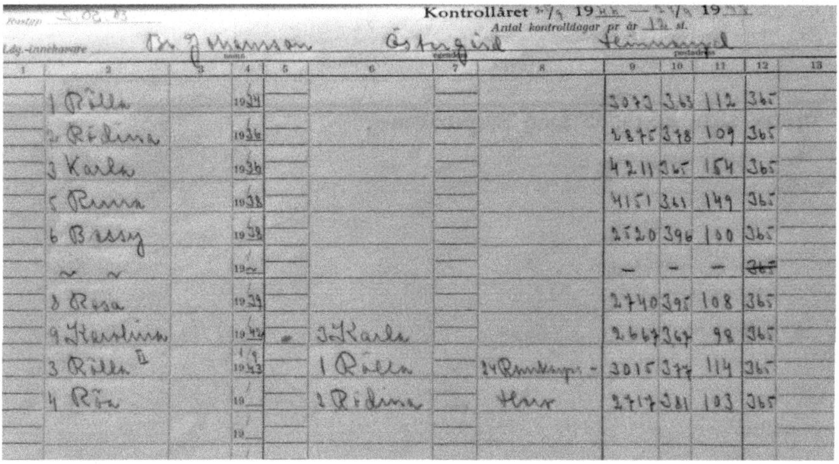

Figure 3.5. Individual yearly milk yield for the 9 cows in the records from *Hinneryd Östergård* 1946/47.

of Sweden). The low average compared to herds in control may be explained by the comparably meagre landscape and modest farm size, but unknown conditions this year may have played a role, too. About 20 years later the average yearly yield had increased to 3108 kilos, over the national average of 2,559 kilo, but still below the national average for dairy cows in milk control (Statistical Yearbook of Sweden). The driving forces behind this increase at *Hinneryd Östergård* are not known, but one possible explanation could be the more regular calvings and accompanying shorter dry periods in 1946/47. The dairy herd does not appear to have changed significantly; it was most probably based of the parent's herd, which is indicated by similar names, 'Rölla', 'Rödina' and 'Röa', in 1946/47 as in 1928, Figure 3.4 and 3.5.

Neither was the feeding substantially changed. According to the notes from 1928 the animals were rather over- than underfed. As shown in Table 3.1 the kinds of feed were changed just a little over time. Hay, straw, oats, barley, and small quantities of imported oil-cakes, were fed in both cases; the only difference was root-beets in 1928 only, and the adding of small quantities of bran and silage in 1946/47. Silage was propagated during WWII, and was found at about 6 per cent of all farms in 1944 (Statistical Yearbook of Sweden 1949: 88, Table 76).

The share of roughage was significant and even increased in the late 1940s whereas the use of feed concentrates was somewhat lower in 1946/47. Sweden's import of feed concentrates was about halved during WWII, when feed concentrates were rationed but reached the pre-war level in 1947/48 (Åmark, 152:292; Historical statistics of Sweden. Part 3: Table 3.3 and 3.5). Hay, straw, oats, barley and root beets were most probably cultivated in the fields of *Hinneryd Östergård*, and exemplifies a common kind of crop circulation on comparably meagre soils.

Table 3.1. Feed used for the dairy cows at *Hinneryd Östergård*

	1928	1946/47
Hay	x	x
Silage		x
Straw, from oats and barley	x	x
Oats	x	x
Barley	x	x
Root beets	x	
Bran, probably purchased		x
Oil-cakes, purchased	x	x

For example, it could be an eight-year circulation plan with: fallow – winter grain – ley (grass and clover) – ley – ley – oats or barley – root beets, potatoes and peas or other leguminous plants) – oats or barley (Ernest, 1939: 341). The crop circulation and dairy farming were obviously planned to be a collaborating whole. The dairy cows were kept grazing from early June to early November in 1928, and until late October in 1946, after which the herd was let out by the end of May 1947. Just a few weeks later, in early June 1947, about 400 kilometres north, the Swedish *Riksdag* was just about to decide about the 1947 agricultural programme, that would come to throw a dark shadow over the grazing 'Rölla', 'Rosa' and 'Karolina'.

II.5. Scaling up and intensification characterized the C form of dairy farming

The previous appreciation of the B form of dairy farming for its time-consuming/job creating character, and supply of dairy products to the whole population, was strongly contrasted by the parliamentary approved agricultural programme of 1947, substantially sharpened in 1967 and valid until the early 1990s. From then on, the authorities were commissioned to reshape farming into a more efficient form that better integrated agriculture into the general Swedish efforts of a more rational use of all factors of production (SOU 1946:42: 16). With a horizon of thirty years, the 1947 programme called for more efficient farming, operated by substantially fewer people at fewer but somewhat larger family farms, while smallholder farming was to be phased out. The total production was to be maintained, sufficient for domestic needs, and supplied at modest, politically regulated consumer prices. In addition to the price regulation on agricultural products the authorities were provided with detailed rules and other tools to work for these structural changes, and rationalization at the farm level.

From 1946 to 1995 as many as 94 per cent of the dairy suppliers and 74 per cent of the dairy cows disappeared, while milk yields rose and herd sizes increased. The processes escalated from around 1960, which marked the beginning of the C form of dairy farming. The 201,000 dairy producers in 1960 declined to 85,000 in 1970 and only 17,000 in 1995 (Statistical Yearbook of Sweden). At the same time the average delivery per farm increased from the previous long lasting 13,000-14,000 kilo per farm and year to 30,000 in 1969, and 188,000 in 1995 (Figure 3). Moreover, the number of dairy cows went down from 1.3 million in 1960 to below 0.5 million in 1995, while the average all dairy cow milk yields was about doubled, from 3,023 to 6,853 kilos, and from 4,395 to 8,033 kilos for dairy cows in official milk control (Figure 1).

Compared to Britain the changes in dairy farming were more radical in Sweden. The number of dairy cows was almost unchanged in Britain between 1960 and 1985, and the decline of dairy suppliers in Britain from 123,000 in 1960 to 41,000 in 1985 was less drastic than the corresponding downturn from 201,000 to 34,000 suppliers. Moreover, the average British dairy cow yielded almost 2,800 kilos in 1960 and 4,100 kilos in 1985, compared to the Swedish increase from 3,000 to 5,700 kilos (Statistical Yearbook of Sweden; Woods, 2014: 298).

The rising Swedish yields per dairy cow did not fully compensate for the many emptied cowsheds, the total production went down from 3.9 to 3.4 million tonnes, or from 525 to 375 kilos per capita per year (Statistical Yearbook of Sweden), which was the level of production when Sweden became member of the European Union. Despite the scaling-up, Swedish dairy farming did not reach an 'industrial' scale during the studied time period, still in 1995 most dairy herds were of the size 10 to 49 dairy cows, far from a level to be called 'industrial' (Figure 3.6).

The agricultural programmes of 1947 and 1967 brought about far-reaching changes, in dairy farming as well as for the entire Swedish society in terms of structural, social and cultural changes, and an increasingly dark, forested landscape. The consequences were far from popular among all dairy farmers. Its' far-reaching consequences in terms of rural exodus were also criticised by significant parts of the total population, of which many of whom had roots in the countryside. The process was met with criticism and debate, although hardly by any kinds of social unrest but rather sights of resignation and talk about missing the disappearing dairy cows, that came to be a symbol for rural change. Neither did the two national based farmers' associations, RLF (*Riksförbundet Landsbygdens Folk)* and SL (*Sveriges Lantbruksförbund)*, appear to have raised considerable objections, at least not in their comments on SOU 1966:30 and 1966:31 that gave the impression that the organization did not want to be seen as negative, or backward looking. Instead, the organizations appeared to have worked to prepare the Swedish farmers for the changing times (Government Bill No 95, 1967). In winter time 1967/68 the RLF arranged a nation-wide study campaign 'My farm in transition' (*Mitt lantbruk i förändring)*. Among the messages in the

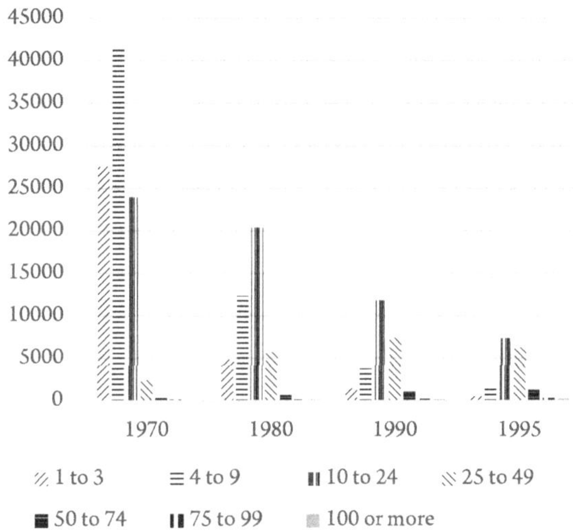

Figure 3.6. Changes in total number of dairy herds, and herds of different size, 1970-1995.
Source: Statistical database, Historical statistics. Swedish Board of Agriculture.

study book was: '...a large share of the current farmers *has to find another job*' (Lejde, 1967: 42).[2]

Returning to the cowsheds, the rising milk yields were no hocus pocus, but a complex interplay of research and farmers' experiences, along with technical innovations and access to investment capital. Among many contributing pieces to this puzzle were more nutritious roughage and more feed concentrate, making it possible for the animals to consume more nutrients, and thus produce more milk (Gustafsson, 1997: 102-04). The feed consumption was further encouraged by increased numbers of feedings per day at many farms, which contrasted with recommendations in the late 1950s to reduce the number of feedings in order to save time (Helmenius-Rydå-Woldmar, 1958: 51, 58-61). As for genetic factors, the artificial insemination technology (further discussed in the chapter by Karl Bruno) gave access to scientifically based animal breeding with potentials for higher milk yields. The use of artificial insemination rose from approximately 40 per cent of all dairy cows in Sweden in 1960 to 70 per cent in 1970 and about 85 per cent in 1990 (Statistical Yearbook of Agriculture, 1991: 150). In addition, the spread of genes for higher milk yields was accelerated by changes in more frequent recruitment, from approximately 15 per cent yearly replacement in the B form, to 25 per cent during the C form, and as much as 37 per

2 In Swedish, with the last words in italics: ... en stor del av de jordbrukare som för närvarande är sysselsatta inom lantbruksnäringen måste *övergå till annan verksamhet*.

cent in 1995 (Table 3.3). A concrete motive to join the milk control was that the establishment of time-saving release systems, through which the milk was transported in glass pipes directly from the udder to a tank, made it difficult to know how much each cow yielded. As earlier said, the C form of dairy farming saw an increased interest in official milk control, from 27% of all dairy cows in 1960/61, to 45 per cent in 1970/71, and a record high 82 per cent in 1995 (Statistical Yearbook of Sweden). The increased association to milk control, and use of artificial insemination contributed to increased and more frequent contacts between dairy farmers and the control/artificial instemination assistants, whose advice most problably contributed, too. At the farm level the economic pressure was of greatest importance, maybe accompanied by elements of competition and status. To have good figures, or even being the best in the parish or the region, may have served as an important driver. Who would not have been proud to have the best yielding dairy herd in the region, or even the highest producing dairy cow in Sweden?

Our focus will now turn specifically to feed and feeding, more precicely to the needs for energy and protein in the previously identified kinds of dairy farming. How about efficiency in the use of feed? For example, did short-lived but high-yielding dairy cattle in 1995 require more or less in total, compared to more long-lived animals that yielded less per year?

III. Differences in energy and protein needs in the A, B and C kinds of dairy farming

An overview of the various forms of dairy farming in this study is given in table 3.2. The first line shows estimated time periods for each of the three main forms, and the second line shows the approximate lowest and highest average milk yield per dairy cow during each of the three time periods. Below that is the average milk yields for dairy cows in official control. The bottom-line shows the volumes that were chosen to be used for the following calculations of energy and protein needs in the various categories. The column to the right was added to include the highest yield level during the studied time period, 8,033 kilos for dairy cows in official control in the year 1995. The comparisons will thus include four different levels of production, A, B, C and the year 1995, and range from 1,400 to 8,000 kilos per dairy cow and year.

The undertaking to compare energy and protein is not only a matter of differences in milk but also about the life cycle of an average dairy cow in each system. We need at least approximate figures about: Life length; time for breeding from birth to the beginning of the first lactation; number of lactations; per cent recruitment;

3 The sum of the quantities on the second line divided by two, e.g. (1020+1785):2=1403.

Table 3.2. Yearly production of milk in the different forms of dairy farming, based on earlier findings in this chapter. (First and last year in brackets.)

	A	B	C	
Time period	– 1913	1920s – 1950s	1960s – 1995	1995
Milk per year, all dairy cows, kilos	1,020 – 1,785 (1,871; 1,913)	2,176 – 2,852 (1927; 1959)	3,023 – 6,853 (1960; 1995)	
Milk per year, controlled dairy cows, kilos	2,928 (1913)	3,362 – 4,398 (1927; 1959)	4,395 – 8,033 (1960; 1995)	8,033 (1995)
Milk yields, feed calculations base, kilos[3]	**1,400**	**2,500**	**5,000**	**8,000**

and body weight as fully grown.[4] The variables are summarized in Table 3.3. The figures should be understood as approximate averages, with individual, socioeconomic and regional variations, and changes over time within the identified time period, in spite of comprehensive research behind.

Tables 3.2 and 3.3 provide the basis for the energy and protein needs presented in Figures 7 to 10, calculated in line with Eriksson et al. (1976). The figures show

4 Potential needs for animal labour are not taken into account. Dairy cows *could* be used as draught animals during the early decades of this study, although to a limited extent and, if so, among the rural poor (Martiin 2005, pp. 262 and 263).

5 Life length: Breeding and number of lactations including two dry moths per year. One calving/year; production until the end of the last lactation.

6 From birth to first calving. Similar birth weight assumed. The breeding times in Table 3.3 are qualified approximates, based on the author's numerous literature studies over the years.

7 Each calving was followed by a lactation period.

8 15 months between each calving, estimated.

9 14 months between each calving. In line with Hinneryd 1928 and 1946/47, and Flach et al., *Sveriges jordbruk vid 1900-talets början*, 1909, Göteborg, p. 188.

10 12 months between each calving. 4 lactations in line with the magazine *Husdjur (Animal husbandry)* 3 October 1980, p. 9.

11 12 months between each calving. 2.7 lactations in line with '*Svensk mjölks lönsamhetskalkyl år 1990-1999*, Appendix 3.

12 The recruitment per cent (replacement frequency) is a logical consequence of the number of lactations. Example from C: In a dairy herd with a 6.2 years average life length the average dairy cow made 4 lactations and thus had to be replaced every fourth year, hence 25 per cent recruitment.

13 Live body weight as a fully grown dairy cow was used for calculations about needs for body maintenance. (Eriksson et al. 1976, p. 49, Table 8.) Needs for 350 kilo body weight is extrapolated from higher weights in the table (400, 450 … 650 kilo).

Table 3.3. Overview, variables of importance for energy and protein needs.

	A	B	C	1995
Milk yields per dairy cow and year, kilos	1,400	2,500	5,000	8,000
Length of life, years[5]	13.0	9.5	6.2	4.9
Breeding, months[6]	36	30	26	26
Number of lactations[7]	8[8]	6[9]	4[10]	2.7[11]
Recruitment, per cent[12]	10	15	25	37
Body weight as fully grown, kilos[13]	350	500	550	600

Sources: See footnotes in the table.

needs for: Body maintenance ('Body weight as fully grown' in Table 3.3); milk yields per dairy cow and year (1.2 Mcal energy and 60 grams digestible crude protein (CDP) per kilo milk [Eriksson et al. 1976: 49]); the three final months of pregnancy; breeding of young heifers as recruitment; additional growth during the first lactation. The additional growth has been assumed as 10 per cent during the first year of lactation, e.g. from 500 to 550 kilos (Widebeck 1997: 88). The additional needs during the final months of pregnancy add comparably marginal needs, taking the differences in calving intervals into account. Needs for recruitment and additional growth are included in the same proportion as the recruitment per cent in each of the four categories. The figures 7 to 10 are calculations based on feeding norms, and do not show what the animals were fed, or what the individual dairy cow may have needed. In the A form, in particular, dairy cattle can be expected to have been temporary and/or permanently underfed. As regards the C form, it is the author's impression that norms and recommendations used to be followed.

It is also important to bear in mind that the bars in the four diagrams were filled with energy and protein of different origin. Form A especially deviates from the others in that the feeding was substantially based on marginal resources, such as by-products from arable farming or various edible stuff found in the forest or elsewhere. The B and C forms had much in common regarding kinds of feed, and rather differed with regard to the quantities fed of: hay, straw, feed grains, bran, roots, and oil cakes. Figure 3.7 synthesizes the calculations of needs for energy, on basis of Tables 3.2 and 3.3.

The differences regarding body maintenance in Figure 3.7, are related to live weight, with the smallest dairy cows in A. The differences are greater when it comes to energy needs for the production of milk, which is explained by the different yield levels in Table 3.2. The differences in needs for recruitment are due to differences in per cent recruitment, such as 15 per cent in B and as much as 37 per cent in the year 1995. We can understand this as costs for higher yielding but more short-lived dairy cattle in 1995. Figure 3.8 is based on the same calculations, shown in per cent.

In the perspective of historical dairy farming, it is interesting to find that as little as about 20 per cent of the total needs were used for production in A, while most of the feed was used for other kinds of needs, such as maintenance of the dairy cow herself. The needs for protein are demonstrated in Figures 3.9 and 3.10.

As shown in Figures 3.8 and 3.10 the production of milk required somewhat smaller shares of the recommended energy intake, but somewhat larger shares of the protein, which is shown in per cent in Table 3.4.

As shown in figures 3.7 to 3.10, and table 3.4, the exchange feed versus milk was improved when the milk yields increased. An answer to the initial question about poor feeding, low yields but long-lived dairy cattle in comparison with high yielding but short-lived cattle in the late 20th century is obviously that the former used proportionally more energy and protein for other needs than the production of milk.

On the other hand, the A form made use of marginal resources that would otherwise have been unused, e.g. leaves and straw, and served as a possible form of dairy forming by that time, efficient or not, and a way to contribute to the survival of the many poor households across the country.

IV. Feed, milk, and society in interaction, a concluding discussion

The earliest identified form of dairy farming, here called A, was characterized by poverty and a scarcity of feed, and in use at least to the mid-1920s, with socioeconomic and regional differences. In parallel, the late 19th and early 20th century saw the development of what is here called an 'Early commercial form' of export oriented, more high-yielding and resources requiring dairy farming, primarily practiced at estates and other big farms. An intermediate form B, was identified as common from the 1920s to 1960. During this period both production and consumption of milk were considered a win-win concept that provided the countryside with jobs and cash income, and the total Swedish population with more nutritious food. The three or four decades that were dominated by the small-scale, family-based, commercial-subsistence directed B, have often come to be seen as typical for the Swedish countryside, and missed ever since. By 1960 a more production and efficiency driven form, C, was clearly distinguishable, characterised by scaling up and intensification, which mirrored official aims for labour-saving efficiency. The final year, 1995, was given special attention. By then yearly milk yields were high, but the average life span of the dairy cows was reduced to a level that makes it relevant to see this as something of a combined milk- and meat-production. Based on the various forms, Part III provides an

Figure 3.7. Energy needs according to feeding norms, in line with the needs for the dairy production forms A, B, and C, and in the year 1995.
Source: Eriksson et al. 1976: 49, 52 (Table 13), 54.

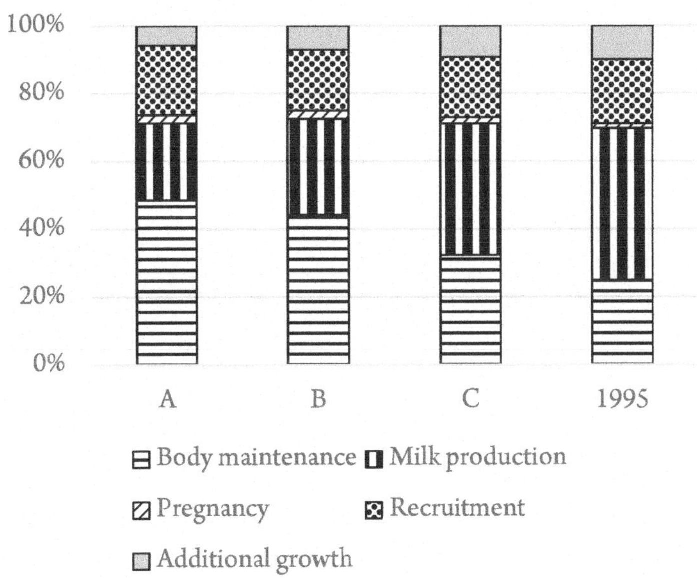

Figure 3.8. Figure 7 recalculated to per cent.

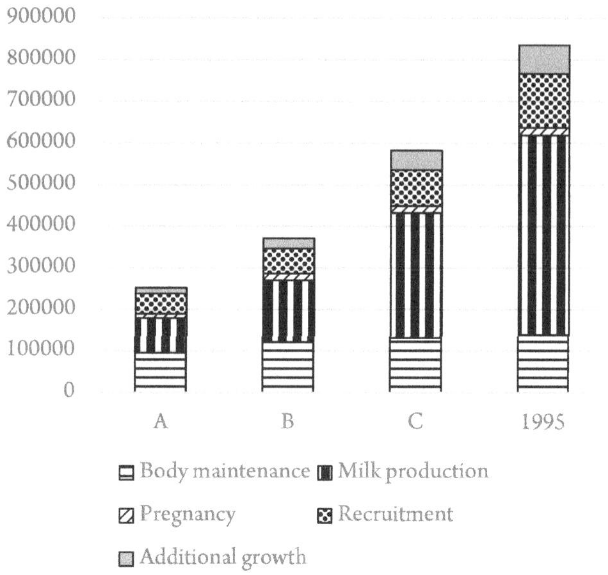

Figure 3.9. Needs for digestible crude protein, according to feeding norms, in line with the needs for the dairy production forms A, B, and C, and in the year 1995.
Source: Eriksson et al. 1976, p. 49.

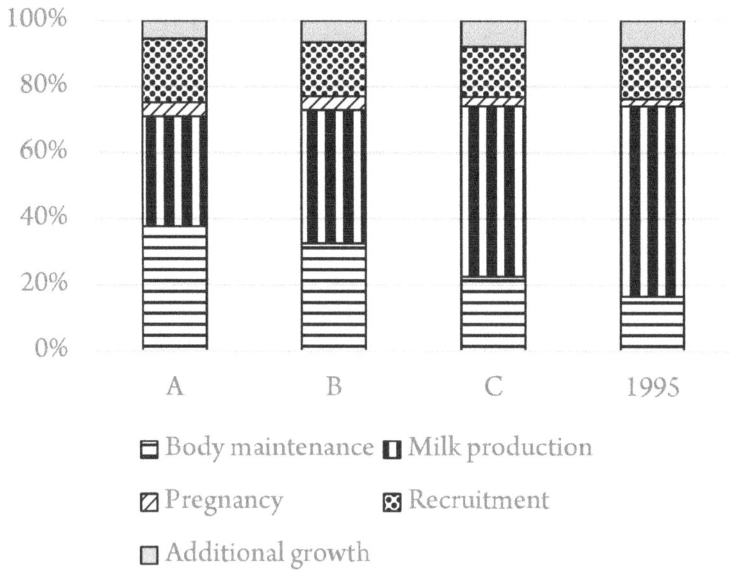

Figure 3.10. Figure 3.8 recalculated to per cent.

Table 3.4. The share of energy and protein in the feed that was used for production of milk. Based on the same calculations as for the previous diagrams.

	A	B	C	1995
Energy in the produced milk vs energy content in the feed, %	23	28	39	43
Protein in the produced milk vs protein content in the feed, %	33	40	52	57

attempt to answer the initial question about how differences in life-span and milk production influenced the total need for feed in the various forms of dairy farming.

It is fascinating to discover how closely large-scale changes in Swedish society were mirrored inside the many cowsheds across the country, where feeding and milking routines were repeated 365 days per year, yet following more or less tangible signs of the times. The methodological approach, to study feed, milk, and society with a point of departure of milk yield per dairy cow, has proved useful to identify different kinds of dairy farming and link this to wider perspectives, such as consumer demand and society at large. In addition, it has reduced the distance to the farmer's perspective, and facilitated the possibility of understanding his/her ways of thinking and working with the animals.

Regarding links to the world outside the barn, the transition from one form of dairy farming to the other was a matter of mutually interacting drivers such as urban demand and access to dairy plants, as well as attitudes towards dairy farming and dairy consumption at political levels. In addition, the importance of trust should be emphasised, trust among producers in becoming dairy suppliers, and trust among consumers regarding hygiene and quality of the perishable fresh milk.

Regarding the feeding of the dairy cattle, the identified categories of Swedish dairy farming indicate an advantageous circle, where increased yields reduced the need for energy and protein per kilo of milk. In form A, related to the late 19th and early 20th centuries, large proportions of the energy and protein were used for the dairy cow herself, while she managed to give only little back per input of energy and protein. Vice versa, the highest yield, 8000 kilo per dairy cow and year in 1995 proved to be the most efficient in terms of energy and protein. However, the calculations are theoretical, and do not take the various kinds of feed into account, for example to what extent were the animals fed by marginal resources, such as leaves; or by on-farm produced feeds such as hay and grains; and/or imported oil cakes. To be short of feed and still manage to have the dairy cattle survive, and have new calves and accompanying lactations, may have required at least as much knowledge and experience as later times' practices with ready-made feed concentrate and various technologies.

Bibliography

Bieleman, J. 'Technological innovation in Dutch cattle breeding and dairy farming, 1850-2000', in *Agricultural History Review*, 2005 Vol. 53.2. pp. 229-50

Brandesten, O. (2005), *Lantbrukarnas organisationer. Agrart och kooperativt 1830-1930*. Skogs- och lantbrukshistoriska meddelanden nr 35. Kungl. Skogs- och lantbruksakademien. Stockholm

Ernest, E. (1939), *Jordbrukslära*. Stockholm: Lantbrukssällskapets Tidskriftsaktiebolag

Eriksson, S; Sanne, S; Thomke, S. (1976), *Fodermedelstabeller och utfodringsrekommendationer*. Stockholm: LTs förlag

Flach, W; Juhlin Dannfelt H; Sundbärg, G. (1909), *Sveriges jordbruk vid 1900-talets början. Statistiskt kartverk*. Göteborg

Gustafsson, A. H. 'Grundläggande utfodringslära', pp. 102-26 in Bergsten, Ch. et al., (1997), *Mjölkkor*. Stockholm: LTs förlag

Helmenius, A; Rydå, K; Woldmar, G. (1958), *Våra Husdjur*. Stockholm: LTs förlag

Henriksen, I. & O'Rourke, K. H, 'Incentives, technology and the shift to year-round dairying in late nineteenth-century Denmark', in *Economic History Review* LVIII, 3 (2005), pp. 520-54

Hernández Adell, I. & Pujol, J., 'Cities and milk consumption in Europe, 1890-1936: the emergence of a new market in Spain', in *Historia Agraria* 73, December 2017, pp. 59-90

Hinneryd Östergård. Milk control accounts for 1928 and 1946/47. (author's possession)

Historical Statistics of Sweden, Part 1. Population. National Central Bureau of Statistics

Historical Statistics of Sweden, Part 3. Foreign trade, Stockholm: National Central Bureau of Statistics

Husdjur, (leading Swedish magazine in animal husbandry, especially dairy farming). 3 October 1980, p. 9

Lampe, M. & Sharp, P. (2014). 'Just Add Milk: A Productivity Analysis of the Revolutionary Changes in Nineteen Century Danish Dairying'. EHES (European Historical Economic Society) Working Papers in Economic History, No 55

Lejde, I. & Jonsson, B. (1967), 'Framtidens lantbruk', in *Mitt lantbruk i omvandling*, Lindström, I. et al. (eds) 1967, Stockholm: LTs förlag

Mejerihantering, Jordbruk med binäringar (1913), Stockholm: National Central Bureau of Statistics

Martiin, C. (2017) 'From farmer to dairy farmer: Swedish dairy farming from the late 1920s to 1990', in *Historia Agraria* 73, December 2017, pp. 7-34

——— (2016), Specialization in Dairying: 'A rapid transformation in Denmark but a drawn-out process in Sweden', pp. 213-32 in Antoine, A. (ed.) *Agricultural specialisation and rural patterns of development*. Rural History in Europe 12. Brepols: Turnhout

———, 'Swedish Milk, a Swedish Duty: Dairy marketing in the 1920s and 1930s' in *Rural History* 21 (2), 2010, pp. 213-32

Martiin (Israelsson), C. (2005), *Kor och människor: Nötkreatursskötsel och besättningsstorlekar på torp och herrgårdar 1850-1914.(In English: Dairy Cattle and People. Cattle husbandry and herd sizes at crofts and estates 1850-1914.)* Published PhD thesis, as Israelsson, C (2005). Hedemora: Gidlunds

Martiin (Israelsson), C. (2002), 'Winterfoderordning för kor', in *De oskäliga kreaturen! Något om synen på lantbrukets djur de senaste 200 åren,* ed. by Cserhalmi, N. Stockholm: The Royal Swedish Academy of Agriculture and Forestry

Martiin (Israelsson), C. (1998), *Frithiof Johanssons dagböcker 1875-1923*. Unpublished essay, Swedish University of Agricultural Sciences, Uppsala

Mjölkhushållning (milk economizing), Questionnaire number 1, ULMA. The Institute for Language and Folklore, Uppsala, Sweden

Myrdal, J. (2013), *Boskapsskötseln under medeltiden.* Nordiska museets förlag: Stockholm

Nyberg, J. (1917), *Mejerihandteringen.* In *Jönköpings läns Kungl. Hushållningssällskaps Historia 1814-1913. Tredje delen,* Jönköping

Peltonen, M. 'The iron cage of dairy farming. Self-sufficiency and specialisation in Finnish peasant farming at the beginning of the twentieth century', in *Agricultural specialisation and rural patterns of development* ed. by Antoine, A., Turnhout: Brepols, 2016, pp. 233-47

Plym Forsell, K. (1997), 'Mjölkkornas hälso- och sjukvård' pp. 161-67 in Bergsten, Ch. et al., (1997) *Mjölkkor.* Stockholm: LTs förlag

Government Bill No 95, Year 1967

Roos, A. 'Kontrollföreningarnas betydelse för jordbrukets utveckling', pp. 221-36, in R. Rydén (Ed.), *Jordbrukets kooperativa föreningar och intresseorganisationer i ett historiskt perspektiv.* 2004, Stockholm: The Royal Academy of Agriculture and Forestry

SOS *Jordbruk och boskapsskötsel* (1949), (Official statistics, arable and animal farming), Stockholm: National Central Bureau of Statistics

SOU 1930:4, *Betänkande angående visa åtgärder för mejerihanteringens och smörexportens främjande.* Stockholm: Ministry of Agriculture

SOU 1936:3, *Förslag om vissa föreskrifter beträffande konsumtionsmjölk.* Stockholm: Ministry of Agriculture

SOU 1938:6, *Betänkande i näringsfrågan, avgivet av Befolkningskommissionen.* Stockholm: Ministry of Health and Social Affairs

SOU 1946:42, *Riktlinjer för den framtida jordbrukspolitiken, Del 1.* Stockholm: Ministry of Agriculture

SOU 1966:30, *Den framtida jordbrukspolitiken A.* Stockholm: Ministry of Agriculture

Statistical Yearbook of Agriculture, Official Statistics of Sweden, Stockholm: National Central Bureau of Statistics, Stockholm

Statistical Yearbook of Sweden, vol. 1914-1997. Official Statistics of Sweden, Stockholm: National Central Bureau of Statistics

Svensk mjölks lönsamhetskalkyl år 1990-1999, Appendix 3. Course material in *Högre kurs i mjölkproduktion* 2000/2001' Swedish University of Agricultural Sciences, Uppsala

Widebeck, L. 'Rekryteringskvigan', pp. 87-101 in Bergsten, Ch. et al., (1997), *Mjölkkor.* Stockholm: LTs förlag

Woods, A, 'Science, disease and dairy production in Britain, 1927 to 1980', in *British Agricultural History Review* 2014, Vol. 62.2, pp. 294-314

Åmark, K, (1952), *Kristidspolitik och kristidshushållning i Sverige under och efter andra världskriget Del 1.* SOU 1952:49. Ministry of Trade: Stockholm

LUCA ANDREONI AND LAURENT HERMENT

4. The devil is always in the details

Good and bad oilseed cakes at the end of the nineteenth century in France and Italy

I. Introduction

At the end of the nineteenth century, European agriculture shifted towards cattle-breeding and dairy production (Koning, 1994). The drop in the price of wheat and vegetal production, including oilseed, makes this evolution clearly visible in northern Europe. This shift caused an intensification in cattle breeding and dairy-farming in a large part of the north of Europe, mainly driven by medium-sized farmers.

To implement this intensification, farmers selected more productive dairy cows or improved local breeds for meat production. The selection of more productive animals was only one step in the effort to increase the yield of milk and meat. In some regions, as in Denmark for example, farmers implemented year-round stabling for dairy-cattle to constantly produce butter of high quality.[1] It was also necessary to feed these more productive animals. In some regions farmers chose to transform arable land into pastures and to improve the production of forage, following and improving the English forage revolution which began during

1 As early as the 18th century, Flemish farmers introduced year-round stabling (De Graef, 2016). On the Danish dairy-industry see, among a vast historiography, Henriksen, 1999; Henriksen-O'Rourke, 2005; Henriksen-Lampe-Sharp, 2011; Henriksen-Hviid-Sharp, 2012; Henriksen-McLaughlin-Sharp, 2015. It is important to point out that these authors considered that the high performance of the Danish dairy industry and the implementation of improved methods were due first and foremost to the adoption of the right institutional framework and the organisation of commercialisation to supply the English market.

Luca Andreoni • Università Politecnica delle Marche, Ancona, Italy: l.andreoni@univpm.it

Laurent Herment • CNRS - EHESS, Centre de Recherches Historiques, Paris, France: laurent.herment@ehess.fr

From Breeding & Feeding to Medicalization, Rural History in Europe, 17 (Turnhout: Brepols, 2023), pp. 77–106
BREPOLS PUBLISHERS 10.1484/M.RURHE-EB.5.135022

the 18th century.[2] In other regions, they also used new feedstuff. Oilseed cakes (a by-product of the oilseed production) became a very important item in this process (Thompson, 1968). Indeed, rapeseed or poppy-seed cakes were used since at least the 18th century in some regions (The Netherlands, French Flanders, and England). However, with the exotic oilseed invasion (peanuts, Indian rapeseed, linseed from Russia, coconut, cottonseed from North America and Egypt, etc.), the quantities of oilseed cakes used by European agriculture became huge.[3] From the middle of the nineteenth century, the increasing use of oilseed cakes and the improvement of forage production caused new issues to arise regarding cattle feeding. At a scientific level, this new field of research, based on chemistry and animal physiology, was developed in German agricultural research stations. In this sense, the laboratories managed by German and Belgian scientists played a crucial role.[4] It appears then that the improvement in cattle-breeding and dairy-farming consisted of a variety of more or less interwoven issues, and that, depending on the issue at stake, were influenced by England, Belgium, Denmark and Germany. Referring to the bibliography on the subject for these countries, when relevant, we would also like to examine the French and Italian cases in this chapter.

It is not possible to apprehend all the issues at stake in understanding the evolution of cattle-breeding and dairy-farming in France and Italy at the end of the 19th century. We shall focus on the issue of cattle feeding and more specifically on the complementary feedstuffs which were promoted by agricultural scientists. Among these materials, oilseed cakes played a crucial role. In discussing the role of oilseed cakes, we do not really intend to study the feeding of livestock as such, from an agronomic point of view. The dual usage as feed and/or fertilizer and accompanying problems relating to the commercialization of oilcakes, such as harmful elements, along with the prescriptions of scientists enable us to devise several issues which were raised for scientists, legislators, industrialists, and farmers.[5] Oilseed cakes are therefore a paradigm through which it is possible to analyze the evolution of two rural societies, grappling with the challenges of the modernization of cattle-breeding and dairy farming. This work is part of what Sheila Jasanoff calls "The idiom of co-production" (Jasanoff 2004a). However, unlike many studies inspired by this multifaceted current (Jasanoff 2004b), we will not focus our analysis on the functioning of research laboratories and their links with the State or public authorities. Indeed, in order to account for the complexity of the economic issues and the debates associated with them, it seems necessary to move away from these places to consider the multiple levels of negotiation

---

2 For the example of Normandy, see Désert, 2007.

3 For France see Daviron, 2014.

4 Regarding German agricultural research stations and their influence in France see: Jas, 2001, and in the United States Finlay, 1988. Petermann, the first director of the Belgian agricultural research station of Gembloux was German. Trained at the Agronomic Institute of Weende, near Göttingen, he became very influential in France. About Petermann see Diser, 2012. For a concise picture of the Italian *stazioni sperimentali*, see Saltini, 2002: 345-47, 350-51.

5 About laboratories in the spreading of scientific knowledge see Latour, 1987.

that led to the commercialisation of oilseed cakes. Therefore, in this chapter we will focus on the roles of companies that provide commercial fertilizers and feed supplements to farmers and their unions. We will mainly rely on the literature produced by industrialists and agricultural organisations.

In the next section we will illustrate the general context of cattle-breeding and dairy-farming in France and Italy. As we will see, these contexts differed greatly from the context of northern Europe. In the third section we shall present the importance of oilseed production in the two countries. In the fourth section we will then show the importance of oilseed cakes for agriculture. Despite the limited use of oilseed cakes, we show that these two countries, first and foremost France, had industry to supply oilseed cakes to farmers to feed cattle, but to also to fertilize. In the fifth section we shall devise the uses of cakes in cattle-breeding and the problems raised by such uses which were partly caused by the different uses of the cakes. In the final section we analyze the problems that science had to overcome to protect farmers against fraud and adulteration, and the industrialist's point of view about this issue.[6] We can then show that the uncertainty surrounding the effect of the use of cakes as animal feed became a central theme which prefigured certain contemporary issues regarding the industrialization of agriculture and breeding.[7]

II. Cattle-breeding and dairy-farming in France and Italy at the end of the 19th century

Although with great caution on the reliability of the available data, the photograph that provided an attentive observer, contemporaneous to the events, such as Ghino Valenti, is at least a testimony of the perception that one had on the eve of WWI about cattle breeding in France and Italy compared to Northern European countries. As figure 4.1 shows, French and Italian livestock was far behind the livestock of North-European countries. It demonstrates that the Italian and French cases were very different to those of Belgium, Germany, or Denmark. To measure the gap between France and Italy and northern Europe it is necessary to illustrate the specificity of the two countries.

National averages are often not very explanatory of the multiplicity of existing situations. In France, the North-Eastern part of the country had embarked on the forage revolution since the 18th century and followed the English/Flemish path consistently. Indeed, some regions of the extreme north of the country (Nord and Pas-de-Calais) developed an advanced agriculture which is comparable to the best

6 The work of Stanziani, 2005, constitutes the reference for fraud and adulteration in France, but he did not study the issue of cattle feedstuff.

7 Unlike Fitzgerald, 2003, we do not consider that mechanisation characterised the process of industrialisation of agriculture. This process, at least in Europe, began with the use of off-farm inputs by farmers (see Gonzáles de Molina et al., 2020: 8).

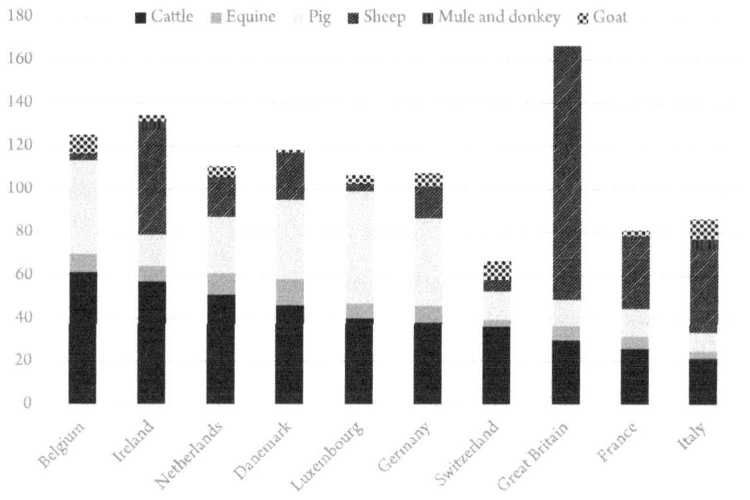

Figure 4.1. Livestock per km² according to Ghino Valenti 1911
Source: Valenti, 1911: 85-87. The source of Valenti's data is Ministero, 1910.

part of England.[8] In this way, cattle-breeding reached a high level of productivity. In the West, the number of livestock was significant, but was mainly made up of little local breeds, except in Lower and Upper Normandy, where farmers usually bred more productive milking cows. Moreover, a large portion of the livestock was in mountainous regions (Jura, Alps, Massif Central, Pyrenees and Vosges) in which it is difficult to reproduce the English/Flemish model. In these regions, cattle-breeding remained extensive. Finally, as for Italy, the Mediterranean zone, which represented more than 10% of the country (in 1914), was ill-suited for livestock, especially bovines.[9] Then, apart from northern part of the country and, to a certain extent, Normandy, the improved breeds were rare, a significant portion of the livestock was fed with poor forage, especially in winter, and year-round stabling was inexistent.

Progress was made in the selection and increase of livestock during the first two decades of the Italian unification, which came about in 1861 (Ministero, 1877: 221-39; Ministero, 1882: II, 200; Valenti, 1911: 84; Corona-Massullo,

8 On agriculture on the Pas-de-Calais see Hubscher, 1979-1980. For the department of North (especially Artois and Flanders) see Delleaux, 2012; Vandewalle, 1994. For an overview, see Herment, 2015.

9 The importance of cheese production (using raw milk) was another significant characteristic of dairy-production.

1992: 379). Nevertheless, the breeding situation in Italy between the end of the nineteenth century and the First World War presented some problematic issues. According to the longstanding and dominant view of agricultural economists and historians, bearing in mind the dominant model of northern Europe, in comparison with other European economies, the average number of cattle was not yet adequate for an advanced economy.[10] As has been pointed out, Italy was in an advanced position as regards the least profitable animal species (mule, donkey, goat) and in pursuit of strategic species for agricultural progress (cattle, horse, pig) (Barsanti, 2002: 113). The country was marked by a profound differentiation between the various rural areas and the central role still played by cereal crops (Valenti, 1911: 5, 83, 99-102; Jacini, 1976: 78; Corona-Massullo, 1992: 356-59; Barsanti, 2002: 107; Tino, 2005; Tino, 2010: 75-76; Tino, 2016). In particular, at two opposite poles, we find the North, which had already embarked on the pathway of zootechnical development;[11] and the South, where cattle struggled to reach a sufficient level.[12] The reception of the Northern European model was not homogeneous in France and Italy, nor could the new husbandry model be applied everywhere.

III. Oilseed and oilseed cake production

During the second part of the nineteenth century, Western Europe's production of oil seed collapsed. As for wheat, Western-Europe was submerged by exotic oilseed and fruit-seed. However, unlike wheat, the flood of exotic oil seeds and fruits caused the elimination of a large part of the European production. The main exotic oil and fruit seeds were sesame seed, cotton seed, peanuts, linseed, rapeseed, coconut, palm, etc. They came from all over the world: French, Dutch and English colonies, South and North America, the Pacific, Eastern Europe, and Russia. They supplied the oil industry, which in turn supplied the industries of

10 For the reconstruction of this debate, see Federico, 1994; Federico, 1996; Federico, 2002: 122-23; the most recent historiography evaluates it in more positive terms, based in particular on the TFP data. See Federico, 2002; Federico, 2009.

11 Albeit with different models: on the one hand, we have the example of a part of Lombardy characterized by a long-standing and efficient irrigation system and by the significant extension of fodder crops (Romani, 1953; Romani 1963; Coppola, 1979: 94-95; Zaninelli, 1979; Bigatti, 1995; Bigatti, 2000; Failla-Fumi 2006), on the other hand, we have that of western Emilia, which had preferred investments in tools (fertilizers, machinery) and in the selection of livestock, being, as it was, unable to count on the capillary irrigation of lower Lombardy (Crainz, 1994: 26, 41, 86-88; Cazzola, 1996: 34-36).

12 Among the reasons, we can mention the uncertain housing process, environmental conditions (soils, climate), crop choices in terms of forage, ownership relations, demographic pressure, long-term rural development paths (Corona-Massullo, 1992: 392-94; Tino, 2005; Tino, 2010; Tino, 2016).

soap, edible oil, oil for lighting, etc.[13] All European countries received these fruits and oilseeds.

Besides the production of oil, the oil industry provided oilseed cakes to farmers. Cakes represented a significant portion of industry turnover. We can analyze the importance of this production with a precise example. We used the archives of Darblay, one of the biggest European millers, which not only milled flour, but also ground oilseed in its plants located in Rouen. A factory accounting book of Bonne Nouvelle for the beginning of the 1880s was preserved. This factory received indigenous and exotic oilseeds. We briefly analyzed one operation. It concerned the grinding of 29,894 kg of rapeseed received at the factory on 26 September and 3 October 1881. The grinding yielded 10,968.50 kg of oil, 17,700 kg of cakes and 1,225.50 kg of refuse. In this case, the cakes represented around 60% of the weight of raw material.[14] In some cases, the proportion is more significant, depending on the nature of the raw material.

The two countries had an oil-industry. In France, at the end of the nineteenth century, plants which processed exotic oilseed were located mainly in Northern departments, in Upper Normandy, and, first and foremost, in Marseille, which gradually became one of the major sites for the commercialization of oil and cakes in Europe (Daumalin, 1992). Throughout the period, French farmers who cropped rapeseed, poppy-seed, linseed, and the like, begged for prohibitive tariffs on exotic oilseeds. However, despite tariffs, indigenous oilseed production had virtually disappeared before WWI, and in any case indigenous production would not have been capable of providing enough materials to the industrialists. The Italian situation was rather more complex. Customs tariffs on seeds and in particular on oilseeds, were designed to protect the production of olive oil, which had gradually been enduring a crisis since the 1880s. Tariffs made the importation of oilseeds rather difficult.[15] In 1891, at the time of the first significant increase in duty taxes, the representatives of the seed-oil industry, which in Italy fed twelve "large factories and other twenty smaller ones, employing several thousand workers", pointed out that the increase in duties would entirely impact edible oils, since industrial oils could not see their price raised, at risk of losing the market entirely to the cheaper oils produced in nearby Marseille (France).[16] This increase would have ended up favouring other low-quality imported oils (American cotton oils), which ultimately also damaged the olive oil market. However, the presence of an artisanal and industrial apparatus for processing olive oil was also useful for processing oilseeds. In Italy, one of the hubs for the production and

13 For margarine and adulteration of butter in France see Stanziani, 2005, and for Belgium and England Deelstra-Burns-Walker, 2014. For lighting see Fressoz, 2020. For the use of linseed oil for painting see MacFadyen, 2018.

14 Archives départementales de l'Essonne, 3/J/174.

15 On the olive oil market see Franciosa, 1940: 72-79, which provides data relating to pressed olive oil only (p. 73); Ramon-Muñoz, 2000; Ramon-Muñoz, 2003; Ramon-Muñoz, 2005; Ramon-Muñoz, 2013.

16 Gismondi et al., 1891: 5 for the quotation. On duty taxes, see Istituto, 1940: 144-45.

commercialisation of oilseed cakes was Genoa. By no coincidence, the important supplier and promoter of oilseed cakes mentioned above (Scerno Gismondi, see n. 16) was located here. In the city an oil industry developed during the same period, alongside fodder and fertilizer industries, which were well connected to the inland and the market of the Po Plain. Furthermore, the variety of uses of these oils, ranging from food to industrial, grew significantly prior to World War I, resulting in a large increase in imports (Giglioli, 1905: 80). Imports reached even more relevant figures in the interwar period (Istituto, 1940: 149). Likewise, the production of cakes increased greatly; unsurprisingly, this supply was not matched by domestic demand.[17] The quantities exported far exceeded the imports (Corona-Massullo, 1992:365; Andreoni, 2021: 134-36).

IV. Oilseed cakes for agriculture and animal husbandry

Oilseed cakes played a dual role in the transformation of agriculture during the nineteenth century. They could be used as fertilizers or as cattle feedstuff. As fertilizers they were in demand for their nitrogen content. Depending on the type of cakes, the nitrogen content ranged from to 2% to 7% (Herment-Mermet, 2018; Andreoni, 2021). At the end of the period being examined, oilseed cakes could provide nitrogenous fertilizers in several French regions, especially in the Mediterranean regions. In these regions, oilseed cakes provide more nitrogen than ammonium sulphate and sodium nitrate to agriculture (Herment 2021, 145-46). The use of cakes as cattle feedstuff is better known. As early as 1967, Collins and Jones underlined the importance of such feedstuff for some English regions.[18]

The use of the two types of oilseed cakes, coming from the residues of seed processing, raised several interwoven questions. First of all, the ration taken by the cattle through the cakes, once the cycle was completed, returned to the fields in the form of manure, which was affected by feedstuff and animal conditions. It is no coincidence that in the texts of the agricultural scientists who promoted the use of cakes, the attention to the digestibility rates of the cakes was often mentioned.[19] Secondly, because the double potential use of cakes raised numerous questions: how to value each type of cake? Did the cakes destined for cattle feed and those destined for fertilizer have the same price? What kinds of cakes should have been devoted to feeding cattle and what kind of cakes could be used as fertilizers? Were such uses alternative or complementary? Could cakes that are

17 The figure is known to the operators of the time and is interpreted as a sign of the backwardness of the livestock system and the insufficient number of cattle. See the *Disegno di legge* 575, presented to the Chamber of Deputies by the then Minister of Agriculture Luigi Rava in 1904 (Atti parlamentari, 1904: 31); Valenti, 1911: 98.

18 See also Van Zanden, 1991, for the Netherlands.

19 Baldassarre, 1889: 16-45; Sansone, 1896: 25-31, who both quoted the work of Emil Theodore von Wolff.

good as fertilizers be good as feedstuff and vice versa? If not, were some types of cakes harmful for animals or plants? Thirdly, how to detect fraud in cakes devoted to fertilisation and cakes devoted to cattle feed? Scientists, first and foremost chemists, claimed that they could help farmers to answer these multiple and interwoven questions. In doing so, they also claimed that their chemical expertise was also a sort of economic expertise.

The simplest questions were the questions about the quality and price of cakes. To determine the quality of a cake as a fertilizer, it was necessary to know its contents in potassium, nitrogen, and phosphates, and what the form of each compound was. In the case of oilseed cakes, nitrogen was the most important compound. From this, it is possible to assess the value of the cake. Table 4.1b, shows that the price of the cake was roughly determined by the proportion of nitrogen.[20] If a specific type of cake moved away from the price determined by the price of the nitrogen compound, farmers were to stop using it. For cakes used as cattle feedstuff, the problem was far more complex but could also be solved thanks to agricultural chemists as is underlined by Antonio Sansone:[21]

> Many farmers now buy the cake as they bought a lot of fertilizer before, being flattered by more or less shrewd propaganda; without thinking that the effect of all feeds, like any fertiliser, depends on the amount of nutrients they contain. [...] There are three useful substances to be taken into consideration when establishing the value of any feed: *albuminoids*, *fatty substances* and *non-nitrogenous extractive materials*, as if it were said that the main components of a fertiliser are: *nitrogen*, *phosphoric acid* and *potash*.[22]

However, unlike what happened with the fertilizing compound, there was an absence of any commercial assessment of the useful substances. Sansone proposed a practical method, based on the nutritional value of the individual components of the cake. A coefficient of 3 was assigned to the most important component, the albuminoids, which are a key component in animal nutrition and which are found in a relatively low percentage in many other kinds of feed; a coefficient of 2 was assigned to fatty substances; a coefficient of 1 was assigned to non-nitrogenous

20 On the price of nitrogen at the end of the 19th century see Page-Herment, 2021.

21 Born in 1866, trained in the Royal Agricultural Institute of Portici (Naples), after extensive research and teaching, Sansone became deputy director of the itinerant chair of agriculture in Parma (Emilia Romagna) in 1894, and then moved two years later to that of Cremona (Lombardy), where he also founded the local *consorzio agrario* (Fumi, 2006: 120). An intense activity of experimentation, research and promotion dates back to these years, in an area such as that of Lower Lombardy, which experienced an extraordinary increase in the number of cattle and in milk production between 1881 and the eve of the Great War (Romani, 1963: 107-10; Cova, 1984; Fumi, 2006: 98-112). Sansone then held other numerous and important administrative and managerial positions: In 1908 he went to head the southern section of the Institute of rustic funds, in 1911 he passed to the Water and Forestry Department of the Ministry of Agriculture, and in 1919 he took over the management of the *Opera Nazionale Combattenti* (Ventura, 1977: 686, 721; Denitto, 1988; Soldani, 2001: 49).

22 Sansone, 1896: 1-2, italics in the source.

extractive materials. By multiplying the individual chemical components by the respective coefficient it was possible to obtain an estimate of the value of the different oilseed cakes per nutritional unit. With a further step, it was possible to determine a relative estimate of the prices of the oilseed cakes available on the market, based on the intrinsic values of the substances that composed them. Starting from the consideration that a quintal of hay stood at 8 *lire* and that the hay was made up of 5.4 parts of albuminoids, 1 of fat, 41 of non-nitrogenous extractive materials, making the appropriate calculations (5.4 * 3 + 1 * 2 + 41 * 1 = 59.2 nutritional units; 8 *lire*: 59.2 = 0.135 lire per nutritional unit) an overall estimate could be reached.[23] According to these calculations, for example, the unshelled peanut cake, which had the highest nutritional value (169.4) should have reached 22.86 *lire* per quintal (Sansone, 1896: 6-11, tav. II-III).

23 The role of hay in establishing nutritional equivalences was widespread during the first half of the 19th century, but still used by Wolff, with modifications; the system of equivalences in hay was abandoned starting from the 1860s: see Jas, 2001: 153. On the attribution of the values 3, 2, 1, already in Wolff, see Poggi, 1891: 8-9.

Table 4.1a. Content in fat and protein, and price (in francs) of cakes sold at Dunkirk in 1897.

Protein matter %		Fat matter %		Oilseed cakes as feedstuff	Available	To be delivered
Min	Max	Min	Max			
27	29	6	7	Egyptian seed cotton	9	9
48	52	7	8	Peanut Rufisque	17,5	-
34	36	12	14	White sesame	13	13,5
34	36	12	14	Gray sesame	12	12
30	31	7	8	Rapeseed to feed	13	-
31	33	7	8	Native rapeseed	13,5	-
31	33	7	8	Œillette du Levant (type of poppy)	11,5	-
36	39	12	14	Œillette blanche de Turquie (type of poppy)	11,5	11,5
31	33	7	8	Linen of 1st quality, small format 2kg	15,5	15,5

Table 4.1b. Content in nitrogen and phosphoric acid, and price (in francs) sold at Dunkirk in 1897.

Nitrogen %		Phosphoric acid %		Oilseed cakes as fertilizer	Available	To be delivered
Min	Max	Min	Max			
4,8	5,25	1,5	2	Rapeseed ravison	7,75	7,75
5,25	5,75	1,5	2	Yellow rapeseed Guzerath	12,5	12,5
5,75	6,5	2,5	3	Poppy, crushed or powdered	11,25	11,25
5,5	6	1,5	2	Rapeseed bigarre	11,5	-
5,5	5,75	1,5	2	Rapeseed from India	11,25	-
5,5	6,25	2	3	Brown sesame	12	-
5,5	6	1,5	3	Yellow rapeseed	12	-

Source of Table 1a and b: *L'Engrais*, 23 April 1897, p. 395.

L'Engrais. Organe spécial Des Fabricants & Marchands d'engrais, de la Grande Exploitation, des Sociétés & Syndicats Agricoles, des Négociants en Grains, des Fabricants de Sucre (herein *L'Engrais*, which means "The fertilizer" in English) was a journal that represented the interests of the fertilizer industry and related industries in France.

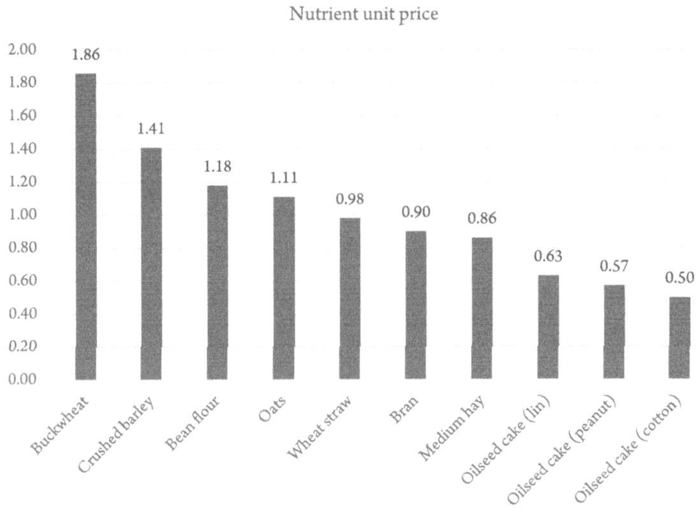

Figure 4.2. Price of fodder (in francs by nutritive unit)
Source: *Bulletin agricole de l'Ouest*, February 1907, p. 25.

Such types of calculations could be found in newspapers published by farmers' unions or by industrialists also in France (Table 4.1a). Depending on the scientific references of the author, or the sources he used, the table and the way in which the feeding value of a cake was calculated could however differ greatly. This was due to the importance that an author placed on the protein content and the fat content and also the improvement in knowledge regarding the feed assimilation process.[24]

Throughout this period, the works of German agricultural scientists played a crucial role in the diffusion of knowledge regarding oilseed cakes. The work of Emil Theodore von Wolff inspired the work of Sansone. Besides Wolff, several other German agronomists and institutions were cited in France or Italy. For example, in 1890, *L'Engrais* published a table to assess the nutritive content of different types of oilseed cakes based on the work of the *Ladw. Centralblatt für Provinz Posen*. In 1904, the *Bourgogne rurale* published an article by Jean Guicherd, who quoted the works of Khun and Kellner.[25] In all of these examples, newspapers published by industrialists or by farmers' unions, more or less accurately

24 About these debates in Germany see Matz, 2015.

25 Jean Guicherd was professor of agriculture in the department of Aube (at the limit of the Burgundy and Champaign). He headed the service of agriculture of the department. He was involved in the farmers' union too (*La Bourgogne rurale*, 20 September 1904, pp. 114-17). Julius Khun headed a department at the University of Halle. He played a significant role in the reorganisation of agricultural studies in Germany. Oscar Kellner was trained at the Hohenheim agronomic station headed by E. von Wolff.

pointed out or referred to *German* methods which were more or less similar to those illustrated by Sansone.

In 1907, the *Bulletin agricole de l'Ouest*, published a table on the price of oilseed cakes as cattle feedstuff, and published a table to compare the price of the nutritive value of different kinds of forage, based on a methodology that is comparable to that illustrated by Sansone. Fig. 4.2, which reproduces the last column of the table, proved that it was not only possible to compare the nutritive efficiency of different kinds of forage, but also their economic efficiency. Beyond the multiple methods illustrated in newspapers, oilseed cakes appeared in all cases as the most valuable cattle feedstuff, and peanuts, sesame, European rapeseed, or cottonseed, were considered as the most nutritive.

Obviously, in scientists' intentions, these calculations were not intended to have an exact value. They were intended to serve exclusively as a guideline for farmers, who could assign a relative value to the single cake chosen, with respect to the set of oilseed cakes available on the market. In this way, he intended to help farmers not to overpay for a cake with a low nutritional value. As Sansone wrote:

> The data exposed, the method itself, therefore, have an approximate value, and can only serve as a guide to avoid committing big mistakes. The exact price may be a few *lire* below or above the figures indicated; but it will not go beyond these limits, which we can call of great tolerance.[26]

V. Good cakes and bad cakes

As underlined above, oilseed cakes were used in cattle breeding in Northern France and the Netherlands from the 18th century onwards. With the increasing use of exotic oilseed, the issue of the harmful nature of the cakes became a crucial one. Not every type of cake could be used to feed cattle. Some cakes are harmful: ricin (castor oil plant), Gujerath (colza from India), *purgère* (*pignon des indes*: Jatropha curcas), etc. On the other hand, shelled peanuts (usually from India), flax, sesame, palm, and cotton, can be used to feed cattle. For indigenous rapeseed, the sources seem contradictory. Finally, for unshelled peanuts (from Senegal) they were sometimes used as fertilisers because of their high cellulose[27] content.

Beyond the issue of the clear definition of what constituted an oilseed cake, and more generally, valuable cattle feedstuff, throughout this period, the double potential uses of cakes, and the great variety of seed used to produce oil, raised numerous questions which are more or less interwoven with issues of fraud and adulteration, which became largely debated, but also with the commercial practices that could be admitted.

---

26 Sansone, 1896: 11. It is of course possible to replace *lire* with *francs* in this quotation.

27 Unshelled peanuts: during the last pressure of peanuts the shells were added to have a better yield. On the use of peanuts in French and Italian agriculture see Andreoni-Herment, *forthcoming*.

The correspondence of Petermann, director of the *station agronomique* de Gembloux, near Namur in Belgium, one of the most influential agricultural scientists in Europe, especially in France, shows the importance of such a question and the need for farmers to have answers to their questions. During the 1870s, the station of Gembloux undertook cake analyses from two points of view. Firstly, Petermann analysed the fat and protein content. Secondly, he analysed the very nature of the seed used to produce cakes.

For example, on 21 December 1875, Petermann informed a farmer that the two samples of cakes he had analysed, contained 11.52% and 10.35% of fat, and 31.49% and 33.75% of protein compounds respectively. Furthermore, he informed the farmer that the sample contained not only rapeseed, but also "ravison" (wild mustard), which contains sulfocyanure d'allyle (Allyl isothiocyanate).[28] As Petermann pointed out: "prolonged use of these cakes presents the (greatest) danger to animal health; significant inflammations of the digestive system, abortions are the disastrous result of its use".[29] This example shows that the calculation of the nutritive content of a cake was not the only issue at stake, it was also necessary to find out the very nature of the seed used to produce oil.

It was first necessary to determine the potential uses of cakes. They depended first and foremost on the very nature of the seed used. On the one hand, cakes could be dangerous, and even lethal for animals, on the other, handsome cakes were not valuable to be used as fertilizers because of their low nitrogen content, or their high fat content which hampers seed germination.[30] On the other hand, a specific cake used as cattle feedstuff could be worthy by itself but could have been, intentionally or unintentionally, adulterated, or could be sold under a false denomination. In any case it was necessary to accurately define the limits between cheating and sales of valuable materials. Finally, as fertilizers benefited from special tariffs for transportation by train, it was necessary to know the exact use of cakes in order to apply one tariff or another. It is not possible to consider every issue in detail here. We shall just consider the two points mentioned first, which are closely linked, and the consequences of such concerns on expertise, and on the law to protect consumers.

The first issue was, at first glance, the most important, but indeed the solution was more straightforward. If a specific type of cake was harmful, it was due to the presence of one or several specific compounds. For example, the castor oil seed contained a harsh purgative substance, and its ingestion by cattle could lead to abortion. The problem with rapeseed was more complex. It was necessary

28 *Ravison* corresponds to *sinapis arvensis,* which is very common in Europe.

29 Archives de la province de Namur, CRA Gembloux, boite 1, CSPA 002.

30 The problem of fat in *organic fertilizers* raised several questions. Degrully evoked the fat content of fertilizers, which could hamper germination (Degrully is cited by *L'Engrais,* 2 January 1891, p. 7). Sometimes, it was said that fat diminished the solubility of phosphoric acid, but Gebeck, a German scientist, did not agree with this idea, see *L'Engrais,* 1 March 1895, pp. 205-06. As for a lot of cases, answers to the concerns of farmers and industrialists came from Germany.

to clearly distinguish between *indigene* (i.e. European) rapeseed cakes, which were used to feed cattle in inland Flanders and the north of France, and the so-called 'Indian rapeseed' cakes (sometimes called Guzerath, sometimes with other denominations), which were harmful.

Beyond the nature of the cakes, the process of oil production could also cause another specific issue. The use of chemical products to improve the rate of oil extraction could transform an edible cake into a harmful cake. In this case, it seems that the knowledge was more difficult to stabilize. For example, some scientists asserted that the use of 'carbon sulphur' was not harmful for cattle, but others did not agree.[31] Antonio Sansone paid particular attention to this issue. The use of chemicals was frequent in the manufacturing process of coconut and palm cakes, but also for other cakes. Quoting a French agricultural scientist, he left open the doubt that it could be toxic and that in particular it could cause abortion.[32] He and others recommended using such types of cakes as fertilizers.

The economic issue was also, of course, very important. Cakes devoted to cattle feeding were usually more expensive than cakes used as fertilizers. It was then very tempting for industrialists to sell cakes of bad quality or adulterated cakes with a low nutritive value seed. This was a big temptation when price of fertilizers fell, or when there was a shortage of fodder. During periods of shortage, the demand for cakes increased, as in 1893 for example.

At the end of the 1880s and during the 1890s numerous instances of adulterated cakes occurred. Scientists were increasingly suspicious about the quality of all kinds of cakes. For example, at the end of 1890s, several publications wondered if cottonseed cakes were harmful or not.[33] Nonetheless, one should bear in mind that intentional adulteration was not necessarily the more problematic issue.

Lastly, industrialists, scientists and farmers had to raise four issues to better assess the problem of fraud and adulteration: the presence of natural impurities in the cakes, which were usually not harmful for cattle; the presence of harmful impurities, which could happen when, as was often the case, a mill used a different kind of oilseed; intentional adulterations without posing any danger (substitution of a good product with a less nourishing substance); intentional adulterations which were harmful or indeed lethal for cattle.

---

31 *L'Engrais,* asserted that the use of "sulfure de carbone" ("carbon disulphide") made cakes unsuitable to feed cattle, unlike the use of "tétrachlorure de carbone" ("tetrachloromethane"). *L'Engrais,* 23 October 1903, p. 1025.

32 'We join Larbalétrier in posing this other question: could the prolonged use of oilseed cakes treated with carbon disulphide in the feeding of dairy cows, not have an influence on the numerous cases of abortion, which have been observed for some years, because even in low doses, does this liquid have abortive properties such as to suggest its use in medicine?' (Sansone, 1896: 86). Albert Larbalétrier was a professor at the agricultural school of the Pas-de-Calais.

33 For example, the work of Cornevin, cited by *L'Engrais,* 29 January 1897, p. 108-09. Cornevin was professor at the veterinary school of Lyon. See also, the notation of a veterinarian of Mayenne, *L'Engrais,* 12 November 1897, p. 1096.

These issues were sometimes more or less interwoven, as underlined by *L'Engrais* in April 1894.[34] It speaks about linseed cakes made with "pavots" (poppies), plaster, barium sulfate (baryte), phosphates, sand. This was clearly adulteration. The newspaper adds:

> Since the prices of linen are very high, these blends have provided good benefits to manufacturers and intermediaries, because chemical analysis can easily reveal mineral matter, but blends of poppies with rice flour, very finely ground peanuts and other organic substances of little value are much more difficult to analyse.[35]

But it added some very important information.

> Adulterations are all the more regrettable as *the linseed cake, by itself, is already sufficiently unclean.* We know that the commercial practices of the London market for the conditioning of grains allow a tolerance of 4% of impure matter, without reduction. This proportion is often exceeded and the seed contains 5, 6 and 7% of other matter.[36]

As is proven by the example drawn from the correspondence of Petermann cited above, this was a very general issue, not only for linseed cakes but also for every type of cake used to feed animals. The presence of *ravison* and *millet* in the cakes analysed by Petermann was not necessarily intentional. Petermann did little more than advise farmers not to use such cakes for long periods.

VI. Science, law, industrialists to rescue farmers

The uncertainties surrounding the quality of cakes raised three questions: How to detect harmful cakes? What was the maximum grade of impurity allowed in a cake? How to protect farmers against bad quality cakes? As for fertilizers and later, the protection of consumers, the problem of the expertise was consubstantial with scientific expertise and the organization of markets (Jas, 2001; Stanziani, 2005; Herment, 2021).

These problems became so significant that several countries passed law to protect farmers against cheating on quality of cakes: England in 1893, and Belgium, in 1895, were pioneers. Italian legislation had not yet intervened regarding this specific point. A rather restrictive bill in this sense would have been proposed by the Minister of Agriculture Luigi Rava in 1904, however, after a tortuous parliamentary journey and the launch of an inquiry involving the various economic actors, carried out by the *Società Italiana degli Agricoltori*, it never became law. Probably, the opposition of a portion of the industrial and commercial figures

---

34 *L'Engrais*, 20 avril 1894, p. 373-74.
35 Ibid.
36 Ibid.

of the time had some involvement in this story (Inchiesta, 1904; Giglioli, 1905; Andreoni-Chiapparino, 2020: 21-23). In France, these laws were linked to the protection of consumers against food adulteration. Every article of the French law of 1905 mentioned both animal feedstuff and human food.[37] In all countries, the problem was then to find a method to determine if a cake was harmful or not.

Not only farmers' unions and scientists wanted law. Industrialists campaigned for a law to condemn cheating, adulteration, fraud, etc. It is not surprising. G.B. Gazzo, a leading representative of the Genoese commercial world, in intervening in a very critical way on some parts of the first version of the law that was presented to the Italian Chamber of Deputies in June 1904, recognized that the first to rejoice about a law that protected farmers from fraud

> will be those who represent honest trade, no less damaged by the frauds of others than the farmers themselves. The frauds in fact take place in a competition based on a discount that serious trade cannot face, and have the effect of discrediting that it unites the good and the sad in the distrust and rejection of the consumer.[38]

The problem, according to Gazzo, was the presence of excessive restrictions:

> Excess is always harmful and when, under the pretext of protecting, intolerable harassment is reached, the result is that either the product increases [...] to the detriment of the consumer, or abandons the market – and the first to abandon it are always those who have the most to lose, that is, earnest traders [...].[39]

As Alessandro Stanziani points out for food adulteration, a closely related issue, at least in the French law of 1905, regulations were meant to make competition safer, not only to protect the consumer. However, the issue of cheating and fraud was not the major concern of industrialists and scientists. This was more a kind of rhetorical figure: for the former to prove their good faith, and for the latter their key role in protecting farmers.

As we have seen above, chemical expertise could be a response to the problem of the value of cakes, but it could not determine whether a cake is edible or not, as noted in Petermann's letter. In this field of expertise, as for water quality, it appears that the expertise of chemists is not valuable, because chemical expertise cannot determine the very nature of the seed used: two different types of cakes can have

---

37 In every article of the French law to protect consumers, it was stated that the law was related to human and animal foodstuff.

38 G.B. Viazzo fu Paolo, 'Sul progetto per la tutela dei concimi', Il Sole, 162 (July 10th, 1904), p. 1, kept in Archivio storico della Camera dei Deputati, Rome, *Disegni e proposte di legge e incarti delle commissioni (1848-1943)*, Legislatura XXI, Sessione II, Esaminati, vol. 806, c. 366r.

39 *Ivi.*

the same amount of fat or protein content.[40] It is then absolutely impossible to infer the harmlessness of a cake from the fat and protein content.[41]

It is necessary to undertake microscopic analysis find out the characteristics of the seed(s) used to produce cakes. Then laws regarding fraud and adulteration were also a means by which to redefine the field of expertise. For cakes, microscopes partly replaced chemical reagents, and botanists complement chemists. From the 1890s onwards one can find advertisements for laboratories specialised in microscopic analyses, and a book which explains how to identify the different kinds of seed. However, as proven by the correspondence of Petermann, this issue was at stake far earlier. For example, on 29 October 1876, Petermann replied to a correspondent who asked for an analysis of linseed cakes:

> Microscopic examination shows that both cakes are made with flax; they are quite pure as regards foreign seeds; are free from seeds that might be harmful to the health of livestock; they do not give off harmful essences such as mustard essence. – On the other hand, N ° S contains a starchy material as a by-product of wheat, bran, barley and it induces a strong starch reaction, from which the pure flax grain is free and which is also lacking in P, which can be designated as a pure cake, but lacking in useful nutritional principles.[42]

As for fertilisers, during the 1870s and 1880s, this new field of expertise triggers new issues regarding not only the way in which to identify fraud and adulteration, but also the limit beyond which it is possible to blame an industrialist.[43] In the case mentioned by Peterman, a pure cake was poor in valuable nutritive compound.

This question raised a last very important issue. What was the limit between a good valuable cake and a bad, inedible, or noxious cake? The limit was linked to the percentage of impurities in the cake, and the content of the noxious compound. As for fertilisers, it was likely that cakes usually contained impurities. However, the problem is more complex than for fertilizers. It is not sufficient to determine if the percentage of valuable compound corresponds to what is asserted by the industrialist, it is also necessary to know, firstly: the content of non-valuable seeds and impurities, and secondly: if such impurities, or non-valuable seeds, were edible or noxious. Moreover, it is necessary to know the long-term effects of non-valuable or noxious elements. Then, the very nature and denomination of the seed used to make cakes was far too insufficient to assess the quality of a specific type of cake.

---

40 For water, see Hamlin, 1990.

41 Sansone, 1896: 2: "Knowledge of the composition is therefore necessary, not only to establish the relative and absolute price of these residues, but also to determine the useful quantity to be given to the animals in the various rations. But the simple chemical determination is not enough: since the oilseed cakes recognized as useful are the subject of skilful sophistication with others or harmful to themselves or simply damaged, so even microscopic examination is necessary".

42 Archives de la province de Namur, CRA Gembloux, boite 1, CSPA 002.

43 For fertilizers, see Jas, 2001; Herment, 2021.

These problems were raised in England, Belgium, and in France as soon as government envisaged enacting a law. We briefly examine the Belgian and French cases. In Belgium, article 2 of the law stated:

> Any delivery of simple or compound materials, containing at least one of the essential nutritional principles (albumin, fat), and intended for feeding farm animals, will, if the delivery exceeds a quantity to be determined by royal decree, be accompanied an invoice certified as accurate by the seller or assignor.
> It will indicate: 1 ° the nature either of the seed, or of the substances or waste from which the delivered materials originate; 2° The minimum quantity per 100 of each of the essential nutritive principles. If the material is sold or transferred as containing, in addition to an essential nutritional principle, a principle favouring animal production, the invoice will provide information on the nature of this principle, its specific property and the proportion in which it is found in the material delivered.[44]

The problem of purity is clearly a central issue. The law did not state that cakes (or more generally feedstuff) had to be pure. It asked that the industrialist provide information about the very nature of the seeds or substances used to produce feedstuff. Nonetheless, the question of the presence of impurities was raised immediately, because it is impossible to guarantee the absence of any kind of adventitious seed, and any kind of waste (sand, soil, etc.). Belgian, but also French industrialists who sold cakes in Belgium, proposed to set limits with the help of scientists. For example, *L'Engrais* published an article about the works of d'Hont, a chemist at Kortrijk (Courtrai), who asserted that a "pure cake" should not contain more than 0.5 to 1% of impurity. To attain such a degree of purity it is necessary to take great care during the industrial process of seed selection. On the other hand, it would be possible to admit a higher percentage of impurities (4%) if any special selection was undertaken.[45] In this last case, the term "raw cake" could be used.

However, even if this distinction were to be accepted, several issues remain. The most significant factor is the fact that some impurities could be noxious. Although, as it was often admitted, to a certain extent, depending on the kind of impurities, if they were not noxious in the short term, it is not possible to assess the long-term effects of such impurities due to a lack of experience.[46]

After the enactment of the French law in 1905, industrialists were very anxious about the decree following the law. Once again, they did not protest against the principle of a law, but they needed to be reassured that the decree would take the uncertainty of the process of producing a good quality cake into account. In 1906, *L'Engrais* took advantage of the Belgian royal decree which specified

44 Law 21 December 1895.

45 *L'Engrais*, 28 May 1897, pp. 520-21.

46 The article refers to the work of Cornevin, 1892.

the law of 1897. The main concerns of industrialists remained the same as those dating back to the middle of the 1890s: how to define a cake of good quality? The Belgian decree offered a solution. Firstly, the cakes which contained more than 12% of impurities could not be sold as pure cakes but as *mélange*. Secondly, some feedstuffs were considered harmful: cakes which contained, either accidentally or naturally, more than 2% of some kinds of cruciferous seeds (white or black mustard, etc.); cakes or substances made with castor oil, purghère, etc.; feedstuffs which contained barium sulphate (*sulfate de barite*), gypsum, chalk, sawdust, etc. and an abnormal proportion of clay or sand.

At the end of the article, to answer to the question raised by a reader, *L'Engrais* provided a table showing the data considered 'normal' for the linseed cake:

Table 4.2. Composition of linseed cake according to *L'Engrais*[47]

	Max	Min	Average
Fat	18,3	5,8	8,6
Albuminoids	43,4	20,5	33,4
Carbohydrate without cellulose	38,4	27,2	31,1
Mineral matter	10,6	4,8	6,4

Source: *L'Engrais*, 29 February 1906, p. 186.

Finally, it was stated that:

> In the absence of regulations in France, it seems to me that the case referred to by our correspondence (15% other matter and 15% water) does not significantly deviate from the tolerance granted by Belgian law; therefore, it would be unreasonable to see a falsification in this case, when one cannot identify only the defect of insufficiently careful manufacturing.[48]

One may be led to believe that such a high percentage of impurity became impunity. Indeed, it is worth bearing in mind that industrialists cannot accurately control the process of production. The most important action, for them, was to forbid noxious cakes and to promote the best-known practices. They cannot offer cakes without impurity. Certainly, they tried to impose the most favourable regulations for them, but the quotation of the works of Belgium agronomic stations, which were very famous, could justify their demand. In their mind, the goal of the law was not to impose pure cakes or pure cattle feedstuffs, but to

47 *L'Engrais*, 29 February 1906, p. 186. The newspaper also noted: "The moisture content gives 12 to 16% water. According to this data taken from the publications of the Belgian food trade adulteration monitoring service, it is certain that a linseed cake can contain 12% of other non-harmful matter and up to 16% water without ceasing to be loyal and marketable".

48 *L'Engrais*, 29 February 1906, p. 187.

impose regulations in a competitive market.[49] Indeed, to impose purity, it would be necessary to have a perfect agricultural process to produce seed crops in which all impurities could be eliminated. How to require perfectly pure cakes with impure materials provided by agriculture?

VII. Conclusion

Oilseed cakes not only participated in the modernisation of agriculture (fertilizers and cattle feedstuffs), but also in the movement of science towards microscopic analyses and concern about the nutritive quality of food and animal feedstuff, and finally led to the definition of certain legal themes that are central to the protection of consumers. It is worth noting that frauds and falsifications regarding animal feedstuffs were comparable to the frauds and falsifications about fertilisers and food. This was a paradox because during the same period, the idea of the need for the industrialisation of agriculture became more and more common among the agricultural elite and agricultural scientists. From this point of view, cakes constituted a good observation point from which to analyse the dual discourse of such an elite. The analysis of the controversy surrounding the quality of cakes reveals that the project of industrial agriculture and cattle breeding was not only an industrialisation of agriculture *per se*, but that farmers and breeders had to accept the state of uncertainty that industrialists had had to face. Industrialists could not guarantee the absolute purity of cakes, but they accepted to conform to more or less restrictive norms to assure fair competition. Then, farmers and breeders, who wanted to be engaged in the industrialisation of their profession, had to accept the same regulations and the same state of uncertainty.

This acceptance was not based on the biological nature of cakes. Indeed, it is possible to show that the same concerns were raised with mineral fertilizers, whose qualities, despite the very strict regulations imposed by law in France and Belgium, were difficult to define and to respect (Herment 2021). One can assume that this state of uncertainty reflected the state of uncertainty of science. It is only a partial explanation for two reasons. Firstly, the uncertainty was due to the impurity delivered to the industrialists with seeds (sand, adventive seeds, etc.). Impurity was thus part of the agricultural process itself. Secondly, as is proven by the possibility to admit 2% of cruciferous seeds in cakes, the state of uncertainty of science regarding the long-term effects of such feedstuffs did not imply that they were forbidden. On the contrary, the state of uncertainty of science became an argument on the basis of which to promote such cakes. From this point of view at this time, agriculture and breeding belonged to the industrial

49 Finally, the law enacted on 18 July 1907 provides: "The lesion of more than a quarter in the purchase of fertilizers or amendments [...] and substances intended for the feeding of farm animals gives the purchaser an action for reduction of the price and for damages"). This is a means by which to indicate a threshold between good and bad cakes close to what industrialists thought.

and scientific logic. Industrialists could plead that an improvement in processing, the development of scientific knowledge should improve the quality of cakes in the future, a very common discourse when industrialists and scientists could not alleviate concerns.[50]

Finally, the importance of international knowledge networks is worth underlining. These networks were not only based on the spread of scientific advancement of German and Belgian agronomic stations, but also on the experience of the implementation of laws to protect farmers and regulate the market in England and Belgium, which inspired French and Italian legislators and allowed for a better grasp of the issues at stake. Moreover, as demonstrated by the work of Antonio Sansone, as professor of agriculture and farmers' union promoter, and the example of French farmers' unions newspapers quoted in the text, often edited by professors of agriculture, these networks aimed to diffuse scientific knowledge at a local level, in particular through the role of agricultural unions. Even if they lagged far behind German, Belgian or, to a certain extent, English famers, far from being isolated, French and Italian farmers were indeed part of such networks.

50 Such a process of argumentation is a common standard in the industry to impose its regulations. Among a vast amount of literature, see: Le Roux, 2011.

Sources and bibliography

Sources

Italy

Archivio storico della Camera dei Deputati, Rome, *Disegni e proposte di legge e incarti delle commissioni (1848-1943)*, Legislatura XXI, Sessione II, Esaminati, vol. 806.

Atti parlamentari (1904), Camera dei Deputati, *Atti parlamentari*, Legislazione XXI, 2° sessione, 1902-1904, Documenti, Disegni di legge e relazioni. Disegno di legge n. 575, presentato dal Ministro Rava di concerto col Ministro dell'Interno Giolitti, June 7th, 1904.

Inchiesta (1904), *Inchiesta della Società degli agricoltori italiani fatta in seguito ad invito della Commissione parlamentare incaricata di riferire sul disegno di legge "Tutela del commercio dei concimi, degli alimenti per bestiame, delle sementi e delle sostanze destinate a prevenire e curare le malattie delle piante agrarie ed a combattere i parassiti"*, Rome, Tipografia dell'Unione cooperativa editrice.

Istituto (1940), Istituto Centrale di Statistica, *Censimento industriale 1937-XVI*, Monografia n. 5, *Industria degli oli vegetali*, Rome, Tipografia Failli.

Ministero (1877), Ministero di Agricoltura, Industria e Commercio, Divisione Agricoltura, *Notizie e studi sull'agricoltura (1876), Relazione al consiglio d'agricoltura*, Rome, Tipografia Eredi Botta.

Ministero (1882), Ministero di Agricoltura, Industria e Commercio, Divisione Agricoltura, *Notizie intorno alle condizioni dell'agricoltura negli anni 1878-1879*, 3 vols, Rome, Stamperia reale, 1881-1882.

Ministero (1910), Ministero di agricoltura, industria e commercio, Direzione generale delle acque e foreste e dei servizi zootecnici, Ispettorato generale dei servizi zootecnici, *Censimento generale del bestiame del 19 marzo 1908. Legge del 14 Luglio 1907, n. 535*, Rome, Stabilimento tipografico G. Civelli.

France

Archives départementales de l'Essonne, 3/J/174.

L'Engrais. Organe spécial Des Fabricants & Marchands d'engrais, de la Grande Exploitation, des Sociétés & Syndicats Agricoles, des Négociants en Grains, des Fabricants de Sucre. Mentioned as *L'Engrais*.

La Bourgogne rurale. Revue mensuelle d'agriculture, de viticulture & d'horticulture

Le Bulletin agricole de l'Ouest

Cornevin, Charles, *Des résidus industriels dans l'alimentation du bétail*, 1892, Paris, Firmin-Didot.

Belgium

Archives de la province de Namur, CRA Gembloux, boite 1, CSPA 002. Correspondence of Petermann

Bibliography

Andreoni, Luca (2021), 'Oilseed Cakes in Italy and France: Opportunities and Difficulties of a Market (late 19th – first half of the 20th Century)', *Jahrbuch für Wirtschaftsgeschichte*, 62, 1, pp. 129-58

——— and Chiapparino, Francesco (2020), 'Fertilizzare la terra', *Proposte e ricerche. Economia e società nella storia dell'Italia centrale*, 84, pp. 9-28

——— and Herment, Laurent (forthcoming), 'Productos globales y agricultura europea: el ejemplo del cacahuate en Francia e Italia', in Alejandro Tortolero Villaseñor (ed.), *Continentes, intercambios e hibridacones. Transferencias técnicas y culturales en la historia rural entre Europa y América, siglos XVI-XX*

Baldassarre, Salvatore (1889), *L'alimentazione intensiva nell'industria del bestiame*, Milan, Tipografia Giovanni Varisco fu Gius

Barsanti, Danilo (2002), 'L'allevamento', in Reginaldo Cianferoni, Zeffiro Ciuffoletti and Leonardo Rombai (eds.), *Storia dell'agricoltura italiana*, 3 vols (2001-2002), vol. III, *L'età contemporanea, 1. Dalle "rivoluzioni agronomiche" alle trasformazioni del Novecento*, Florence, Accademia dei Georgofili – Edizioni Polistampa, pp. 95-128

Bigatti, Giorgio (1995), *La provincia delle acque. Ambiente, istituzioni e tecnici in Lombardia tra Sette e Ottocento*, Milan, Franco Angeli

——— (ed.) (2000), *Uomini e acque. Il territorio lodigiano tra passato e presente*, s.l, Giona

Cazzola, Franco (1996), *Storia delle campagne padane dall'Ottocento a oggi*, Milan, Bruno Mondadori

Collins, Edward John T. and Jones, Eric L. (1967), 'Sectoral Advance in English Agriculture, 1850-80', *Agricultural History Review*, 15, 2, pp. 65-81

Coppola, Gauro (1979), *Il mais nell'economia agricola lombarda*, Bologna: Il Mulino

Corona, Gabriella and Massullo, Gino (1992²), 'La terra e le tecniche. Innovazioni produttive e lavoro agricolo nei secoli XIX e XX', in Piero Bevilacqua (ed.), *Storia dell'agricoltura italiana in età contemporanea*, 3 vols, Venice, Marsilio, 1989-1991, vol. I, *Spazi e paesaggi*, pp. 353-449

Cornevin, Charles (1892), *Des résidus industriels dans l'alimentation du bétail*, Paris, Firmin-Didot

Cova, Alberto (1984), 'Problemi dell'agricoltura cremonese negli anni della prima guerra mondiale (1914-1920)', in Peter Hertner and Giorgio Mori (eds.), *La transizione dall'economia di guerra all'economia di pace in Italia e in Germania dopo la prima guerra mondiale*, Bologna, Il Mulino, pp. 153-78

Crainz, Guido (1994), *Padania. Il mondo dei braccianti dall'Ottocento alla fuga dalle campagne*, Rome, Donzelli

Daumalin, Xavier (1992), *Marseille et l'ouest africain, l'outre-mer des industriels (1841-1956)*, Marseille, CCIMP

Daviron, Benoît (2014), 'Turning Organic Economies into Mineral Economies: Two Centuries of Oil and Fat Use and Supply in France', Paper presented at the European Social Science History Conference (Wien, April 23-26)

Deelstra, Hendrik, Burns, D. Thorburn and Walker, Michael J. (2014), 'The adulteration of food, lessons from the past, with reference to butter, margarine and fraud', *European Food Research and Technology*, 239, pp. 725-44

De Graef, Pieter (2016), *Urbs in Rure. Urban Manure and Fertiliser Improvement in 18th Century Flemish Farming*, Unpublished doctoral thesis, Universiteit Antwerpen

Denitto, Anna Lucia (1988), 'Cultura agraria e investimenti a confronto. Antonio Sansone e la direzione del patrimonio meridionale dell'Italia di Fondi Rustici (1907-1912)', *Itinerari di ricerca storica*, 2, pp. 37-77

Désert, Gabriel (2007^2), *Les paysans du Calvados, 1815-1895*, Caen, CRHQ

Delleaux, Fulgence (2012), *Les censiers et les mutations des campagnes du Hainaut français. La formation originale d'une structure socio-économique (fin XVIIe-début XIXe siècle)*, Namur, Presses universitaires de Namur

Diser, Lyvia (2012), 'Laboratory versus Farm: The Triumph of Laboratory Science in Belgian Agriculture at the End of the Nineteenth Century', *Agricultural History*, 86, 1, pp. 31-64

Failla, Osvaldo and Fumi, Gianpiero (eds.) (2006), *Gli agronomi in Lombardia: dalle cattedre ambulanti ad oggi*, Milan, Franco Angeli

Federico, Giovanni (1994), 'Agricoltura e sviluppo (1820-1950): verso una reinterpretazione?', in Pierluigi Ciocca (ed.), *Il progresso economico dell'Italia. Permanenze, discontinuità, limiti*, Bologna, Il Mulino, pp. 81-107

——— (1996), 'Italy, 1860-1940: A Little-Known Success Story', *The Economic History Review*, 49, 4, pp. 764-86

——— (2002), 'L'agricoltura italiana: successo o fallimento?', in Pierluigi Ciocca and Gianni Toniolo (eds.), *Storia economica d'Italia*, 3 vols (1999-2003), 3, *Industrie, mercati, istituzioni, 1, Le strutture dell'economia*, Milan and Rome-Bari, Intesa-Laterza, pp. 99-136

——— (2009), 'Agriculture and economic development in Italy', in Pedro Lains and Vicente Pinilla (eds.), *Agricultural and Economic Development in Europe Since 1870*, London & New York, Routledge, pp. 234-54

Finlay, Mark R. (1988), 'The German Agricultural Experiment Stations and the Beginnings of American Agricultural Research', *Agricultural History*, 62, 2, pp. 41-50

Fitzgerald, Deborah (2003), *Every Farm a factory. The Industrial Ideal in American Agriculture*, New haven and London, Yale University Press

Forni, Gaetano (2004), *La formazione scientifico-culturale dell'agronomo da fine '700 al '900. Un'analisi critica*, in Giuliana Biagioli and Rossano Pazzagli (eds.), *Agricoltura come manifattura. Istruzione agraria, professionalizzazione e sviluppo agricolo nell'Ottocento*, 2 vols, Florence, Leo S. Olschki, vol. I, pp. 157-69

Franciosa, Luchino (1940), *L'olivo nella economia italiana. Parte prima*, Rome, Tipografia Failli

Fressoz, Jean-Baptiste (2020), 'Une histoire matérielle de la lumière', in François Jarrige and Alexis Vrignon (eds.), *Face à la puissance. Une histoire des énergies alternatives et renouvelables à l'âge industriel*, Paris, La Découverte, pp. 84-99

Fumi, Gianpiero (2006), 'Divulgazione e sperimentazione nell'opera della Cattedra ambulante di agricoltura di Cremona da Sansone a De Carolis (1896-1935)', in Osvaldo Failla and Gianpiero Fumi (eds.), *Gli agronomi in Lombardia: dalle cattedre ambulanti ad oggi*, Milan, Franco Angeli, pp. 87-128

Giglioli, Italo (1905), *Concimi, mangimi, sementi e sostanze antiparassitarie. Commercio, frodi e repressioni delle frodi, specialmente in Italia*, Rome, Tip. Della Camera Dei Deputati («Annali di agricoltura», n. 236)

Gismondi, Scerno & C., Rossi, Davidde & Figlio, Gianolo, Morteo and Società Comm.le, I.G. Schmidt (1891), *Petizione al Parlamento Nazionale in ordine ai dazi sui semi oleosi e sugli oli*, 30 novembre 1891, Rome, Tipografia Ciotola

González de Molina, M., Guzmán Casado, D. Infante-Amate, J., Aguilera-Fernandez, E., Vila Traver, J. and Garcia Ruiz, R. (2020), *The Social Metabolism of Spanish Agriculture, 1900-2008. The Mediterranean Way Towards Industrialization*, Heidelberg, SpringerOpen

Hamlin, Christopher (1990), *A Science of Impurity. Water analysis in Nineteenth Century Britain*, Berkeley and Los Angeles, University of California Press

Henriksen, Igrind (1999), 'Avoiding Lock-In: Cooperative Creameries in Denmark, 1882-1903', *European Review of Economic History*, 3, 1, pp. 57-78

——— and O'Rourke, Kevin (2005), 'Incentives, technology and the shift to year-round dairying in late nineteenth-century Denmark', *The Economic History Review*, 58, 3, pp. 520-54

Henriksen, Ingrid, Lampe, Markus and Sharp, Paul (2011), 'The Role of Technology and Institutions for Growth: Danish Creameries in the Late Nineteenth Century', *European Review of Economic History*, 15, 3, pp. 475-93

———, Hviid, Morten and Sharp, Paul (2012), 'Law and Peace: Contracts and the Success of the Danish Dairy Cooperatives', *The Journal of Economic History*, 72, 1, pp. 197-224

———, McLaughlin, Eoin and Sharp, Paul (2015), 'Contracts and Cooperation: the Relative Failure of the Irish Dairy Industry in the Late Nineteenth Century Reconsidered', *European Review of Economic History*, 19, 4, pp. 412-31

Herment, Laurent (2015), 'Northern France, 1750-2000', in Eric Thoen and Tim Soens (eds.), *Rural Economy and Society in North-Western Europe, 500-2000*, 4 vols (2010-2015), vol. 4, *Struggling with The Environment. Land Use and Productivity*, Turnhout, Brepols, pp. 184-217

——— (2021), *La plus belle firme du monde ne peut offrir que ce qu'elle a. L'émergence des filières agroindustrielles en France : Les engrais commerciaux 1860-1913*, Mémoire d'habilitation à diriger des recherches, 2 tomes, Paris, Université de Paris 1 Panthéon-Sorbonne

——— and Mermet, Eric (2018), 'L'aire de diffusion des engrais commerciaux en France au milieu du XIXe siècle', *Histoire et Mesure*, 33, 1, pp. 111-47

Hubscher, Ronald (1979-1980), *L'agriculture et la Société rurale dans le Pas-de-Calais du milieu du XIXe siècle à 1914*, 2 vols, Arras, Mémoires de la Commission Départementales de Monuments Historiques du Pas-de-Calais, Tome XX

Jacini, Stefano (1976), *I risultati della Inchiesta agraria (1884). La situazione dell'agricoltura e dei contadini italiani dopo l'Unità*, ed. by Giacomina Nenci, Turin, Einaudi

Jas, Nathalie (2001), *Au carrefour de la chimie et de l'agriculture. Les sciences agronomiques en France et en Allemagne 1840-1914*, Paris, Éditions des Archives contemporaines

Jasanoff, Sheila (2004a), 'The idiom of co-production', in Sheila Jasanoff (ed.), *State of knowledge. The co-production of science and social* order, London – New York, Routledge, pp. 1-12

——— (2004b), 'Ordering knowledge, ordering society', in Sheila Jasanoff (ed.), *State of knowledge. The co-production of science and social order*, London – New York, Routledge, pp. 13-45

Koning, Niek (1994), *The Failure of Agrarian Capitalism. Agrarian politics in the UK, Germany, the Netherlands, and the USA, 1846-1919*, London and New-York, Routledge

Latour, Bruno (1987), *Science in Action. How to Follow Scientists and Engineers through Society*, Cambridge, MA., Harvard University Press

Le Roux, Thomas (2011), Le laboratoire des pollutions industrielles. Paris, 1770-1830, Paris, Albin Michel

MacFadyen, Joshua (2018), *Flax Americana. A History of the Fibre and Oil That Covered a Continent*, Montréal, McGill-Queen's University Press

Matz, Brendan (2015), 'Nutrition Science and the Practice of Animal Feeding in Germany, 1850-1880', in Denise Phillips and Sharon Kingsland (eds.), *New Perspectives on the History of Life Sciences and Agriculture*, Heidelberg, Springer, pp. 163-81

Page, Arnaud, Herment, Laurent (2021), 'The Price of Nitrogen at the End of the Nineteenth Century', *Jahrbuch für Wirtchaftsgeschichte*, 62, 1, pp. 49-70

Poggi, Tito (1891), *Valutazione chimica dei foraggi*, Turin, Francesco Casanova

Ramon-Muñoz, Ramon (2000), 'Specialization in the international market for olive oil before World War II', in Sevket Pamuk and Jeffrey G. Williamson (eds.), *The Mediterranean response to globalization before 1950*, London, Routledge, pp. 159-98

——— (2003), 'El comercio exteriordel aceite de olive en Italia y España, 1850-1936', in Carlos Barciela Lopez and Antonio Di Vittorio (eds.), *Las industrias agroalimentarias en Italia y Espana durante los siglos XIX y XX*, Alicante, Universidad de Alicante, pp. 497-555

——— (2005), 'La producción y el comercio de aceite de oliva en los países del Mediterráneo (1850-1938): competencia y especialización', *Mediterráneo Económico*, 7, 329-54

——— (2013), 'Modernizing the Mediterranean olive-oil industry, 1850s-1930s', in Alain Drouard and Derek J. Oddy (eds.), *The Food Industries of Europe in the Nineteenth and Twentieth Centuries*, London, Routledge, pp. 71-88

Romani, Mario (1953), *L'agricoltura in Lombardia dal periodo delle riforme al 1859. Struttura, organizzazione sociale e tecnica*, Milan, Vita e Pensiero

——— (1963), *Un secolo di vita agricola in Lombardia (1861-1961)*, Milan, Giuffrè

Saltini, Antonio (2002), 'Il Sapere agronomico. L'agronomia italiana tra Ottocento e Novecento: dal divorzio all'aggiornamento ai moduli europei', in Reginaldo Cianferoni, Zeffiro Ciuffoletti and Leonardo Rombai (eds.), *Storia dell'agricoltura italiana*, 3 vols (2001-2002), vol. III, *L'età contemporanea, 1. Dalle "rivoluzioni agronomiche" alle trasformazioni del Novecento*, Florence, Accademia dei Georgofili – Edizioni Polistampa, pp. 333-67

Sansone, Antonio (1896), *I panelli di semi oleosi nell'alimentazione del bestiame*, Parma, Stabilimento tipografico Luigi Battei, Pubblicazioni della Cattedra ambulante di agricoltura per la provincia di Parma n. 3

Soldani, Simonetta (2001), 'A scuola di agricoltura', in Anna Pia Bidolli and Simonetta Soldani (eds.), *Fonti per la storia della scuola*, 7 vols (1994-2005), vol. VI, *L'istruzione agraria, 1861-1928*, Rome, Ministero per i beni culturali e ambientali, Ufficio centrale per i beni archivistici, pp. 27-68 (Pubblicazioni degli Archivi di Stato, Fonti n. 17)

Stanziani, Alessandro (2005), *Histoire de la qualité alimentaire. XIXe-XXe siècle*, Paris, Seuil

Thomson, F.M.L. (1968), 'The second agricultural revolution 1815-1880', *The Economic History Review*, 21, 1, pp. 62-77

Tino, Pietro (2005), 'Tra maggese, letame e rifiuti urbani. La reintegrazione della fertilità della terra nell'Italia meridionale tra Settecento e Novecento', *I frutti di Demetra*, 5, pp. 19-28

——— (2016), 'Il rapporto tra agricoltura e allevamento nel Mezzogiorno del Novecento', *Mélanges de l'École française de Rome – Antiquité*, 128, 2, pp. 349-63, consulted on 29/05/2020, https://journals.openedition.org/mefra/268

——— (2010), *Le radici della vita. Storia della fertilità della terra nel Mezzogiorno (secoli XIX-XX)*, Rome, Edizioni XL

Valenti, Ghino (1911), 'L'Italia agricola dal 1861 al 1911', in *Cinquanta anni di storia italiana*, R. Accademia dei Lincei (ed.), Milan: Hoepli, vol. II, monografia VII

Vandewalle, Paul (1994), *Quatre siècles d'agriculture dans la région de Dunkerque, 1590-1990. Une étude statistique*, Gand, Centre Belge d'Histoire Rurale (Publication n° 110)

Van Zanden, Jan Luiten (1991), 'The First Green Revolution: The Growth of Production and Productivity in European Agriculture, 1870-1914', *The Economic History Review*, 44, 2, pp. 215-39

Vaquero Piñeiro, Manuel and Giommi, Francesca (2020), 'Semi e panelli oleosi. La nascita dell'industria mangimistica in Italia (1881-1945)', *Rivista di storia dell'agricoltura*, 60, 2, pp. 71-88

Ventura, Angelo (1977), 'La Federconsorzi dall'età liberale al fascismo: ascesa e capitolazione della borghesia agraria, 1892-1932', *Quaderni storici*, 36, pp. 683-737

Wille, Robert-Jan (2015), 'The Coproduction of Station Morphology and Agricultural Management in the Tropics: Transformations in Botany at the Botanical Garden at Buitenzorg, Java 1880-1904', in ed. Denise Phillips and Sharon Kingsland (eds.), *New Perspectives on the History of Life Sciences and Agriculture*, Heidelberg, Springer, pp. 253-75

Zaninelli, Sergio (1979), 'L'agricoltura lombarda dal Settecento all'unità', in *Atti del Convegno Nazionale di Studi sul rilancio dell'Agricoltura italiana, sotto l'alto patronato del Presidente della Repubblica (Siena, 15-16 dicembre 1977)*, 2 vols, Siena, Monte dei Paschi di Siena, pp. 191-207

TELMO OTERO RODRÍGUEZ AND DANIEL LANERO TÁBOAS

5. From dairy innovation to dairy specialization

Disruption of cattle improvement and milk commercialization in Galicia (1900-1975)

I. Introduction

Since the 1960s, intense dairy specialization has been occurring in Galicia and many other regions of the world. Climate conditions in this northwest region of Spain link it to other Atlantic agricultures, especially those of Bretagne and Ireland. During the 1960s, the Francoist authorities promoted the implantation of specialized cattle and technologies for obtaining milk in a productivist sense, following the logic of the Green Revolution.[1] Late specialization – compared to other European regions or even other northern regions of the Iberian Peninsula – has not hindered the consolidation of the dairy sector. It has shared all the successes of the most advanced agricultures in terms of technification, high-level production on farms and a dairy chain that connects it with industry and distribution, along with all the deficiencies related to hyper-intensification and its ecological footprint.

Before specialization entered the discussion, the production and sale of milk conformed to the multi-purpose logic of local cattle, as did calves destined for the meat market, which has been the focus of agrarian literature generally. In fact, the first studies on this issue came from economic history, which attempted to analyse early-twentieth-century agricultures based on elements that connected them to

1 As understood in the agrarian literature: agriculture of purely capitalist logic, based on production in growth. Reproduction depends on achieving it in the most econonomically efficient way possible. See Picado (2011: 26-35); Moser and Varley (2013: 20-27).

Telmo Otero Rodríguez • University of Santiago de Compostela: telmo.otero.rodriguez@usc.es

Daniel Lanero Táboas • University of Santiago de Compostela. CISPAC: daniel.lanero@usc.es

From Breeding & Feeding to Medicalization, Rural History in Europe, 17 (Turnhout: Brepols, 2023), pp. 107–134
BREPOLS PUBLISHERS 10.1484/M.RURHE-EB.5.135023

the rest of the system. Accordingly, it emphasized the existence of a market begun in the late nineteenth century for exporting calves by sea to England. With the crisis at the end of that century, the market was re-oriented towards supplying diverse nuclei on the peninsula by rail. This functioned until the Spanish Civil War (Villares, 1982: 371-82; Carmona and De la Puente, 1988: 184-98; Carmona, 2000: 315-39; Martínez, 1995: 143-207). The question of cattle improvement also emerged in the early twentieth century and ended up dividing technical experts and agronomists. An official line was established for improvement through cross-breeding with imported Simmental breed. These multi-purpose cattle, for milk, meat and traction were brought from Switzerland with the idea of increasing the weight of the descendants. This created conflict with other agricultural technical experts who defended time-honoured, farm-based 'pure selection' practices (Alenda, 1983: 367-70; Fernández, 1992: 259-71; Conde Gómez, 2015: 77-92).

The scarce research on milk also comes from economic history but arrived at opposite conclusions to studies on meat. They attribute the general increase in Galician milk production during the first third of the twentieth century to the region's substantially higher number of milk cows, compared to others on the peninsula, rather than to their increased production. Milk was not yet economically relevant: cooperatives or processing industries had not managed to establish themselves, so there was nothing resembling a 'dairy sector' in today's sense of the word (Martínez, 1991: 33-39; 1995: 236-50). From an economic perspective, the search for qualitative, larger-scale improvement and commercialization of beef cattle generated the understanding that meat should be the preponderant productive orientation. Thus, Galicia was assumed to be a place for meat specialization, to the detriment of any indication of dairy specialization. This notion is also very present in more general references to the region in Spanish agrarian literature (Calcedo, 1996: 233-35).

From our point of view, this interpretation needs revision for several reasons. The agrarian processes discussed here have been understood from positions more closely linked to orthodox economics. Accordingly, they centre on how agriculture relates to the rest of the economic system: in other words, its capacity to provide resources to the rest of the economy. This is measured with parameters such as efficiency, work productivity, production or commercialization. Parameter variability depends directly on how much technology has been incorporated into a given area of agriculture. In the nineteenth century a modernizing current began to spread in agriculture, but it was deeply rooted in the Enlightenment period (Fernández, Pan-Montojo and Cabo, 2014: 28-31). Its representatives – always from outside the peasantry – sought to make agriculture more productive by adopting technologies that brought it closer to emerging industrialization models. The challenge was initially assumed by a 'learned agriculture' that tried to translate new technologies from other regions to local spheres. Later, after competition from American production and the crisis at the end of the century, the states themselves became more aware of the need to develop regional agriculture for greater productivity. They took on the task of transmitting technologies by training new technical experts and creating a network of research centres for this purpose. So, we see how peasant agriculture has long been under this modernizing influence and how its representatives conditioned the peasantry towards transformation through the incorporation of technology. The type of actors involved – the state, technical experts, farmers – were key in this process, as they defined the results and their implications.

Thus, the conclusions derived from most of the research relevant to our case indicate very little improvement in agricultural output. This suggests that prior to the Green Revolution, farmers generally did not embrace technology for the productive improvement and commercialization of milk. This may have been slightly less the case for meat, which led to the understanding that Galicia was specializing in beef. However, some agricultural technologies were adopted in other regions, resulting in more relevant production and more visible advances. In the northern regions of Cantabria and Asturias, for example, Friesian cattle had been introduced in the nineteenth century and dairy processing industries were abundant (De la Puente, 1992: 151-95; Langreo, 1995: 52-104). Significant dairy specialization was also occurring in other regions of Europe, raising the question as to why it had not occurred in Galicia. The generally vague explanations habitually refer to scarce genetic selection, deficient cattle nutrition or the small size and poor innovation capacity of existing industries. This interpretation certainly fits with explanations of agricultural backwardness in Galicia, compared to other regions of the northern Iberian Peninsula or Europe. Seen through a teleological lens, the situation persisted until the 1960s, when the modernizing impulse of the state gave rise to new specialized dairy or beef operations that clearly improved the efficiency and production parameters of Galician agriculture.

Accordingly, the modernizing actions of the 1960s and all their consequences were inescapable as they largely involved perfecting the path to specialization and productive increase, to finally achieve what was understood as authentic agricultural progress. This vision managed to legitimize the Green Revolution by transmitting the idea of a single and inevitable path for agricultural development.

Studies criticizing this productivist agricultural model appeared very early on, pointing out its inherent environmental costs – pollution, fires, etc. – along with socioeconomic costs in the form of agricultural concentration and hyper-intensification or rural abandonment.[2] Other visions emerged to question whether the modernizing process need have progressed as it did or if the consequences were the necessary, inevitable and only way to improve agriculture. This also led to thinking about alternative development processes that were thwarted by the modernizing paradigm. Research focusing on these agricultures and their farmer-driven innovation capacity – before being abruptly modernized – has demonstrated that state-promoted Green Revolution modernization was not the only or necessary solution for achieving agricultural progress.[3]

---

2 See Naredo J.M. (2004 [1974]: 215-59), González de Molina and Alier (2001: 7-31).

3 Concerning this debate see also: Pujol et alii (2001: 97-108), Simpson (1997: 27-41).

In applying this premise to the Galician case (Fernández, 1992) discovered an innovation framework in the first third of the twentieth century by studying diverse spheres such as cattle, fertilizers, crops, and mechanization. It was emerging in similar societies across Europe from the imbrication of strong, organized farming with technical experts and state research organisms. Later, Conde (2015: 35-48) demonstrated how cattle improvement during that period and the configuration of the multi-purpouse local breed *Rubia Galega* in the following decades depended more on family-based selection of cattle with better qualities than on official improvement tactics.

Studying the dairy question should further our understanding along this line. On one end, state-driven modernization gave rise to sudden dairy specialization and an agro-industrial conglomerate, the strengths and weaknesses of which have been amply studied from the economic perspective (López and Valdes, 2020: 219-51). On the other, given the lack of research on what happened prior to that, we find room for new interpretations. Authors such as Conde (2015) are reaching beyond the perspective of final production data to focus on understanding the dynamics of change that emerged from the symbiosis between endogenous innovation and external technology, as observed from their corresponding historical dimension.

Exploring Galician milk issues offers a long-term perspective that reflects a line of work intended to demonstrate the existence of innovation processes that were crushed by the intense modernizing push, along with the consequences for agriculture and societies (Moser and Varley, 2013: 16-26; Fernández, Pan-Montojo and Lanero, 2017: 7-13; Fernández, Pan – Montojo and Lanero, 2018: 9-14). Fernández and Lanero (2020: 30-32) have also identified the need to further develop this perspective and pose new questions about Galician milk, as an opportunity to adjust the historical lens in relation to dairy progress. Our objective in this text is to continue developing the hypothesis that state modernization projects created a break with the innovation dynamic that was emerging in family-based agriculture. We will do so by studying cattle improvement in Galician dairy production and milk commercialization strategies, tracking state-driven and farmer-based approaches in two contiguous time periods that nonetheless had different logics and processes of change.

We have indications that milk-related cattle improvement and commercialization were happening prior to the Green Revolution, since at least the early twentieth century. The question for us is how to interpret these processes and gain insight concerning what we know happened in the 1960s. Peasant families were commercializing milk from the early twentieth century – and surely before – in a way coherent with their awareness of increasing demand. In this context, agricultural technical experts began to implement the first projects for improving milk production. In a state with scarce modernizing capacity, solutions arose that were more aligned with the needs of rural families. They were interested in selling more milk because they perceived the demand, but they also needed to maintain other cattle aptitudes which were vital for the reproduction of the peasant family.

This connection catalysed an endogenous innovation process that fit the logic and needs of farming families. Our hypothesis is that this fledgling dynamic was disrupted by the Civil War and then crushed in the wake of state-driven modernization projects. This is why so little remains of what happened prior to state modernization and why our results might easily be overlooked. Our first objective is to trace this earlier dynamic and seek deeper knowledge of the innovative capacity of family-based agriculture, which was generally obstructed, leaving the state to promote innovation.

After the Spanish Civil War (1936-39), milk production in Galicia continued to be a relevant issue for the Francoist regime. Innovation in cattle improvement and milk commercialization would be mediated by two non-agricultural agents: the state and industry. Based on their productivity criteria, these actors defined the forms and pace at which smallholding agriculture would be transformed, then took the actions required to achieve it. The Francoist New State*(Nuevo Estado)*, which prized centralization and authoritarianism, took charge of cattle improvement while newly created dairy processing industries monopolized commercialization. Unlike what had occurred before, the state acquired a strong modernizing will at this time and created new transmission pathways to accomplish it, such as the corporativist framework and the agricultural extension agencies. Industries also deployed their own formulas to this end and agrarian change took place under new conditions. From our point of view, this provoked a rupture with the prior dynamic that had new implications for the rural context.

Innovation derived from smallholding agriculture, which responded to family-based needs and logics, yielded to agricultural modernization promoted by the state and industry, to meet their needs.

Our second objective was to study the agricultural modernization measures promoted by the state and industry. However, all this took time in Galicia. Though the idea of dairy specialization appeared in technical literature in the 1940s, nothing of great transcendence materialized until the 1960s. At that time, the state and industry organized campaigns to import Friesian cows, expand pastures and increase fodder corn crops, silos and other technologies. The resulting model of intensive, specialized production facilities – for milk, along with meat, wine or potatoes in Galicia – altered family-based agriculture to fit state productive objectives.

We seek to overcome the interpretation of scarce dairy specialization based on economic end-production data, which identifies the early milk-related processes as antecedents to what occurred in the 1960s. In doing so, we want to better understand the tangible changes that occurred by looking at how state-implanted modernization destroyed earlier agricultural innovation capacity.

II. There was always milk in the house and it was never wasted: milk and innovation in Galician agriculture (1900-1936)

Milk is undoubtedly a key food, and one that best demonstrates nutritional change in Western societies during the twentieth century. Spain was no exception to the European dynamic of milk consumption (Nicolau, Pujol and Hernández, 2010: 148-51; Hernández, 2012: 222-29; Hernández and Pujol, 2016: 187-212; Hernández and Pujol, 2017: 65-68; Hernández, Muñoz and Pujol, 2019: 28-36). As in most other European regions (Valenze, 2011: 260-67; Atkins, 2017: 37; Martiin, 2010: 213-22; Martiin, 2017: 23-29) perceptions about milk had been changing since the late nineteenth century and consumption was increasing at every level of society.

In Europe and in Spain, towns and cities led the process, because that was where scientific ideas arrived first and where more people heard about them (Hernández, Nicolau and Pujol, 2007: 304-09; Hernández, 2012: 229-60; Hernández and Pujol, 2017: 64-66). The best example of this was the need for local administrations to regulate and test the milk that came in. At this time, hygiene and public health was becoming a concern for municipal authorities, especially those with larger populations. It led to municipal initiatives to install laboratories (Sanz, 2014: 86-90; Lorda and Rúa, 2020: 38-47). Given that most of the products consumed in the cities were directly supplied by nearby rural areas, one of the main tasks of the laboratories was to analyse food before it reached the consumer. Milk was one they tried hardest to control because it was so easy to adulterate.[4] We see a constant effort at milk testing in Madrid and Barcelona during the first third of the century, but even more so from 1930 on (Hernández and Pujol, 2017: 69-73).

Similar processes were underway in Galicia. However, unlike other cities with predominately urban cattle or dairy facilities, here, dairy women – called *leiteiras* – would bring milk in large jugs carried on burros or carts from the nearby rural areas into the urban centres. This task was primarily done by women. Habitually, one or two women in each parish would collect the milk from all the nearby families and sell it at a fixed location or deliver it directly to the homes of their clients. Later, as they had access to motorized vehicles, they expanded their radius of collection and the quantities they managed. In this way, they became established intermediaries in the milk business.

We can see the relevance of milk consumption in Galicia – and the northern provinces of the peninsula generally – in the statistics compiled in 1925 by the National Association of Cattle Farmers of the Realm (*Asociación Nacional de Ganaderos del Reino*) who were the first to address the question statewide. Until then, no private or public organism had considered the need to analyse the milk

4 The archive material shows a larger quantity of milk testing than another kind of food.

consumption of the population. The attempt to quantify it at that time is itself an indication of its increasing importance in general terms. The study of the presence of milk women allows us to better understand how milk was supplied for this increasing consumption and the dynamics it generated in the agricultural supply.

Table 5.1. Milk consumption in Galicia for the year 1925

Galician Provinces	Total milk production (litres)	Total milk consumption (litres)	% of milk consumed from total produced per year	Consumption per resident (l./year)
Coruña	43,576,557	31,993,199	73	73
Lugo	57,128,135	35,270,194	61	72
Ourense	24,599,196	21,406,964	87	51
Pontevedra	22,124,103	17,557,424	79	32

Source: *Leche, queso y manteca: Estadística de la producción en España*. Asociación General de Ganaderos del Reino, 1925

Milkwomen increased in number from 1910 on, as did concern for public health. The municipalities found it necessary to test this product as urban consumption grew. We do not know when milkwomen began to supply the demand for milk in urban populations; but milk analyses by laboratories provide the first indications of their presence. This study is based on preserved documentation from the municipal laboratories of Santiago de Compostela and A Coruña.[5]

The first recorded analyses of the Municipal Laboratory of Santiago de Compostela date from 1907. It is actually a list of sanctions applied to merchants for poor-quality food that year, eighteen of which were fines for watered-down milk. However, there was no thorough attempt to monitor the milk supply until 1921. At that point, the city government saw the need to establish a general registry of all the milkwomen who came into Santiago, assigning a number to each jug. In A Coruña, the first analyses were done by a pharmacy in 1911, but the city did not consider installing a laboratory until 1928. The documentation reveals the urgency with which it addressed the task at that point, because the facility was functioning by 1930. Until then, the city did not register milkwomen or systematically analyse milk.

5 The two most important Galician cities at the time. Santiago de Compostela had 24,120 inhabitants in 1900 and 38,270 in 1930. A Coruña had 43,971 inhabitants in 1900 and by 1930 had reached 74,132. The Santiago de Compostela laboratory, founded in 1891, was one of the first in Spain. The A Coruña city hall did not make the decision to install a laboratory until 1928.

Table 5.2. Number of monthly milk analyses in Santiago de Compostela and A Coruña

Santiago de Compostela											A Coruña	
	1907	1910	1913	1914	1916	1925	1929	1930	1931	1932	1933	1930	1935
January			1	89						10	45		17
February			12								32		19
March			139								41		12
April										63	36		19
May		92								49	37	11	23
June	18							13		36	21		19
July							24		47	12	36		22
August			345		180	12			66				14
September			45							11		37	3
October										30		59	14
November			2			12				32		15	2
December										60		12	16

Sources: For Santiago de Compostela: Carpeta 'Multas por leche aguada', Caja A.M. 2192. Fondo Sanidad. Archivo Histórico Universitario de Santiago de Compostela; Carpeta '1925-1933', Caja A.M. 2193. Fondo Sanidad. Archivo Histórico Universitario de Santiago de Compostela. For La Coruña: 'Expedientes sobre reconocimiento por el farmacéutico municipal de varios jarros de leche' Caja 7545. Fondo Sanidad. Archivo Municipal de A Coruña, Legajo 00 'Laboratorio Municipal – Partes de Inspección de Alimentos' Caja 5153, Fondo Sanidad. Archivo Municipal de A Coruña. (Blank cells mean "zero analyses")

Table 5.3. Number of milk analyses and results by decade

Santiago de Compostela						A Coruña				
Quality	Good	Adequate	Poor	Total	%Poor	Good	Adequate	Poor	Total	%Poor
1901-10	41	40	29	110	26.3					
1911-20	651	75	23	749	3.2			9		
1921-30	20		17	37	45.9					
1931-35	547		141	715	19.7	216	41	58	310	18.7

Sources *1901-10*: Carpetas 'Multas por leche aguada', Caja A.M. 2192. Fondo Sanidad. Archivo Histórico Universitario de Santiago de Compostela. *1911-1920*: Carpeta '1925-1933', Caja A.M. 2193. Idem., and 'Expedientes sobre reconocimiento por el farmacéutico municipal de varios jarros de leche -1911' Caja 7545. Fondo Sanidad. Archivo Municipal de A Coruña. *1921-30 and 1931-35*: Carpeta '1925-1933', Caja A.M. 2193. Fondo Sanidad. Archivo Histórico Universitario de Santiago de Compostela, Legajo 00 'Laboratorio Municipal – Partes de Inspección de Alimentos' Caja 5153, Fondo Sanidad. Archivo Municipal de A Coruña. (Blank cells mean 'zero analyses')

Thus, we can see that from the beginning of the century, incoming milk began to be a municipal concern in these cities, and timid attempts – even for that time – were made to regulate this product. In Santiago de Compostela, public health technicians began recording all samples taken, the name of each milkwoman and analysis results in 1910. The preserved bulletins show that during that decade, analyses were not done regularly or throughout the year. In Table 1, we see that 100 samples were taken annually, but concentrated into a single month. However, 1913 constitutes an exception and is the only year for which we have a complete annual record. A few samples were taken throughout the year, accompanied by high numbers of analyses in March and especially August. A Coruña first addressed milk quality in 1911, when a local chemist was commissioned to do nine analyses in August, when the heat most endangered milk quality, as was described in the results.[6] Another step was taken in 1913, when a veterinary corps was organized to inspect milk by districts.[7]

By 1930, efforts at regulation became more methodical and stable. In Santiago de Compostela, a registry established in 1921 marked an important step for the city administration. It reflected the need to act more effectively, given the increasing numbers of milkwomen entering the city. However, the sparse documentation indicates that analyses during this decade maintained their prior pace, and samples were only taken in one or two months of each year. This changed in 1932, when health technicians began taking more samples systematically, every two weeks throughout the year, averaging 40-45 analyses monthly. In A Coruña, we find the same tendency. Once the laboratory was operative, analyses were done weekly throughout the year, but until 1935 fewer samples were taken than in Santiago (average of 16 per month). From the earliest official registries, each analysis was identified with the name of the milkwoman and the number of the jug from which the sample was taken. In the absence of a complete registry that would indicate exactly how many milkwomen supplied these cities, we estimated this data using the highest number that appears in the bulletins.

In Santiago de Compostela, we know that at least 923 milkwomen were registered in 1925 and the number had increased to 1,420 by 1931. In more populous A Coruña, there were at least 3,009 registered milkwomen in 1935. Neither city had urban cattle facilities, nor was milk brought in from other regions by train. Thus, supply depended completely on the milkwomen, and their increasing numbers would reflect increased demand.

In the early twentieth century, the liberal Spanish state did not actively intervene in the milk supply or in promoting consumption, as occurred in the second half of the century. This depended more on local agricultural providers and city governments that wanted to ensure the quality of the milk supply. In

6 'Expedientes sobre reconocimiento por el farmacéutico municipal de varios jarros de leche'. Caja 7545. Fondo Sanidad. Archivo Municipal de A Coruña.

7 'Expediente sobre reconocimiento de leche que se expende a la población a fin de decomisar la que se encuentra alterada'. Caja 7545. Fondo Sanidad. Archivo Municipal de A Coruña.

the 1920s increasing consumption created a deficit of milk and dairy products in Spain.[8] National associations such as the National Association of Cattle Farmers of the Realm responded by demanding greater state implication in fostering milk production.[9] During that liberal era, however, the state did not promote 'productivist' projects or have the capacity to do so, as it did in the second half of the century (Fernández, 2007: 327-44). Instead, initiatives to increase milk production emerged from family-based agriculture: not from a state-driven productivist dynamic, but from attempts to take advantage of increasing milk demand. Farming families were aware of what was happening around them and responded to increasing consumption by sending more milkwomen to the urban centres, where demand was highest. In the 1920s, this growing tendency led to new formulas for commercialization, appearing the first cooperatives. It also implicated new actors, such as technical experts and agrarian organisms for improving dairy production.

First of all, Galicia had an important cheese tradition in specific areas. It was habitual to make cheese for self-consumption in remote areas, as one more product that could be obtained from milk. From this, many families' cheese-making enterprises originated that by the 1920s were reaching markets across the peninsula and were present in the trade fairs for state agricultural products.[10]

Along with this, four dairy cooperatives were established in Galicia in the 1920s. After gathering milk from local producers, cooperatives distributed it in the nearby villages. Peasants themselves founded these cooperatives and rapidly adopted technological innovations that clearly demonstrated their capacity for growth. For instance, the Laíño Cooperative of Producers (*Cooperativa de productores de Laíño*) was able to process 1,500 litres of pasteurized milk every day, which was sold in nearby villages and in the town of Santiago de Compostela.[11]

In the field of cattle improvement, in the early twentieth century, cattle contests organized by agrarian societies to accelerate cattle improvement became increasingly commonplace throughout Galicia.[12] In the 1920s, agricultural researcher and geneticist Cruz Gallastegui, director of the Biological Mission of Galicia (*Misión Biológica de Galicia*) – one of the main research organisms in Galicia – made strides in this area by promoting dairy production contests

8 *Estadística General de Comercio Exterior de España* (1926-1927, 1935). Dirección General de Aduanas. Madrid.

9 *Leche, queso y manteca: Estadística de la producción en España*. Asociación General de Ganaderos. Madrid. 1925.

10 Rof, J. *La producción de la leche y sus derivados en Galicia*. Cátedra de la Producción Pecuaria de Galicia. 1953 and "El queso gallego" by Juan Rof Codina in biweekly newspaper *El Faro Villalvés* N° 13. 1932.

11 Laíño is a parish that belongs to the municipality of Dodro, located in the South of the province of A Coruña. Rof, J. *Cooperativismo*. Talleres Tipográficos Ruíz de Lara, Cuenca. 1932 and different news between 1927 and 1933 in the daily newspapers *El compostelano* and *El Pueblo Gallego*.

12 Cajas 61/442 and 61/443, Dirección General de Ganadería (Libro 1.1), Fondo Ministerio de Agricultura (11), Archivo General de la Administración. Madrid. See also Conde (2015: 49-57).

(Fernández, 1992: 297-98; Conde, 2015: 95-96). The objective was to identify the best milk cows of local breed *Rubia Galega* and monitor their offspring, using the novel system of Genealogical Record Books. Implementation of this project depended largely on the public organism that financed and organized the contests: the provincial government of Pontevedra, where the Biological Mission facilities were located. It monitored dairy production from 1923 to 1929.

At another major agricultural research centre in the region, the Experimental Farm (*Granja Experimental*) in A Coruña province, director Leopoldo Hernández Robredo began work in 1933 on a programme to improve dairy production (Fernández, 1992: 293-94). Unlike Gallastegui, he directly crossed local cattle with Schwyz cattle, another multi-purpose Swiss breed but with better milk production than the local breeds.

Records from this organism indicate that after importing several head of Schwyz cattle, the programme remained active until 1939, resulting in a significant number of cross-bred descendants.

These improvement programmes were autonomously initiated by local researchers, unaffiliated with any state projects or stipulations from the Ministry of Agriculture. The decentralized research framework enabled the directors of the two centres to promote projects adapted to the dynamics and possibilities of each area (Fernández, 1992: 105-61; Fernandez, 2007: 109-54). The agrarian experts at that time understood farmers as innovators and sought improvements adapted to productive needs as well as to the reproductive circumstances of farming families. The dairy improvement projects discussed here fit perfectly with this premise. While seeking to improve dairy production so that farming families could take advantage of increasing demand, the formulas proposed by Gallastegui and Hernandez Robredo also prioritized maintaining multifunctional capacities in cross-bred descendants. The ideal: to increase milk production in cows that were also robust for work and had good offspring. Otherwise, farmers would not adopt external technological improvements. Neither agronomist pursued increased output at any cost. Accordingly, neither made recourse to the Friesian breed, the clear choice for an exclusively productivist objective. In fact, there were no Friesian cows in the region during the first three decades of the twentieth century.[13] They appeared in the 1940 as the first indicator of a change of course in agrarian transformation (Fernández, 1992: 300).

The facts presented thus far lead us to attribute a much more relevant role to milk in Galicia than was previously assumed. Far from being overshadowed by meat production, an innovation dynamic was emerging that combined endogenous productive improvement with cooperative formulas for commercialization.

---

13 From the 1920s on, merchants and private enterprises based in Madrid and Barcelona were importing Friesian cattle for urban dairies in those cities and dairy regions in the north of the Peninsula (Cajas 61/676-678-685-694-696-702-705, D.G.G. (Libro 1.28-1.30), Fondo Mº de Agricultura (11), A.G.A. Madrid. For Galicia, we only know of one cattle importer in the northeast who advertised Friesian cattle (Fernández, 1992: 299).

Though very embryonic – with only a handful of cooperatives and early experiments in dairy improvement – it invites us to ask what might have happened if the Civil War had not interrupted it. Associated lines of research and cooperatives disappeared in the wake of that conflict, but the issue of dairy improvement did not. It was taken up afresh with new formulas that aligned with quite different objectives.

III. Innovation demolished: Agrarian modernization projects in early Francoism (1940-1955)

The Spanish Civil War put an end to all prior research lines and marginalized most of the practising agrarian technical experts. The entire innovation structure that had been emerging and spreading – the integrated effect of family-based agriculture, agrarian technical experts and research centres – was dismantled (Fernández 1988: 244-46; Fernández, 1992: 463; Fernández 2007: 217-31). In this new regime born of a coup d'état, Francoist victory and the construction of the New State also entailed a project for agrarian renovation based on a logic of breaking with the past (Lanero, 2011: 432). It established the productive objectives of post-war agricultural policy along with the modernization and industrialization deemed necessary to achieve them. In reality, the new policy was nothing more than an attempt to fulfil the decades-old aspirations of 'educated agriculture'. These were embodied in the National Association of Cattle Farmers of the Realm, for example, or the earlier engineers who drafted the first statistical records on cattle in 1891.[14]

In Galicia, the modernizing intent of the Francoist agrarian authorities was on display at the 1944 Agricultural Congress of Galicia (*Congreso Agrícola de Galicia*), a publicity event to present the lines of action for modernizing the agrarian sphere (Bernárdez, 1999: 85-95; Lanero, 2011: 433-38). The agrarian leaders in this new project made milk a central issue, which clearly indicates the relevance it had been acquiring in prior decades. The difference, with respect to the former innovation pathway, however, was the intent to improve dairy production by following strictly productive criteria. For the first time, the concept of *dairy specialization* of cattle appeared in Galicia, but at this moment in the Galician countryside the conditions were not in place for directly implanting the Friesian breed. Peasant families could not use them for work, nor could they manage to feed them adequately to get the best production.

Improvement efforts would focus on enhancing the dairy aptitudes of the local *Rubia Galega* breed, using Gallastegui's earlier experimental method of selecting cows in dairy contests and monitoring production of offspring.

---

14 *La ganadería en España: Avance sobre la riqueza pecuaria en 1891.* Junta Consultiva Agronómica.

Monitoring of dairy production in the four Galician provinces had begun in 1940, four years prior to that Congress, and could be considered the first state intervention in dairy cattle improvement. In 1933, the recently created General Livestock Directorate (*Dirección General de Ganadería*) had officially established a Dairy Yield Verification Service (*Servicio de Comprobación de Rendimiento Lácteo*) in the northern provinces of the peninsula that functioned in the same way as the Genealogical Record Books implemented by Gallastegui in the 1920s.[15] Provincial Livestock Development Boards (*Juntas de Fomento Pecuario*) had been established in Galicia and tasked with registering livestock. Official veterinarians from these organisms tried to get farmers to identify the best dairy cows.[16] After the 1944 Congress, dairy registry became a fundamental line of work, along with expansion of meadows and cattle fodder crops.

The project announced in the Agricultural Congress materialized as the Agricultural Plan for Galicia (*Plan Agrícola de Galicia*). It included the Bovine Selection Service (*Servicio de Selección de Ganado Vacuno*), which handled dairy registry in the four Galician provinces. The funding difficulties, the turf wars between organism directors and the feeble results of the initial project have been studied extensively (Bernárdez, 1999:117-42; Fernández, 2007: 247-51; Lanero, 2011: 473-79). The Dairy Yield Verification Service, though independent from the Agricultural Plan for Galicia, faced similar problems and followed a parallel trajectory. The veterinarians responsible for reporting results emphasized the difficulties in doing the verification work. This was largely due to personnel shortages and the unwillingness of farmers to register their cattle for fear of confiscation, which had been commonplace during the war and early post-war years.[17]

Dairy improvement of traditional cattle was not at all hegemonic. Contrary to the aims of official improvement policy, husbandry stations for crossbreeding with Simmental bulls persisted across Galicia.[18] Meanwhile, actions for modernizing dairy improvement came nowhere near fulfilling their objectives. Thus, not only had the work of pre-war agrarian experts been interrupted in this sphere, but the new top-down project also failed.

Furthermore, as part of its production intervention strategy, the state attempted to organise the milk supply to cities by creating a National Plan for Milk Distribution Centres (Langreo 1995: 158-63; Calcedo, 1996: 236-37). But implementation was very slow and in Galicia, these Milk Plants were not even operative until the mid-1960s. This favoured the growth of private milk companies, some of

15 The *Dirección General de Ganadería* belonging to the *Ministerio de Agricultura* had been created in 1931 as the maximum state authority on livestock issues.

16 Caja 61/414 – Legajo 390, Dirección General de Ganadería (Libro 1.1). Fondo Ministerio de Agricultura (11), Archivo General de la Administración, Madrid.

17 Idem.

18 Significantly, the Civil Government of A Coruña did not agree to eliminate official and private breeding stations with Simmental bulls until 1952. Carpeta 'Datos estaciones pecuarias y paradas para las memorias de la D.G.G. en 1966', Caja 55/24214, D.G.G. (Libro 1.28-1.30), Fondo Mº de Agricultura (11), A.G.A. Madrid.

which had been established in the 1940s. The first dairy transformation industry, *Ilepsa*, was founded in 1939 as part of Nestlé, to produce condensed milk. Next, the *Larsa* cheese factory was set up by local entrepreneurs. After that came *M.G.*, which belonged to a cheese industry in the neighbouring region of Asturias. These factories monopolised milk supply and transformation, expelling local families of breeders from this activity. For unknown reasons, we have lost all trace of the few cooperatives that existed prior to the Civil War. The new cooperatives had to integrate into the corporative structures of the regime, such as U.T.E.C.O. (*Unión Territorial de Cooperativas del Campo* – Territorial Union of Rural Cooperatives). The most prominent case is the Leyma cooperative which owned the Central Lechera of Coruña, the Milk Distribution Centre that supplied that city. There were other small attempts to start cooperatives with state support, but most of them could not withstand the strong competition in milk collection from the large companies just named.

In 1955, the Agricultural Plan for Galicia stopped working with dairy registries for local cattle to implement new diary improvement formulas. At the Agrarian Congress organized by the Official Agrarian Union Chamber of A Coruña (*Cámara Oficial Sindical Agraria de A Coruña*) in 1955, it was clear that the 1940s defence of local cattle selection had been abandoned to pursue crossbreeding with foreign varieties.[19] We should understand recovery of this formula as an attempt to remedy the deficiencies of the prior project. Production criteria remained the same, but the focus shifted to accelerating cattle improvement to meet production needs more quickly. Curiously, it had almost been forecasted ten years earlier by César Fernández Quintanilla, the agrarian engineer responsible for the cattle improvement project of the 1940s, in this interesting quote:

> If tomorrow we find that the local cow did not give everything we expected, it will not have done any irreparable damage (…) the crossbreeding that will have to be assessed at that point would be quicker and much more effective.[20]

Economic history has interpreted the autarky period as a shutdown of economic development, describing it as 'twenty lost years for agriculture and cattle farming in productive terms' (Calcedo, 1996; Martínez, 2000; Domínguez 2001a: 57-62; Domínguez, 2001b: 42-43). Measures taken in the late 1950s began the recovery from this parenthesis. However, and in agreement with Fernández (2007: 344-48), we should not understand the Francoist agricultural policies of the late 1950s as the reactivation of a pre-Civil-War process. Rather, modernizing measures were enhanced at this time to achieve productive objectives based on actions more aligned with Green Revolution criteria. Milk industries also became active agents in the modernization process and played a role equivalent to that

---

19 Gómez De Azcárate, J.; and Gómez, J. Ponencia *Ganado Vacuno*. Cámara Oficial Sindical Agraria de La Coruña – Plan Agrario de 1955. Centro de Investigaciones Agrarias de Mabegondo, 1955.

20 Fernández, C., *Apuntes para un estudio de la mejora vacuna en Galicia* (Borrador). Carpeta 'Mejora Vacuna en Galicia 1939-1959'. Centro de Investigaciones Agrarias de Mabegondo.

of the state in transmitting agrarian technologies to farmers. Ilepsa implemented such strategies from the beginning by subsidizing breeders who were more receptive to the purchase of cattle or to expanded grazing (Iturra, 1988: 38-48).[21] From the end of the 1950s, both Ilepsa and Larsa actively participated in cattle improvement projects. They funded two centres for monitoring milk production and promoted cattle contests.[22] After they were dismantled in the 1940s, the new model only increased the distance from pre-war innovation dynamics.

Accordingly, in 1956 the General Livestock Directorate began to import cattle breeds-for experimentation: the Schwyz (imported before the war) and South Devon – both rustic, mixed-aptitude milk and meat breeds – and the more demanding Dutch Friesian with exclusive dairy aptitudes.[23] The rather different characteristics of these cattle reflected two improvement proposals. The first South Devon head were sent to a research centre in the Galician interior, which had a dry, harsh climate and less meadowland. The objective: extensive breeding with local cattle to improve milk and meat production. However, only a few bulls were sent to the artificial insemination centres in the Galician interior.[24] In the end, the South Devon and Schwyz improvement options were more of a temporary solution to increase production without compromising the rustic features and work capacity still needed by farming families. As funds and conditions permitted, the Francoist authorities worked to spread Friesian cattle across the region, through direct distribution or breeding with traditional cattle. Massive implantation of this breed marked the definitive break with the innovative logic of family-based agriculture and the advent of a new agricultural model that answered to the productive rhythms and needs established by the processing industry or the state.

IV. Promoting the dairy breed to 'modernize the countryside': consolidation of dairy specialization (1960-1975)

By the 1960s, Friesians were a familiar breed in Galicia. In fact, the Local Board for Livestock development in Santiago de Compostela had used Friesian bulls in

21 Interview with Francisco Díaz Santalla, technical adviser of Ilepsa's Department of Agrarian Development (1972-2012). Recorded by Telmo Otero Rodríguez (Padrón, 17-12-2020).

22 'El Primer campeonato Regional de rendimiento lechero de la raza vacuna gallega Rubia ' (Interview with Benito Fernández García Fierro, Provincial Chief of Livestock) in daily newspaper *La Noche* (4/02/1957) and 'Lalín' (cattle contest report) in daily newspaper *La Noche* (10/10/1959).

23 Carpeta ' Importaciones frisona', Caja 55/24214, D.G.G. (Libro 1.28-1.30),Fondo Mº de Agricultura (11), A.G.A. Madrid.

24 Fernández, C., *Boletín Español de la raza South Devon*. Centro de Investigaciones Agrarias de Mabegondo. 1958.

its breeding stations during the 40s.[25] Another report from 1944 indicated that 11% of the cattle in a municipality in southwest Galicia were crossed with this breed.[26] These small manifestations reflect Friesian crossbreeding based on the innovative logic of farming families. In the 1960s, however, the Francoist authorities attempted the rapid and massive implantation of this breed. In this section, we will follow the actions involved in this undertaking, using documentation from state organisms such as the General Livestock Directorate, and organisms in the corporative framework tasked with local implantation. These include the Official Agrarian Union Chambers (*Cámaras Sindicales Oficiales Agrarias*, C.O.S.A.) at the provincial level and the Union Brotherhoods of Land and Cattle Farmers (*Hermandades Sindicales de Labradores y Ganaderos*, H.S.L.G.) at the municipal level.

In 1959, documents sent from C.O.S.A. to the H.S.L.G., informed of the plan to import Dutch Friesians, 'as the breed with highest milk production and therefore highest yield to the farmer'.[27] It intended to import 'five or six thousand Dutch head and even more if demand is greater. The Ministry of Agriculture has great interest in establishing 'a strong dairy cattle nucleus in our region'.[28]

The cattle had to be requested by the peasant families themselves; the state paid for half the price of each head and its importation. Before the cattle was handed over to the farmers, veterinarians would verify that the petitioners met the necessary conditions for the survival of the animal. Because these cattle were brought in as an experimental project, the contract conditions were very demanding for the farmers, especially concerning animal nutrition and hygiene.

This is not surprising if we understand Friesian cattle as an exogenous technology being implanted with all haste in an adverse environment. The Friesian cattle came from damp, cool environments with more food available year-round and required special care to maintain their productive qualities. Thus, modernizing actions to increase feed capacity and adapt living conditions for a more demanding breed were considered necessary complements to importation.

> Importing select cattle and distributing it in the province without coordinating this activity with others to develop an effective cattle improvement plan, might show few results or even become a negative factor. For this, we understand that the importation of select cattle should be accompanied by these other improvement actions.[29]

---

25 I.U. 061/002 Carpeta *Junta Local de Fomento Pecuario. 1932-1943*. Archivo Histórico Universitario de Santiago de Compostela.

26 Cited in Fernández, C., *Apuntes para un estudio de la mejora vacuna en Galicia* (Borrador). Carpeta 'Mejora Vacuna en Galicia 1939-1959'. Centro de Investigaciones Agrarias de Mabegondo.

27 Communiqué 20/60 of the Cámara Oficial Sindical Agraria de la Provincia de la Coruña to the Hermandades Sindicales de Labradores y Ganaderos. Arquivo do Reino de Galicia, A Coruña, 1959.

28 Idem.

29 Communiqué 8/60 from C.O.S.A. to H.S.L.G., A.R.G. A Coruña, 1960.

To this end, many state actions were carried out across Galicia. In the province of A Coruña, for example, a specific modernization plan was implemented in 1960: Plan for Agricultural Expansion of A Coruña. This *comprehensive agricultural modernization package* used subsidies to promote the acquisition of Friesian cattle and the adaptation of farms to create adequate conditions for establishing them in rural Galicia. To address nutrition, land was tilled to make artificial meadows and increase fodder corn crops, the staple food of the new breed. Pesticides and fertilizers were promoted to treat these crops, silos were built to store grass, and new machinery was acquired to process grass and corn. The plan also supported the construction of new stables with trenches for processing excrement into liquid manure. Milk companies had an active role in achieving these objectives by allocating significant resources to fund research projects to increase the productivity of pastureland. During the 1960s, Ilepsa financed two demonstration fields in strategic areas along with research on grazing lands, conducted by the agronomist Valeriano Yepes.[30] Cattle nutrition was a primordial issue for dairy transformation companies. They – especially Nestlé – dedicated ample human and economic resources to increasing production from peasant families. Factories also invested significantly in refrigeration technologies. As with other technologies in the early 1970s, private industry collaborated with state funding to finance a network of community refrigeration sites in parishes where significant amounts of milk were already being collected or where a production increase was expected.[31] This milestone technology forever transformed dairy systems and collection capacity in the rural context.

Returning to the topic of imports, documentation from COSA and the General Livestock Directorate indicate that cattle arrived throughout the 1960s, to the extent that 606 head of Friesian cattle had been directly imported and ceded to farm families by 1969.[32] However, we will see that the spread of Friesian cattle depended more on the local efforts of research organisms and imports by private entities.

In this experimental venture, imported cattle was requested by interested farmers under an assignment contract. Alongside the farmer's obligation to provide adequate care, the state would subsidize half the cost of each head of cattle. In return, the farmer would give the first female calf to the research organisms when it reached nine months of age. The goal was to recover calves and create select Friesian cattle centres. For every imported head, the state received another without cost which had been raised by the farmer, which could be ceded under

30 Interview with Francisco Díaz Santalla, technical adviser of Ilepsa's Department of Agrarian Development (1972-2012). Recorded by Telmo Otero Rodríguez (Padrón, 17-12-2020).

31 Interview with Francisco Díaz Santalla, technical adviser of Ilepsa's Department of Agrarian Development (1972-2012). Recorded by Telmo Otero Rodríguez (Padrón, 17-12-2020).

32 Carpetas 'Partes de cesiones y recuperaciones. 1969-1972', Caja 55/24233 a 55/24238, Dirección General de Ganadería (Libro 1.28-1.30), Fondo Ministerio de Agricultura (11), Archivo General de la Administración. Madrid.

Table 5.4. Friesian cattle registered in the genealogical books

	1964	1965	1966	1967	1968	1969	1970
Collaborating farmers	385	424	445	472	487	504	531
Recorded in the final registry for females	75	149	194	238	292	364	507
Recorded in the final registry for males	24	33	42	63	76	99	125
Recorded in the birth registry	665	1214	1214	1567	2359	3451	5138
Recorded in the auxiliary Registry (cattle recorded after the first registry was closed)	778	1204	1561	1919	2303	2013	3031
Total registered	1542	2380	3001	3787	5030	6527	8801

Source: Author's own based on data from the file labeled 'Memorias del Servicio de Libros Genealógicos y comprobación de rendimiento del ganado' Dirección General de Ganadería. Archivo General de la Administración. Madrid.

the same contract conditions to another farmer (and at fourteen months it could give birth).

The research organisms also controlled lineage through genealogical records and by monitoring milk production. In this way, they could propagate controlled, select bloodlines on the farms that had initially acquired the cattle and eventually on other farms throughout the province.

The following data from A Coruña province illustrate how select cattle multiplied under the control of the agrarian organisms.[33]

The number of registered animals and participating cattle farms grew as the decade progressed. Meanwhile, the agrarian organisms increased their livestock from 1,542 select animals registered in the Genealogical Book in 1964 to 8,801 in 1970

We must consider that these data refer only to heads imported by the state and that their destination was mainly the Protected Family Farms *(Explotaciones*

33 Carpeta 'Memorias del Servicio de Libros Genealógicos y comprobación de rendimiento del ganado', Caja 61/12837, D.G.G. (Libro 1.03-1.05), Fondo Mº de Agricultura (11) A.G.A. Madrid.

Familiares Protegidas).[34] This special category had been created by the state to promote farms where officials had identified good feeding and hygienic conditions, to place imported cattle there first. These owners fit well with the model of innovative smallholders who maintained close ties with the local elites and were well informed about favourable loan conditions for the acquisition of new agricultural technologies or cattle. Francoist authorities were particularly interested in promoting such farms and using them as models to encourage the spread of new technologies.[35]

However, the number of heads imported and managed by the state agrarian institutions was really small compared to the total number of imported heads. Private importers were actually more responsible for the spreading of the Friesian breed. In the following years, most breeder families gradually bought and incorporated imported heads, based on their own economic reproduction strategies and the profitability of milk.[36]

Along with the dissemination of these select individuals, the expansion of cattle with improved milk aptitude depended also on crossbreeding of Friesian bulls with local cows. We used General Livestock Directorate surveys of those in charge of the artificial insemination centres throughout the region from the mid-1950s on to assess the expansion of crossbred cattle.[37] The surveys were done in 1970 but contained information from the moment the centres were opened concerning the number of inseminations (unfortunately the breed inseminated was not indicated), the technical means available and the predominant breeds in each municipality. In 1970, pure Friesian cattle outnumbered the local breed in 14 municipal districts of A Coruña where nuclei for select individuals had been established. Crossbreeds and pure Friesians were distributed more evenly in the rest of the province.

Family farms began to replace their multifunctional native cattle *Rubia Galega* with dairy cows to increase milk production as a family reproduction strategy for the future. This transformation happened at a different pace for each family farm and the 'modern' family farms that emerged were more determined to adopt new technologies. The spread of refrigeration technology from the 1970s on – first through community refrigeration sites and later with individual cold storage tanks – accelerated this dynamic. As businesses increased their milk collection capacity, farmers sought to produce more. This drove dairy specialization among rural families and further concentrated the process in specific areas.

---

34 'Informe resumen de la labor desarrollada por la Junta Coordinadora de Mejora Ganadera en el año 1959', Caja 61/3694 – Legajo 3010/3, D.G.G. (Libro 1.07), Fondo Mº de Agricultura (11) A.G.A. Madrid.

35 Caja 5013, Archivo de la Diputación Provincial de A Coruña. A Coruña. 1956.

36 Interview with Francisco Díaz Santalla, technical adviser of Ilepsa's Department of Agrarian Development (1972-2012). Recorded by Telmo Otero Rodríguez (Padrón, 17-12-2020).

37 Carpeta 'Informes sobre las visitas realizadas a los Centros de Inseminación', Caja 55/24217-18, D.G.G. (Libro 1.28-1.30), Fondo Mº de Agricultura (11), A.G.A, Madrid, 1970.

It explains why milk specialization occurred mainly in areas under the direct influence of the milk factories during the following decades. The state and private milk companies managed to reach their modernization objectives by occupying the void left by the breakdown of the peasant innovation model decades earlier.

V. Conclusions

Rather than being relatively insignificant compared to meat, milk was driving change from the early twentieth century on. This was implemented by the farmers themselves, though they are often characterized as reticent to change. Farming families demonstrated their capacity to respond to increasing milk consumption, first by sending their women to where they could sell it – long before there were registries, surely – and later by looking for new commercialization formulas, such as cooperatives. Technical experts assisted the trend by seeking to improve milk production in a way that prioritized adaptation to the needs of farming families over purely productivist ends.

The research structure understood farmers as innovators and sought to introduce endogenous improvements that fit with their logic. From our perspective, this confluence of farming families and research was creating an innovation dynamic that included both productive improvement of cattle and commercialization strategies to improve supply.

This dynamic was truncated in 1940s, when the state tried to mediate the process of agricultural change by defining from above the productive objectives, the pace for achieving them and the modernizing actions they entailed. However, we have seen that the state projects for improving milk production in the 1940s did not have the desired impact. Also, the implementation of a centralized supply system came late, at a moment when even the state was leaning more toward private business initiatives. Whereas the pre-war innovation dynamic emerged from the imbrication of farmers and research, now it was occurring at the confluence of the state and recently established private industries. This enabled modernizing objectives to be met from the late 1950s on, giving dairy companies an active role in state modernizing projects. They accelerated implementation of new cattle formulas with imported, exclusively dairy breeds and the transmission of new technologies for crops, machinery, refrigeration… all aimed at achieving the highest yield from these new cattle. Implantation of this complete modernizing package sought to increase a surplus intended exclusively for new processing industries, generating dynamics that would increase the scale of dairy farms.

We can see that from very early on, agriculture in Galicia and other regions was subjected to an external modernizing influence that conditioned farmers to transform by adopting technology, based on what the rest of society required of them. The results and the implications of those agrarian change processes depended on the circumstances surrounding their adoption in relation to the state and industry as driving forces: with official organisms, research centres and technical experts as transmitters, and peasant and farmer associations as mediators and recipients. In examining the case of milk in Galicia, we distinguished two models of agrarian development. Though we found features of gradual transition between them, we do not think it discredits the idea of a rupture, due to the dramatically different circumstances in which they developed.

The state in the early twentieth century was only beginning to demonstrate its modernizing will and create pathways to transmit it. This was the function of the research centres with their agrarian technical experts. At that time, associations gave rural society mediating capacity to contend with external influences. This conditioned the agrarian experts and research organisms to create technologies adjusted to the needs of farmers interested in obtaining benefits – in this case from the demand for milk – but whose reproduction depended on the limitations of the rural context. The rupture came when these circumstances changed.

From the 1940s on, the state had clearer modernizing lines and projects. The old peasant associations lost their mediating role as the Francoist regime implemented new transmission pathways, such as the H.S.L.G. and the Agrarian Chambers, while also increasing the capacity for agricultural technology transfer. However, dairy transformation companies and private actors were actually more successful in achieving this objective. Dairy companies fostered technology transfer to adjust peasant production to their productive requirements through dairy specialization. Peasant families interested in the benefits offered by industry adopted these new technologies. This allowed them to overcome the limitations of the context and exploit it to achieve higher production. At that point, agriculture became subordinated to private industry, which provided production technology, and the dairy processing industry. This relates to current concerns over low agricultural prices and the survival of dairies.

Though we have identified two distinct processes, there were also continuities, as the modernizing influence was always driving change. Thus, when peasant families of the first period selected cattle that produced more milk, crossbred such cattle with selected heads or acquired specialized cattle, they were adopting a technology that caused them to see themselves as modernizers. When peasant families of the second period decided to specialize in milk production, it could be said that this innovation originated from within the families. Similarly, when the experts tried to improve the milk aptitude of traditional cattle or import breeds that would make it possible to preserve multi-functionality during early Francoism, it might seem that such improvements were mindful of the needs of families. What marks the difference, from our perspective, is that in the first case, this modernizing improvement originated in the peasant context and responded to their needs, interests or logic. Thus, these changes were not so much modernizing as 'innovative'. In the second case, dairy specialization and the adoption of technologies on that scale may have been an innovation that families chose, but it was really induced by the state or industry, which leads us to understand it as an external or 'modernizing' change. The same occurred with the introduction of multifunctional breeds. Experts during early Francoism still valued these cattle because they adapted to the needs of the families at that time, but this also met the productive objectives of the state and industry. Clearly there was a modernizing dynamic that continued throughout the twentieth century, but the circumstances in which it was applied gave rise to different transformations and implications: from maintaining peasant logic and relationship with the environment to ignoring or disregarding such logic and exploiting the environment. As we can see, studying the past has shed light on alternative development paths for agriculture that are not so far removed from the questions we ask today about sustainable food production or the demise of rural life.

Bibliography

Alenda, R. (1983) 'La preocupación por la mejora del ganado vacuno gallego: su historia y metodología', *I Xornadas de Estudo sobor do Tema Recursos Básicos da Agricultura Galega*, Sada, pp. 351-81

Atkins, P. J. (2017) 'The Long British genealogy of quality in the British drinking-milk sector', *Historia Agraria*, 73, pp. 35-58

Bernárdez, A. (1999) *Planificación agraria na Galicia da Autarquía (1939-1955)*, Santiago de Compostela

Calcedo, V. (1996) 'Crisis, evolución y cambio en la ganadería de vacuno de leche de la España húmeda. (1950-2000)', Domínguez, R. (ed.), *La vocación ganadera del norte de España. Del modelo tradicional a los desafíos del mercado mundial*, Madrid, pp. 205-86

Carmona Badía, X. and De la Puente, L. (1988), 'Crisis agraria y vías de evolución ganadera en Galicia y Cantabria', Garrabou, R. (ed.), *La crisis agraria del siglo XIX*, Barcelona, pp. 181-211

——— (2000) 'Sobre as orixes da orientación exportadora na producción bovina galega: as exportacións a Inglaterra na segunda metade do século XIX', Fernández, L. (ed.), *Terra e progreso: Historia agraria da Galicia contemporánea*, Vigo, pp. 305-52

Conde, D. (2015) *¿Canto vale unha vaca? Da cuestión agraria á cuestión pecuaria*, A Coruña

De la Puente, L. (1992) *Transformaciones Agrarias en Cantabria, 1860-1930. Especialización vacuna y construcción del espacio agrario*, Santander

Domínguez, R. (2001a) 'Las transformaciones del sector ganadero en España (1940-1985)', *Ager*, 1, pp. 47-83

——— (2001b) 'La ganadería española: del franquismo a la CEE. Balance de un sector olvidado', *Historia Agraria*, 23, pp. 39-52

Fernández, L. (1988) 'Agricultura, gandería e economía de guerra: novas orientacións de política agropecuaria para Galicia: 1936-1939', *Grial*, 100, pp. 233-47

——— (1992) *Labregos con ciencia. Estado sociedade e innovación tecnolóxica na agricultura galega (1850 -1939)*, Vigo

——— (2007) *El apagón tecnológico del franquismo. Estado e innovación en la agricultura española del siglo XX*, Valencia

———, Pan – Montojo, J. and Cabo, M. (eds.) (2014) *Agriculture in the age of fascism: authoritarian technocracy and rural modernization, 1922-1945*, Turnhout

——— and Lanero, D. (2020) *Leche y lecheras en el siglo XX. De la fusión innovadora orgánica a la Revolución Verde*, Zaragoza

González de Molina, M. and Martínez, J. (eds.) (2001) *Naturaleza transformada*, Barcelona

Hernández, I. (2012) *La difusión de un nuevo alimento: Producción y consumo de leche en España 1865-1936* (unpublished PhD University of Barcelona), Barcelona

——— and Pujol, J. (2016) 'Economic growth and biological innovation: the development of the European dairy sector, 1865-1940', *Rural History: Economy, Society, Culture*, 27 (2), pp. 187-212

——— and ——— (2017) 'Cities and milk consumption in Europe, 1890 -1936: the emergence of a new market in Spain', *Historia Agraria,* 73, pp. 59-90

———, Muñoz, F. and Pujol, J. (2019) 'A New Statistical Methodology for Evaluating the Diffusion of Milk in the Spanish Population: Consumer Groups and Milk Consumption, 1865-1981', *Investigaciones de Historia Económica,* 15 (1), pp. 23-37

Langreo, A. (1996) *Historia de la industria láctea española: una aplicación a Asturias,* Madrid

Lanero, D. (2011) *Historia dun ermo asociativo. Labregos sindicatos verticais e políticas agrarias en Galicia baixo o franquismo,* A Coruña

López, E. and Valdes,P. (2020) 'A dinámica do sector lácteo galego desde a integración na Unión Europea; perspectivas e retos de futuro nun mercado liberalizado', Fernández, L. and Lanero, D. (eds.), *Leche y lecheras en el siglo XX. De la fusión innovadora orgánica a la Revolución Verde,* Zaragoza, pp, 219-51

Martiin, C. (2010) 'Swedish Milk, a Swedish Duty: Dairy Marketing in the 1920s and the 1930s', *Rural History,* 21 (2), pp. 213-22

——— (2017) 'From farmer to dairy farmer: Swedish dairy farming from the late 1920s to 1990', *Historia Agraria,* 73, pp. 7-34

Martínez, A. (1991) 'Antecedentes del sector lácteo gallego 1890-1935', *Agricultura y Sociedad,* 59, pp. 9-40

——— (1995) *Cooperativismo y transformaciones agrarias en Galicia (1886-1943),* Madrid

——— (2000) 'La ganadería gallega durante el primer franquismo: crónica de un tiempo perdido (1936-1960)', *Historia Agraria,* 20, pp. 197-224

Moser, P. and Varley, T. (eds.) (2013) *Integration through subordination: the politics of agricultural modernisation in industrial Europe,* Turnhout

——— and ——— (2013) 'The State and agricultural modernisation in the nineteenth and twentieth centuries in Europe', Moser, P. and Varley, T. (eds.), *Integration through subordination: the politics of agricultural modernisation in industrial Europe,* Turnhout, pp. 13-39

Naredo, J. M. (2004 [1974]) *La evolución de la agricultura en España,* Granada

Nicolau, R., Pujol, J. and Hernandez, I. (2010) 'Milk, Social Acceptance of a New Food in Europe: Catalonia, XIXth-XXth', *Dynamis,* 30, pp. 145-65

Pan-Montojo, J., Fernández, L. and Lanero, D. (2017) *From the new peasant to the modern agriculturalist, 1918-1960: a comparison of paradigm shifts between Spain and Portugal,* unpublished paper, Rural History Conference, Leuven

———, ——— and ——— (2018) *From peasants with techno-science to modernised farmers: the role of agents in the twentieth-century technological models of the Iberian Peninsula,* unpublished paper, Transrural History, Santiago de Compostela

Picado, W. (2011), 'Breve historia semántica de la Revolución Verde', Lanero, D. and Freire, D. (eds.) *Agriculturas e innovación tecnológica en la Península Ibérica (1946-1975),* Madrid, pp. 25-50

Pujol, J. (2001) *El pozo de todos los males. Sobre el atraso en la agricultura española contemporánea,* Barcelona

———, Nicolau, R. and Hernández, I. (2007) 'El consumo de leche fresca en Cataluña entre mediados del S.XIX y 1935: la difusión de un nuevo alimento', *Historia Agraria*, 42, pp. 303-25

Sanz, G. (2013) 'Health, local taxation and market competition: municipal laboratories and state food control in Spain, 1887-1914' Moser, P. and Varley, T. (eds.), *Integration through subordination: the politics of agricultural modernisation in industrial Europe*, Turnhout, pp. 85-107

Simón, D. and Rúa, M. L. (2020) 'Beneficencia, higiene alimentaria y salud pública (Ourense1910-1936): El Laboratorio Municipal, la Gota de Leche y el Instituto provincial de Higiene' Fernández, L. and Lanero, D. (eds.) *Leche y lecheras en el siglo XX. De la fusión innovadora orgánica a la Revolución Verde*, Zaragoza, pp. 37-65

Simpson, J. (1997) *La agricultura española (1765-1965): la larga siesta*, Madrid

Valenze, D. (2011) *Milk: a local and Global History*, New Haven

Villares, R. (1982), *La propiedad de la tierra en Galicia*, Madrid.

MARCO MARIGLIANO

6. Innovations and associations in Italian dairy cattle breeding

The case study of the Frisona Italiana *breed*

1. Introduction

The aim of this analysis is to understand if and how the appearance of associations, which were increasingly important among breeders and experts in the field from the turn of the century to the early 1980s, was influenced by the increasing need to innovate livestock farming from the perspective of both productivity and profitability. We believe that without the role played by these associations in bringing together people with common interests and providing managerial expertise, the sector would not have been able to develop or even adopt the various innovations implemented in this period.

To confirm this hypothesis, we will refer to the Italian Holstein-Friesian breed of cattle (*Frisona Italiana*) – known primarily for its high milk yield – which became predominant in the 1960s, surpassing other breeds that were previously more widespread in Italy. In fact, after the Second World War, many Italian farmers replaced their dual- or triple-purpose cows with Holstein-Friesians, a highly specialized American dairy breed. This changeover, due to several factors, changed these farmers' mindset. They could no longer consider breeding as a sort of craft and, at the same time, they could not master all the crucial aspects of a new 'scientific' breeding.

One of the reasons for this was clearly financial: with the end of WWII and, above all, due to the intervention of international institutions (UNRRA and OECE), domestic demand for dairy products rose. The various breeds spread all over Italy: in particular, the Brown Swiss *(Bruna Alpina)* were characterized by low average milk yield, by far too low to fully meet the new demand in the country. However, despite this, the Brown Swiss remained the most widespread breed in Italy until 1962 and it took quite a long time before the Holstein-Friesian breed – universally known for its high milk production – was introduced. There are several different explanations for this, but one of the most pertinent is the initial lack of technical innovations – especially the absence of a national herd book – coupled with the need to improve breeding techniques, such as artificial insemination using frozen semen. Moreover, another fundamental issue should be

From Breeding & Feeding to Medicalization, Rural History in Europe, 17 (Turnhout: Brepols, 2023), pp. 135–156
BREPOLS PUBLISHERS 10.1484/M.RURHE-EB.5.135024

added: as soon as the technical bottleneck was eliminated, farmers were forced to group together in order to face the new challenges of modern breeding. Therefore, considering these two factors, this paper can be used to draw comparisons with other top livestock breeding nations, both in Europe and North America, especially as regards the reasons that led farmers to set up new associations.[1]

II. Herd books or herd book? The situation in Italy from the Unification to the Eve of WWII

Indeed, there were two main interlinked issues. First of all, the American Holstein-Friesian breed could not initially be imported into Italy because this would have entailed very high expenses for farmers. Secondly, about 80% of Italian milk was used for cheese production; therefore, milk had to contain large amounts of fat and protein, a characteristic that Holstein-Friesians did not have. Hence, the ideal breed that could have met Italian requirements did not exist. The only option was to improve the existing breeds, making them more suitable to meet the country's demands. However, Italy's situation was not easy: the Unification of the country led to the creation of a socio-political structure that only partially affected agriculture. The State juridical structure was still that existing under the House of Savoy, updated by the Minister of Justice and the Interior Umberto Rattazzi in October-November 1859. In a context of reorganisation of codes, public services, justice, and the juridical body, there was no immediate interest in reforming Italian agriculture.[2] Likely, this interest lacked during the centralization process that took place after the Unification of Italy since the government did not conduct a proper statistical investigation.[3]

---

1 For the Dutch case, one of the most interesting at European level because the Friesian breed originated in the Netherlands, please refer, among others, Theunissen (2008: 637-76 and 2012: 278-309); information on North American associations can be found in Derry (2015. As regards the Italian case, given that the northern part of the country – and in particular the Po Valley – was the most suited and well developed for cattle breeding, it is not surprising that references to the establishment and development of associations are mainly traceable in studies relating to these areas. In particular: Romani (1963); Cazzola (1993); Tedeschi & Stranieri (2011). On the situation in the other European countries, see Grothe (1993). The American Holstein was introduced in Europe for the first time in 1929, starting precisely from farms in the surroundings of Rome and Cremona, in Italy. That marked the beginning in Italy of the "holsteinization" process, which, however, did not affect all the Italian population and ended only some years ago. On the contrary, in Europe selections continued to prefer dual purpose cows, that is for milk and meat. Only in the second half of the 1980s they started the insemination of European Friesians using Holstein bulls' semen. For a more in-depht analysis, see Fusco (1990: 230).

2 On this topic, please refer, among others, to Capone (1981: 24).

3 There were not even overall data on the Italian animal husbandry assets. Only with the royal decree n. 1035, on 2nd June 1927, when the National Institute of Statistics was charged with agrarian statistics, the quantity and richness of the data led to sufficiently consistent and complete information. The first official cattle census was planned in 1928 and carried out only in 1930. Nevertheless, Cesare Correnti and Pietro Maestri published in the 1864 an estimate, difficult to

Also, the contribution from production and processing factories was not sufficient. Since these factories were mainly located and developed in the North of Italy, this led to increased milk and dairy production, mostly in the Po Valley. The factories could not get rid of their overproduction unless reducing prices and trying to sell it in foreign markets.[4] Since the mid-1870s, the sectors underwent a severe crisis due to overproduction and negative consumers' opinions. Frequently consumers were not satisfied with the quality of much cheese, produced by industrial technicians using old processing systems (Fascetti, 1926: 578 and seq.; Cavazzoli, 2002: 139 and seq.).

However, in that period some voluntary organizations called Itinerant Teachers of Agriculture (*Cattedre Ambulanti dell'Agricoltura*) appeared.[5] These Teachers were created by agricultural associations and individual farmers as voluntary organizations to provide some help in the field of agriculture, and they were formed by agronomists and were well-evaluated by the most developed Italian farmers.[6] The

---

verify, of 1.391.731 cows. Cf. *L'Italia in 150 anni: sommario di statistiche storiche 1861-2010*, ed. by ISTAT (Rome: ISTAT, 2010); Barsanti (2002: 105-06). During the Itinerant Teachers of Agriculture (see footnote 6) Congress that took place in Milan on 23rd September 1906, Bartolomeo Moreschi, the then chief inspector for the zootechnical service of the Ministry of Agriculture, referred to the 1881 cow asset census. However, Moreschi himself said that those data were not sufficient to understand the exact amount of cow heads bred in Italy.

4 For example, in 1870 Polenghi, an industrial entrepreneur from Codogno, exported Lombard butter to London using refrigerated wagons. Cf. G. Fascetti, 'Sulle vicende del caseificio in Italia', *L'Italia agricola*, Year 63, no. 12, December 1926, pp. 578 and seq.

5 The first hint of this structure that strongly contributed to the development of Italian agriculture can be found in Pisa in 1839, then in Marche and in the Po Valley. However, they were short-duration initiatives and were linked mainly to the professors that had established them. The first real Itinerant Teacher was created in Rovigo by Pergentino Doni, a professor who established an "Itinerant School of Agriculture", then "Itinerant Teacher of Agriculture". This reform was adopted since 1892 by many local structures like the one in Rovigo, mainly in the North (starting from Parma, then Bologna, Ferrara, Mantova, Novara, and Piacenza) and in the Centre of Italy. The number of these new structures increased significantly, from 37 in 1900 to around 100 in 1915. At the beginning, each Itinerant Teacher was administered by autonomous commissions financed mainly by local administration and to a lesser degree by the Ministry of Agriculture. These organizations were mainly directed by agriculture graduates, but there were also people appointed despite of not being expert in the field. Since 1901 the Ministry of Agriculture had the exclusive right of appointing their directors and in 1906 it established the rules about the composition of vigilance commissions. However, the official establishment of the Itinerant Teachers took place with the law n. 513 on 13rd July 1907, which marked their first legal recognition with the creation of voluntary consortia between the State and local authorities and bodies. Moreover, the new law entrusted the functioning of each Itinerant Teacher to the local structures and farmers which used its services. During WWI, Itinerant Teachers were charged by the Ministry of Agriculture with several official tasks, even if they remained substantially private bodies. Fileni (1954: 349-53); Zucchini (1970). On the actions of the Itinerant Teachers, especially on that in the North of Italy – the most active – refers to Carera (2018); Rognoni (2006); Mazzotti (2017); Corti; Camozzini & Buzzoni: 90).

6 Initially, the interest on the part of the State was scarce. Only the 27th November 1919 decree guaranteed the constant functioning of the Itinerant Teachers through permanent contributions from the government and local bodies. The progressive nationalization of the Itinerant Teachers went further thanks to a 1927 decree establishing that the by-right president of each vigilance

Itinerant Teachers of Agriculture held meetings, conferences, practical lessons, and lessons in several experimental areas, as well as other activities in the field of teaching and propaganda. These organizations, in the first post-war period, contributed to improve the whole agricultural sector. It is important to point out, in fact, that the work of the Itinerant Teachers of Agriculture was not only related to breeding activities; several innovations involving the entire agricultural sector were developed and used by farmers thanks to these Teachers.

For example, Alessandro Romani, as long ago as 1845, introduced in a farm belonging to Counts Jacini the Ladino clover – a spontaneous forage crop in multi-species meadows – which, up to that time, had never been used in monoculture (Fusco, 1990: 20 and seq.). However, it was only at the end of the 19th century that farmers in Cremona began to use it under the guidance of Antonio Sansone, the director of the newly founded local Itinerant Teacher of Agriculture organization (Fumi, 2006: 99 and seq.). And in 1919 Vincenzo De Carolis, who had been the director of Cremona's Itinerant Teacher of Agriculture structure since 1908, ensured the foundation of the *Robbiani* institute. The aim of this institute – subsequently called *Porcellasco* – was to scientifically evaluate which cows had the best milk yield-feed consumption ratio so that they could be selected for cattle breeders in the area.[7]

Furthermore, Jacini's innovation persuaded Antonio Zanelli, a breeder from Reggio Emilia, to set up an experimental center in 1874, with the aim of improving breeding, feeding and fattening methods, as well as to assist and train breeders through theoretical and practical courses.[8]

Focusing on Zanelli and De Carolis, it can be seen that these innovators had two main objectives. On the one hand, they tried to improve the sector by recommending new techniques and systems, and they introduced the first Friesian cows from the Netherlands, demonstrating their commitment to experimentation, which was uncommon at that time. On the other, nevertheless, it was considered fundamental to interpret the needs of agriculture through ongoing dialogue with the livestock farming players – from individual farmers to associations – at territorial level. In addition, they focused heavily on basic training and technical assistance to farmers in the area. The various courses promoted by the two institutes – together with those offered by the various agricultural schools that were being set up, first and foremost the dairy school in Lodi – demonstrate this new interest. The *avant-garde* potential of most of the innovations introduced in this period was not immediately grasped and these 'pioneers' understood that,

---

commission should be a representative of the Minister of Economics. The process of nationalization of the Itinerant Teachers concluded in June 1934 with the establishment of "Local Royal Agriculture Inspectorates", which replaced the Itinerant Teachers, officially suppressed by the 13rd July 1935 ministerial decree n.112.

7 For a discussion on the history of the institute, its role in the improvement of the Friesian breed in Italy, and the introduction of the first Holstein head, refer to Marigliano (2018).

8 On the subject of teaching in agriculture, refer to Saltini (2002). Regarding the figure of Zanelli, refer to Fusco (1990: 21); Cavazzoli (2002: 156).

without an adequate cognitive substrate, no novelty would have spread, either in the Po Valley or in the rest of Italy. Many of the above-mentioned Itinerant Teachers of Agriculture were also fundamental in promoting and disseminating the new discoveries, as well as in training farmers and breeders in the Po Valley, but their role went beyond this. In fact, the technicians of the Itinerant Teachers of Agriculture organizations actively supported companies in financial difficulties and fostered the creation of local cooperatives. This last point is certainly interesting because, thanks to this choice, an attempt was made also in the livestock farming field to combat the isolation of farmers as associations were considered an important driver of innovation in agronomy. In this regard for example, we should remember De Carolis' desire to select the best breeding animals, not only for the *Porcellasco* institute, but for all local breeders.

All these efforts from Itinerant Teachers of Agriculture also led to the creation of several local herd books. As defined by Telesforo Bonadonna (a well-known Italian agronomist), a herd book is:

> The register of records required to ascertain the origin, date of birth, ancestry (paternal and maternal), descent and financial value of purebred animals, belonging to a specific breed and corresponding to morphological and functional standards established for admission.
> When this selection of animal populations (cattle, horses or pigs) was first introduced, the genealogical registration was limited to listing the known ancestors and descendants of a particular animal.
> [...]
> Information needed to establish the ability to transfer the animals' morpho-functional characteristics to descendants was not included.[9]

Certainly, the creation of a herd book provided advantages to both breeders and buyers. First of all, it was possible to establish the value of the animal in a fairly uniform and public manner, providing guarantees to the buyer. At the same time, this guarantee also led to protecting the ownership of the animal.

In Italy, however, a national herd book – a fundamental tool for improving the sector – was sadly lacking. Although important, we can say that with local herd books several problems were encountered. Directors and inspectors decided independently which animals to include in the local herd book. In fact, they could therefore restrict imports into their territory of competence only to animals with characteristics that they considered appropriate, or to animals of particular breeds.

Below is an account of what occurred at Albertini's farm (one of the most important in Italy before the Second World War) in Fiumicino (Rome), where an attempt was made to sell some daughters of Friesian cattle that had been imported from the Netherlands:

9 Bonadonna (1969: 901-02).

> The first difficulties arose from the Herd Book of the Province of Milan which, in December 1933, prevented Count Mapelli – the owner of a farm in that province – from registering a bull that had been booked at the *Bonifica di Torre in Pietra* because only pure breeds were accepted, meaning that only animals of Dutch origin were admitted and not those which had been crossbred with other American bulls for breeding.[10]

The Albertini family, therefore, had to accept to sell their bulls only in those areas where they could register them in the local herd books. Thus it is clear, looking at history from this point of view, why Luigi Albertini – the owner of the *Torre in Pietra* farm in Fiumicino – decided, in the 1930s, to establish close business relations with *Porcellasco*, giving it also some very expensive young Holstein bulls, which had been brought directly from the United States. However, Albertini's estate, which was certainly top level, needed to liaise with a territory other than that of the Roman countryside, where no breeders or livestock farmers could appreciate and benefit from the innovations that had been introduced in the farm. Lombardy, and especially the area near Cremona, was instead much more responsive, thanks to innovation and training activities that had allowed developing the full potential of its territory.

III. First attempts to establish associations: *AIA* and *Associazione Nazionale degli Allevatori di Pezzata Nera* (1944-1945)

The situation in the second 20th century post-war period did not improve at all; it was actually even more confused.

The second major point of this article is associations in the cattle breeding field. At a time of great changes – as occurred at the end of the Second World War – the absence of institutional guidance and proper management of provincial herd books in Italy led to the farmers' desire to group together. This resulted in the foundation of the Italian Breeders Association (*Associazione Italiana Allevatori – AIA*) in Rome on 28 August 1944, when Italy was still divided. The Association was established with the aim of bringing together the breeders of the entire country, even though at first AIA could not represent all members: economies and cultures were still regionalized and very different environments, especially in terms of farmers' entrepreneurial visions. Although AIA was present in a different manner on the entire Italian territory, in the second half of the 1940s, it was the

---

10 The same situation also applied to the introduction of Dutch animals. For example, Salvatore Muzio, President of a consortium for the reclamation of Chillivari (Sassari, Sardinia), ironically said that "whoever is in command (i.e. the powers that be) doesn't allow the Sardinian soil to be contaminated, even by the very modest import of a few Dutch animals on an experimental basis", because the introduction of a Dutch bull, to be used for cross-breeding tests with Brown-Swiss animals, was denied. Albertini (2001: 135).

reference point for all farmers in the liberated Italian areas. Its main function was to provide technical assistance that had previously been the responsibility of the Agricultural Inspectorates (Fusco-Fusco, 1994: 11 and seq.).

AIA, however, had to assist the entire livestock sector. The same lack of central support and the fact that the country was still divided led the breeders of black-and-white cattle, almost entirely concentrated in the Po Valley, to form their own specific association. It was 'sketched out' in a meeting held on 20 March 1945 in the small village of Corte Grande di Quistro (north of Cremona), after the withdrawal of the forces occupying the territories of Cremona. However, the association called National Association of Black-and-White Cattle Breeders (*Associazione Nazionale degli Allevatori di Pezzata Nera*) was still numerically too small to have a say at national level. In 1946, with the appointment of Antonio Segni as Minister of Agriculture, the Ministry once again took over responsibility for monitoring the selection process and entrusted the Provincial Agricultural Inspectorates with carrying out checks on animal productivity and keeping herd books. This meant a return to the concept of provincial herd books, in which each inspectorate adopted its own selective line of the breed by authorizing importations of the same, therefore there were once again two different propensities.

The Association of Black-and-White Cattle Breeders survived for ten years, but it gained prominence at national level only when the custom of setting up umbrella associations became fundamental for improving the breed, that is when artificial insemination using frozen semen gained ground.

IV. Innovations in techniques: artificial insemination and frozen semen after WWII

The practice of inseminating cows using artificial instruments, rather than adopting natural mating, was already known worldwide in the early 1930s (Derry, 2015: 95 and seq.; Herman, 1980: 3). The breeders, however, did not immediately use this new methodology, which aimed to select an animal with specific characteristics in order to achieve livestock improvement. The two main problems that needed to be overcome by using artificial insemination were the low fertility rates of natural mating and, above all, diseases transmitted by sexual contact. It is possible to understand the decision that was taken not to use the new technique in order to make the entire breed more productive if we consider that, until the 1950s, the semen collected was sold fresh and lost its effectiveness in a matter of hours, thus it was impossible to fertilize most of the cows in the area with just a few sires because they were perhaps too far from the cowsheds. Despite this, some breeders began to try and use animals from the United States and Canada that could improve their breeds, although, practically speaking, this was impossible if they did not belong to some sort of association.

On 13 February 1946, during a meeting in Melegnano, the breeders of the province of Milan set up a Bull Consortium (*Consorzio Tori*) section within the

Figure 6.1. Photo of the plane leaving Seattle
Source: Particular in the magazine *Bianco Nero*, August 1947, p. 1.

Association of Black-and-White Cattle Breeders and created a stud bull center in the province of Milan in order to use artificial fertilization thanks to the semen collected from the bulls that had been purchased by the Association. Only a few months later, four other sections of the Consortium were established (Piacenza, Modena, Ravenna and Rome), whose members were keenly interested in using the seminal fluid of American sires that would be purchased (Unknown author, 1946: 7). After numerous postponements and delays, due to red tape and financial difficulties, on 7 June 1947, an airplane arrived at Milan's Forlanini airport from Seattle with four bulls that had been purchased by the important Carnation Milk Farm and they were to be used by the Association (Unknown author, 1947: 1). However, the semen collected from these sires could not be used for a real improvement of the breed on a national scale for two reasons: the above-mentioned provincial herd books and due to the fact that the semen could be stored only for a short time after it had been collected.

However, the invention of frozen semen in 1952 brought drastic changes. This technique was endorsed at the Second International Congress on Animal Reproduction Physiology and Artificial Fertilization held in Copenhagen in 1952, where researchers demonstrated the possibility of freezing sperm with liquid nitrogen, contributing to the final success of this practice (Bonadonna, 1947: 1, 2, 7). Indeed, from that moment on, every breeder could potentially choose any bull in the world to fertilize his cows. Moreover, it was possible to preserve a sire's seminal fluid even for years after its death.

This led to a huge leap forward in improving selection practices. Once again, breeders decided that it was a good idea to group together to purchase foreign breeding bulls in order to improve Italian cattle breeds. At the Italian Dairy Cattle Show in Cremona in 1953, the Association presented its breeders with a proposal to purchase high-ranking bulls, selected for artificial insemination and that would

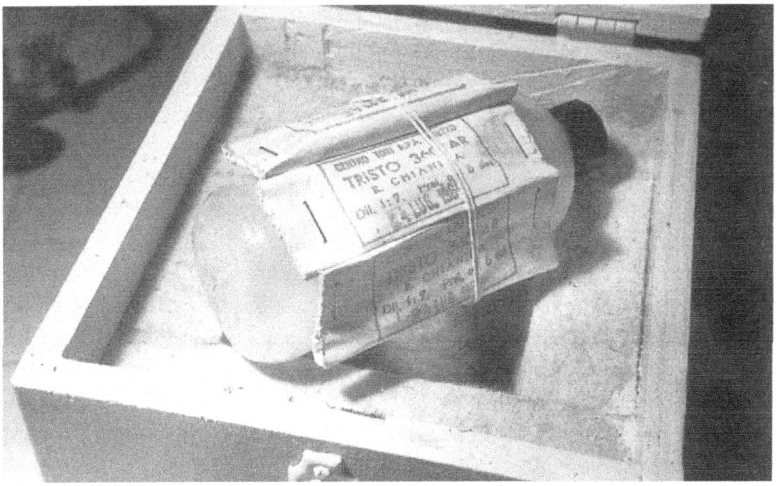

Figure 6.2. Dilution of seminal fluid
Source: Fusco-Fusco, *Aia è...*, p. 222.

be included in the stud centers. This proposition arose from an analysis of what was occurring in the United States, where experiments were underway on deep-frozen storage of seminal fluid from the best American and Canadian selected animals. On an experimental basis, the semen was stored for long periods, even for several months, and served to fertilize tens of thousands of cows throughout North America. Therefore, it was necessary to repeat what had already been done in previous years, i.e. to set up a consortium of breeders which were able to overcome the kind of problems that single breeders were unable to tackle on their own, such as the very high costs of finding suitable animals and the need for the presence of highly specialized structures and operators.

V. The interrelation between innovation and founding associations: the creation of a National herd book and the foundation of *ANAFI* (1956)

The initiative proposed by the Black-and-White Cattle Breeders Association aimed to increase the number of bulls that were able to produce in new generations of cows a significant increase in milk yield, as well as fat and protein content. However, the point was still the same: in order to attain this objective, it was necessary to solve the issue of creating a national genealogical book and its management, but also forming associations of breeders became more pressing. For this reason, the Departmental Inspectorate of Milan – which was entrusted with supervising livestock activities in the area – on 22 May 1956, brought together the Departmental Livestock Commission, the Provincial Agricultural Inspectors with their livestock workers, the directors of livestock stations in the region, veterinarians and managers of livestock associations in order to collectively examine and discuss the situation. On that occasion, the topic of unifying the genealogical books – which were still of a provincial nature – and the existing relations with the associations of breeders at national level were analysed. As it had been impossible to trade certified animals from one province to another in previous years, due to the different approaches of the books, it was agreed that the national associations should be supported and assisted in coordinating the work of the national selection of cattle, although under the supervision of the Inspectorates, which remained responsible for the activity laid down by law. However, it was already clear that the small Association of Black-and-White Cattle Breeders was no longer only a reference point for the breeders of the province of Cremona; it was becoming a place where all black-and-white cattle breeders in Italy could gather together and it was now ready to face national problems.

On 24 June 1956, the national herd book of the Italian Holstein-Friesian cattle was created. On this date, in fact, the Ministry of Agriculture approved the regulations for herd books and the functional controls of the two major breeds in Italy: *Frisona Italiana* and *Bruna Alpina*. The new national books were subject to the supervision of the Managerial Technical Commissions (then still directly controlled by the Ministry) which operated together until December 1961. However, the cattle-breeders association now played a prominent role in taking decisions on the matter and they wanted to be recognized as being more important than before.

The complexity of the sector – which had become progressively more scientific and whose practices were increasingly expensive – no longer allowed individual breeders to act independently. It is therefore not surprising that ANAFI was established just after the creation of the herd book. Following the Unification of the books, on 23 July 1957, in the office of the notary Libero Manfredi (in Cremona), the members resolved to replace the existing Association of Black-and-White Cattle Breeders, founded in 1945, with the new national Association of *Frisona Italiana* Cattle Breeders (*Associazione Nazionale Allevatori di Bovini di Razza Frisona Italiana – ANAFI*). The newly formed association was legally recognized by Presidential Decree no. 1290 of 26 May 1959, published in the Official Gazette no. 35 of 11 February 1960 and the charter of the association, already approved on 3rd March 1956, was notarized. The first objective that ANAFI members wanted to achieve was to obtain the right to manage the herd book independently, albeit under the supervision of the Ministry of Agriculture. This meant having control over the selection choices of the breed and not being subjected to the interference exerted until then by the livestock specialists. Moreover, farmers would no longer have to 'put up with' the intervention of public officials in the choice of animals to be imported: the selection procedure, the choice of bulls or semen and the homogeneity of milk production controls were now the responsibility of the breeders and not of individual civil service employees, who worked in different provinces and regions and therefore could not provide any support to a technical instrument such as the national herd book.

The introduction of the book stepped up the selection process that was required to compete with Northern Europe and North America, however, this was possible only because the individual breeders belonging to ANAFI decided to commit themselves unstintingly within the operational setup of the new associations.

The joint meeting of the Technical Steering Committee of 20 December 1961 is a good example of this. Here, the members – bearing in mind the changes underway in the organization of the activities set out in the national herd book – decided to include in the Technical Steering Committee the representatives of the breeders. Thus, a special Committee, composed of the presidents of the Commissions and the two presidents of the Associations (ANAFI and *Bruna Alpina*), was appointed to draw up the final text. At the same meeting, the transfer of the management of the *Frisona Italiana* herd book from the Departmental Inspectorate of Milan to the ANAFI headquarters in Cremona was also proposed. Both decisions were sent to the Ministry of Agriculture for approval, which was granted by means of two regulations issued the following year, one on 10 February and the other on 30 May, which resolved the transfer of the herd book to the association and the amendment of the regulations. It was also decided that the funding of activities relating to the management of the herd book and the performance of functional checks, hitherto carried out and managed with funds from the livestock initiatives referred to in Article 3 of Law no. 1366 of 29 June 1929, would also be taken care of by the farmers.[11]

We can say that the major accomplishment, the national herd book, the first innovation discussed in this contribution, was completed in that period. The Association then began to focus on promoting new selection trends, the primary one being the 'frozen semen' issue, but also to provide effective guidance to member breeders. The message conveyed to farmers addressed the factors involved in the selection of cattle for their farms: not only in terms of milk production, on which the economic performance of the farm was calculated, but also as regards morphology, longevity, fat and protein content, etc. In particular, it advised members to separately register each cow on their farm, indicating its genealogy, date of birth, present and past milk production career, as well as the data of daily production, control of fat content every fortnight and weight control on a monthly basis. It also strongly recommended controlling sires, indicating the lines of action in choosing the seminal fluid. On the one hand, this aimed to ensure an increase in the quantity and quality of milk production and, on the other, to have animals that were morphologically more in line with high-level cattle farming (Monestiroli, 1962: 16-18).

11 *Nuovo assetto organizzativo della C.T.D. e trasferimento di gestione del libro genealogico all'Associazione Nazionale di razza*, art. 3 and 6, pp. 29, 31.

VI. Expected results: the improvement of the *Frisona Italiana* until the 1960s

The creation of the national association marked a new important turning point. Thanks to new controls on imported cows, there was a further benefit for the breed: a steady reduction in Northern European imports, especially of cattle without a pedigree (Il Comitato Direttivo, 1963: 11-17). In their place, semen from sires was imported, especially from North America, to improve the productivity and quantity of fat and protein in milk of Italian cows.

Table 6.1. Import of Friesian cattle, 1959-1962

Year	Total	The Netherlands	Denmark	West Germany	Sweden	USA	Canada
1959	16,585	15,298 (92.2%)	550 (3.3%)	98 (0.6%)	625 (3.7%)	6 (0.04%)	8 (0.05%)
1960	18,508	17,446 (94.3%)	288 (1.5%)	123 (0.7%)	640 (3.4%)	5 (0.03%)	6 (0.03%)
1961	5,365	5,024 (93.6%)	150 (2.8%)	144 (2.7%)	45 (0.8%)	2 (0.04%)	-
1962	5,885[12]	5,056 (85.9%)	380 (6.4%)	113 (1.9%)	211 (3.6%)	40 (0.7%)	27 (0.4%)

Source: Il Comitato Direttivo, 'La relazione del Comitato Direttivo della Frisona Italiana nell'Assemblea Generale ordinaria del 7 maggio 1963'.

In those years, imports were mainly made by *ad hoc* bodies and cooperatives, which performed these operations with the aim of pursuing their business interests. The largest imports of cattle were dispatched to three provinces: Cremona, Milan and Rome. Therefore, they were also the most interested in regulating import flows. In fact, this led the Provincial Section of Cremona of the national herd book to ask the Technical Commission for clarifications and instructions on two specific topics:

1 if animals born in Italy with the aid of semen imported with a regular authorization issued by the Ministries of Agriculture and Health, pursuant to Article 19 of Presidential Decree no. 1256 of 28 January 1958, were eligible to be registered in the herd book;[13]
2 if necessary, it was also asked if the registration in the book would be subject or not to the approval of the Provincial Technical Commission.

This may seem a technical issue, but in the breeding system that was until then dominated by sellers and inspectors, the subject was of utmost importance.

Following the request of the Cremona section, the Ministry of Agriculture, after consulting the Technical Steering Committee and ANAFI, established that the number of doses to be imported should be proportional to the possibilities of using them on the farms concerned.

In response to the second request, the Ministry decided that the sires, even if they were not imported, should be authorized by the Provincial Technical Commissions, in accordance with Article 16 of the Decree.[14] With the application of the ministerial directives, it was therefore possible to control the import flows, which until then had been in the hands of traders who had a financial interest, not a zootechnical one (Unknown author, 1965: 17-23).

However, the improvement in selection was also due to the approval of Law no. 126 of 3 February 1963, whereby the Ministry of Agriculture endorsed two fundamental principles for the future development of dairy farming: the obligation to register the sires in the herd book and the transfer of the management of herd books to the Italian Breeders Association, or AIA (Camera di Commercio,

---

12 The total of 5,855 head of cattle is given by the import of 48 animals from England and 10 from France.

13 Art. 19: The use of foreign seminal material from animals of proven livestock interest is subject to specific authorizations. [...]. Importation is allowed under the following conditions: that the semen comes from breeding stock whose technical requirements are higher than those provided for in Article 15; that the material is accompanied by a certificate of origin and certificate of good health issued by the veterinary authorities of the country of origin; that the tubes containing the seminal fluid are sealed and indicate, on a special label, clear and irremovable indications of the origin and the reproducer [...]. Presidential Decree of 23 January 1958 (no. 1256), 'Norme di attuazione della legge 25 luglio 1952, n. 1009, sulla fecondazione artificiale degli animali', in *Unione Operatori Fecondazione Artificiale Animale* (Pavia: UOFAA, [unknown date]), pp. 22-23.

14 Art. 16: [...] morpho-genotypical requirements of the sires used for artificial insemination are laid down by the Technical Commissions of the Herd Book [...]. In *Ivi*, p. 20.

Industria e Agricoltura di Firenze, 1963: 5). The law streamlined and regulated the complex issue of cattle breeding and provided a guarantee that the bulls, after approval by the Provincial Livestock Commissions (Articles 3 and 4), would be considered of excellent genealogy.

Immediately after the approval of Law no. 126, during a meeting in March 1963, the Technical Commission Executive Committee of ANAFI modified articles 10 and 11 of the Regulation on the herd book of the breed, establishing new national guidelines for registration of cattle in the book:

> Art. 10: The Book is composed of two fundamental documents:
> 1) Young Cattle Register, in which male and female calves, born from parents that are both registered in the Herd Book, are registered at birth [...];
> (2) Register of bulls and cows [...].
>
> Art. 11: [...] for bulls used for the insemination of cow herds, the requirements are:
> a) official pedigree records referring to at least three generations must be available;
> b) the minimum milk and fat production required by the standard, referring to the subject's mother;
> c) morphological score prescribed by standards.[15]

Through the processing of data made after the introduction of Law no. 126 and the new regulations, it was possible to compare mothers' and daughters' productivity and, later, for the first time in Italy, to perform analytical comparisons between daughters of different bulls.[16]

The ratification of the new law immediately led to a substantial increase in the number of animals registered in the herd book and subject to functional controls.

At the end of 1963, out of a total of 905,611 cows, *Frisona Italiana* cows under production control amounted to 106,441, distributed in 3,247 farms, equal to 11% of the entire population. The cattle registered in the book totaled 165,665 in 3,247 farms (51 animals per farm), of which 3,443 bulls, 70,842 cows (about 74% of

15 [Unknown author], 'La riunione della Commissione Tecnica Direttiva della Frisona Italiana', *Bianco Nero*, No 3, March 1963a, pp. 10-11.

16 It is important to underline, however, that if on the one hand Law no. 126 recognized the commitment and work of the National Associations of Breeders, primarily the Association of *Frisona Italiana* Breeders, on the other hand it accelerated the decrease in the consistency of the cattle of the small breeds which, for financial reasons, were not able to establish autonomous herd books, and therefore could not have good quality bulls for breeding. In the late 1970s, the National Research Council and the Ministry of Agriculture and Forestry carried out a "Project aimed at Safeguarding the Genetic Resources of Animal Populations", with the aim of preserving sparsely distributed animal breeds and species. As a result of this, *AIA* established the "Register of Sparsely Distributed Animals" (*Registro Anagrafico delle Popolazioni a Limitata diffusione*) in order to "study, safeguard and improve these animals for mankind over time". *Atlante etnografico delle popolazioni bovine allevate in Italia (1983)*, ed. by C.N.R., (Milan: C.N.R., 1983), p. 3.

cows under production control) and 91,380 heifers and calves (Unknown author, 1963b: 10-22; 39-41).

VII. The leading role of *ANAFI* in the improvements achieved

Once again, the setting up of associations made it possible to achieve results that were unattainable for individual farmers. For instance, we can consider the case of the 'Tests on a statistical basis' program. In order to comply with the provisions of Article 1 of Law no. 126, AIA, in agreement with ANAFI, set up the aforementioned program relating to controls and, in particular, the average production of all primiparae of the same age and reared in the same environment compared to the average of the daughters of the bull being tested (Consolini, 1964: 10-28). Every year, progeny testing was conducted on 20 young bulls. This choice was made on the recommendation of the provincial commissions concerned and the bulls were taken to the two most important structures of artificial insemination at the time: the Mantova Centre and the Lombardy Livestock Development Authority (*Ente Lombardo per il Potenziamento Zootecnico*), which was entrusted with the task of distributing the seminal material for more than 40,000 cows in the chosen areas of Cremona, Mantova and Milan. The experiment was conducted under the direction and control of a committee set up within the Managerial Technical Committee. At the end of the first stage of the experiment, it was finally possible to equip artificial insemination centers with bulls from national breeders (Il Comitato Direttivo, 1965: 12-29).

But without a clear indication of the best sires in the world, of how many doses of semen were available and at what price, individual breeders were still unable to improve their animals. In 1969, therefore, ANAFI began publishing two genealogical registers: Young Cattle and Bulls. With the first one, thanks to the coordinated work of the breed inspectors, it was possible to have, for the first time, uniform results throughout Italy, while the second one listed the genealogies of all the bulls registered in the herd book and in the artificial insemination centers (Unknown author, 1970: 18-41).

In this regard, the need to spread the practice of artificial insemination as a means of improving production led ANAFI to produce a publication, *F.A. '69* (*Fecondazione Artificale '69*), which collected data on the implementation of artificial insemination of *Frisona Italiana* throughout the country. In addition to the indications that were required to update and steer breeders towards the most suitable solutions for their own breeding needs, the volume included a series of synoptic tables from which all people working in the field of artificial insemination could obtain information for the purpose of their activities.[17]

17 The synoptic tables concerned: the list of production facilities of semen; the distribution system of semen; the number of artificially inseminated cows; the origin of sires; the year of birth of sires;

VIII. Outcomes: the symbiotic bond between innovations and associations

In conclusion, what emerges in the history of Italian dairy cattle breeding – and in Friesian cattle breeding in particular – is the preeminent role played by innovations and associations, and it is also clear that innovations strongly influenced the livestock sector only after national associations were established and developed. It is particularly striking that, for decades, the improvement in milk quality and quantity, the techniques to preserve animal health and the training of farmers were not within the remit of the central government, but of local private individuals, in a kind of 'bottom-up' model. There was no clear willingness on the part of the Ministry of Agriculture to improve the agricultural sector as a whole; rather, this objective, which was not clearly specified, was in any case pursued by individual farmers and Itinerant Teachers, who acted independently. Each director generally concentrated his efforts on the most developed agricultural sector of the Province under his responsibility, resulting in an effective advancement of agronomic knowledge, limited however to specific sectors and within its district boundaries.

In this first phase, which lasted up to the end of the Second World War, the approach of agricultural technicians was clearly heuristic. The 'fluid' legal nature of the Teachers, which for some time were autonomous from the central administration, and the limited financial resources – a feature shared with the sector's schools and institutes – made it unfeasible to update the sector's agenda to bring it in line with predefined objectives and systems. This is exemplified, in the case of zootechnics, specifically in the case of Local Herd Books, which were first managed by the Itinerant Teachers. In actual fact, each director decided independently which animals could be inserted in the book he was responsible for and, as a result, he exercised significant influence on the trade of animals, both at local level and with foreign sellers. Therefore, the features of the breeding bulls were different depending on the book, a bias that prevented any consistent production improvement between the various territories.

It was the keen interest of Sansone and De Carolis, in particular in breeding, as well as the fact that the countryside near Cremona was already specialized in milk production, that permitted this area to develop its potential and form a cooperative consciousness which was effective, even nationally, after World War II. Indeed, it was at this stage that both innovations and associations played a decisive role in the development of Italian dairy farming. Firstly, an important change over the previous period was the increase in domestic demand for both milk and cheese, initially largely due to the financial aid provided by international

---

the list of sires. This information was compared with previous data to understand the evolution of the centers to produce seminal material, the numerical increase of sires and their origin. [Unknown author], 'F.A. '69', *Bianco Nero*, No 1, January 1971, p. 44.

organisations.[18] The fact that the yield of various breeds in Italy was too low to satisfy this demand made the issue of improvements in productivity – and therefore also selection a problem of national interest. The various laws enacted during the period are examples of a desire to organize and manage the matter at central level so as to formulate clear objectives and a common agenda for the entire national territory.[19]

Despite this, the Brown Swiss, although with lower production levels than the Friesian breed and therefore not fully able to meet the new Italian needs, remained the most appreciated and widespread breed in the country until the early 1960s. The reason for this apparent contradiction lies in the fact that, until then, Italy still lacked the technical conditions that would have allowed a real improvement of the breed with the American semen, at that time the best from the point of view of milk yields. With the introduction of the practice of frozen semen insemination the situation changed dramatically.

It was only at this point, thanks to the new central interest that had led to the creation of the National Herd Book and with the appropriate knowledge and technical skills, that the spread of associations, which were highly desired by the technicians of the Itinerant Teacher system, became truly fundamental.

In a sector in which the State increasingly intervened and where the exchange of improved genetic material between States and not between companies became fundamental, the action of single innovators became less important and, not only the know-how of the various operators, but also human capital became essential. In the light of this consideration, it is clear how associations played a crucial role. Indeed, human capital can be effective if inserted into a structure that guides and urges it to achieve continuous improvement.

---

18 On the new trends in meat consumption, that were increasing because of the economic boom and of the idea that eating meat every day was an indicator of prosperity, and on the evolution of milk and dairy production in Italy, more and more oriented towards higher quality production such as cheese and dairy products like yoghurt and creams in a context of reduced milk consumption (essentially limited to children), Cavazza (2010: 53-78); Stranieri & Tedeschi (2017: 95-111). In Italy, stable and cow breeding renewal took place based on the experience of other European countries also because the European integration process tended to make breeding methods uniform for hygienic and health reasons. In ERP documents is not possible to find any references to courses for farmers aimed at explaining USA practices, while there were courses for people working in the industrial sector. As for the agricultural sector, there were aids for farmers buying agricultural machines, hybrid seeds and fertilizers, as well as funds for reclaiming lands and renovating stables and other buildings of the farm. On the CAP in Italy, cf. among others Fanfani (1998: 75-351); Canali (2011: 71-170).

19 Law no. 1009 of 25 July 1952 on the central control of insemination, so that it could only be carried out by veterinarians authorized by the Ministry of Health; the decree of July 1954 laying down rules for the duty-free import of pure-bred animals for breeding thanks to a certificate issued by experts appointed by the Ministry of Agriculture; the fundamental promulgation at legal level, on 24 June 1956, of the National Herd Book of the Friesian and Alpine Brown Swiss breeds, which went beyond the already mentioned limits of provincial books.

Table 6.2. *Frisona* and *Bruna Alpina* heads from 1942 to 1987

Year	Frisona	Bruna Alpina	Total
1942	189.000	2.205.000	2.394.000
1956	1.086.072	1.845.278	2.931.350
1962	1.836.000	1.876.000	3.712.000
1970	2.958.000	1.479.000	4.437.000
1979	3.505.750	1.143.914	4.649.664
1987	3.119.966	857.286	4.177.252

Source: Fusco, *La Frisona Italiana*, p. 62.

Bibliography

Albertini, L. (2001), *La bonifica del Senatore Albertini, 1926-1945. Storia dei primi anni della Bonifica di Torre in Pietra*, Fiumicino: Graffiti

Barsanti, D., (2002) 'L'allevamento', in *Storia dell'agricoltura italiana. III L'età contemporanea 1. Dalle «Rivoluzioni agronomiche» alle trasformazioni del Novecento*, ed. by R. Cianferoni, Z.Ciuffoletti, L. Rombai, Florence, Polistampa

Bonadonna, T. (1969), 'Libro genealogico', in *Enclopedia Agaria Italiana, vol. VI*, Rome: REDA

——— (September-October-November 1947), 'Programmi e speranze per il 1° Congresso Internazionale di Fisio-patologia della Riproduzione Animale e di Fecondazione Artificiale. Milano, 23-30 giugno 1948', *Bianco Nero*

Camera di Commercio Industria e Agricoltura di Firenze, Ispettorato Provinciale dell'Agricoltura di Firenze (1963), *La nuova disciplina della riproduzione bovina*, Florence: Bruno Coppini & C.

Capone, A. (1981), 'Destra e Sinistra da Cavour a Crispi', in *Storia d'Italia, Volume XX*, ed. By G. Galasso, Turin: UTET

Canali, M. (2011), 'L'Italia e l'integrazione delle agricolture europee (1960-1990)', in, *L'agricoltura e gli economisti agrari in Italia dall'Ottocento al Novecento*, ed. by M. Canali, G. Di Sandro, B. Farolfi, M. Fornasari, Milan, Franco Angeli

Carera, A. (2018), *Agricultural Resilience in Lombardy. How it supported Economic Growth in the 19th and 20th Century*, Milan, Vita e Pensiero

Cavazza, S. (2010), 'Dal consumo desiderato al consumo realizzato: l'avvento della società dei consumi nell'Italia postbellica', in *La rivoluzione dei consumi. Società di massa e benessere in Europa 1945-2000*, ed. by S. Cavazza, E. Scarpellini, Bologna, Il Mulino

Cavazzoli, L. (2002), 'Ricerca e formazione nella "filiera del latte" fra Otto e Novecento', in *Oro bianco. Il settore lattiero-caseario in Val Padana tra Otto e Novecento*, ed. by P. Battilani – G. Bigatti, Lodi, Giona

Cazzola, F. (1993), 'L'agricoltura nello sviluppo di una grande regione industriale italiana: la valle del Po', in *Studi sull'agricoltura italiana. Società rurale e modernizzazione*, ed. by P. P. D'Attorre – A. De Bernardi, Milan, Feltrinelli

Il Comitato Direttivo (May 1963), 'La relazione del Comitato Direttivo della Frisona Italiana nell'Assemblea Generale ordinaria del 7 maggio 1963', *Bianco Nero*, no. 5

Il Comitato Direttivo (May 1965), 'L'Assemblea Generale ordinaria dell'Associazione Nazionale Allevatori Bovini della Razza Frisona Italiana. 4 maggio 1965', *Bianco Nero*, no. 5

Consolini, A. (May 1964), 'Importante riunione della Commissione Tecnica Direttiva del Libro Genealogico Nazionale della Frisona Italiana', *Bianco Nero*, No 5

Corti, M., Camozzini, G., Buzzoni, P. (2015), *Arte casearia e zootecnia. Tradizioni da leggenda in Valsassina*, Missaglia, Bellavite

Derry, M. (2015), *Masterminding nature. The breeding of animals*, 1750-2010 Toronto, University of Toronto Press

Fanfani, R. (1998), *Lo sviluppo della Politica agricola comunitaria*, Rome, Carocci

Fascetti, G. (December 1926), 'Sulle vicende del caseificio in Italia', *L'Italia agricola*, Year 63, no. 12

Fileni, E. (1954), 'Cattedre Ambulanti di Agricoltura', in *Enciclopedia Agraria Italiana vol. II*, Rome, REDA

Fumi, G. (2006), 'Divulgazione e sperimentazione nell'opera della Cattedra ambulante di agricoltura di Cremona da Sansone a De Carolis (1896-1935)', in *Gli agronomi in Lombardia: dalle cattedre ambulanti ad oggi*, ed. by O. Failla, G. Fumi, Milan, Franco Angeli

Fusco, R. (1990), *La Frisona Italiana. Evoluzione, lotte e traguardi di cinque generazioni di allevatori*, Rome, ANAFI edizioni agricole

Fusco E., Fusco R. (1994), *1944 1994. Cinquant'anni di progresso. AIA è...*, Rome, AIA

Grothe, P. (1995), *Holstein Friesian. A Global Breed*, Doetinchem, Misser

Marigliano, M. (2018), 'Il problema dell'innovazione in zootecnia in prospettiva storica: il caso della Frisona Italiana' (unpublished doctoral thesis, Università Cattolica del Sacro Cuore, a.y. 2017/2018)

Mazzotti, O. (2017), *"Istruire dalla cattedra, istruire coll'esempio!". Conoscenze agrarie e capitale umano in Romanga tra Otto e Novecento*, Bologna, Il Mulino

Monestiroli, C. (November-December 1962), 'Come pianificare una selezione aziendale. Nulla si può additare al caso, ma tutto deve essere regolamentato e deciso da una volontà che sa che cosa vuole ottenere', *Bianco Nero*, no. 6

Rognoni, G. (2006), 'Le cattedre ambulanti per la zootecnia', in *Gli agronomi in Lombardia: dalle cattedre ambulanti ad oggi*, ed. by O. Failla, G. Fumi, Milan, Franco Angeli

Romani, M. (1963), *Un secolo di vita agricola in Lombardia (1861-1961)*, Milan, Giuffrè

Saltini, A. (2002), 'Il Sapere agronomico. L'agronomia italiana tra Ottocento e Novecento: dal divorzio all'aggiornamento ai moduli europei', in *Storia dell'agricoltura italiana. III L'età contemporanea 1. Dalle «Rivoluzioni agronomiche» alle trasformazioni del Novecento*, ed. by. R. Cianferoni and others, Florence, Polistampa

Stranieri, S., Tedeschi, P. (2017), 'The Influence of the European Institutions and Policies on the Italian Dairy Sector', in *Cheese manufacturing in the 20th century: the Italian experience in the international context*, ed. by C. Besana, R. D'Errico. R. Ghezzi, Brussels, Peter Lang

Tedeschi, P., Stranieri, S. (2011), 'L'evoluzione del settore lattiero-caseario lombardo dall'Ottocento al Duemila', in *La civiltà del latte. Fonti, simboli e prodotti dal Tardoantico al Novecento*, ed. by G. Archetti – A. Baronio, Brescia, Fondazione civiltà bresciana

Theunissen, B. (2008), 'Breeding Without Mendelism: Theory and Practice of Dairy Cattle Breeding in the Netherlands (1900-1950)', *Journal of the History of Biology*, n°41, p. 637-76

——— (2012), 'Breeding for Nobility or for Production? Cultures of Dairy Cattle Breeding in the Netherlands, 1945-1995', *Isis*, n°103, p. 278-309

Unione Operatori Fecondazione Artificiale Animale, *Raccolta di Leggi sulla Riproduzione Animale* (Pavia: UOFAA, [unknown date])

Zucchini, M. (1970), *Le Cattedre Ambulanti di Agricoltura*, Rome, Giovanni Volpe Editore

Atlante etnografico delle popolazioni bovine allevate in Italia (1983), ed. by C.N.R., Milan, C.N.R.

L'Italia in 150 anni: sommario di statistiche storiche 1861-2010 (2010), ed. by ISTAT, Rome, ISTAT

[Unknown author], 'Assemblea Generale Ordinaria della Associazione Nazionale Allevatori Bovini della razza Frisona Italiana. 12 marzo 1970', *Bianco Nero*, no. 3, March 1970

[Unknown author] 'Assemblea straordinaria dei soci', *Bianco Nero*, August 1946

[Unknown author], 'F.A. '69', *Bianco Nero*, no. 1, January 1971

[Unknown author], 'La riunione della Commissione Tecnica Direttiva della Frisona Italiana', *Bianco Nero*, no. 3, March 1963a

[Unknown author], 'L'assemblea generale ordinaria della Associazione Nazionale Allevatori bovini di Frisona Italiana. 24 marzo 1964', *Bianco Nero*, no. 3, March 1963b

[Unknown author], 'Libro Genealogico Nazionale della Razza Frisona Italiana – notiziario n. 28', *Bianco Nero*, no. 1, January 1965

[Unknown author], 'Seattle-Milano', *Bianco Nero*, August 1947

RONALD PLANTINGA, MERIJN KNIBBE AND MARIJN MOLEMA

7. Fair value?

The price of milk and the intricate path of bio-based productivity growth in Dutch dairying, 1950-1980

I. Introduction

The post-war period between 1950 and 1980 is widely recognized as a period of tremendous development in agricultural production. A combination of political ideologies, economic and political stimuli and scientific and technological developments led, in the whole of Western Europe, to an interrelated set of growth stimuli such as mechanization, land reclamation, lower trade barriers, standardized trade regulations, public subsidies, and the more effective protection of animals and crops from disease (Brassley, Martiin & Pan-Montojo, 2016). As a result, agricultural production increased tremendously. The extensive dimension (more land, fertilizer, pesticides, machines, animals) of this growth process is well documented (Thoen, Soens & Brassley, 2015). But we can also look at intensive growth (better land, fertilizer, pesticides, machines, animals). Current historiography is paying more and more attention to the influx of knowledge into primary production systems and the intermediary role of institutions who produce, disseminate, and apply agricultural know-how (Auderset & Moser, 2018; Zwarts, 2021) - a process which often succeeded by paying attention to local details.

This chapter adds to this literature by concentrating on the knowledge-driven, intensive growth of dairy farming. Since the rise of industrialized dairy processing in the late nineteenth century, expertise on milk has been developed by farmers as well as by various organizations. The translation of theoretical knowledge into more practical 'useful knowledge' was crucial, as it enabled farmers and factories to use the production factors land, labour, and capital more efficiently (Mokyr, 2002). Over the last 10 years, important works by agricultural historians have been published that develop a social-cultural perspective on knowledge production and dissemination. According to these studies, multiple actors such as farmers, entrepreneurs, scientists, and politicians established networks and, despite occasional discord and conflicting interests, were involved in these processes (Uekötter, 2012; Auderset & Moser, 2018; Zwarts, 2021). Another strand

From Breeding & Feeding to Medicalization, Rural History in Europe, 17 (Turnhout: Brepols, 2023), pp. 157–182
BREPOLS PUBLISHERS 10.1484/M.RURHE-EB.5.135025

of literature has specifically analysed the adaptation of animals and plants – in short, biological innovation – with the explicit goal of improving productivity in agriculture (Stine & Tarr, 1998; Gorman & Mendelsohn, 2010). Literature on dairy cattle breeding has argued that, apart from the availability of knowledge, social and cultural factors were important as well (Hernández-Adell & Pujol, 2016; Orland, 2016; Theunissen, 2008). Also, as plants and animals are part of an ecosystem, biological innovation is more complicated than 'technical' innovation. Progress can be set back by the emergence of new pathogens (a change in the ecosystem) while genetic changes that increase potential yield or growth or any other single biological factor may come at a cost as it might influence the health and stamina of animals, meaning that progress might have to be measured by more than just the increase in physiological yields (Olmstead & Rhode, 2002).

In this chapter, we focus on the (lack of?) contribution of pricing systems to biological innovation. More specifically, we study the introduction of multi-component pricing (MCP) of milk in the Netherlands (which, as far as we know, pioneered MCP) and its influence on the behaviour of dairy farmers with regard to breeding. MCP means that the farm gate price of milk is based not just on fat content but also on other factors, such as protein content. In 1958 it was introduced fast and, surely in the province of Friesland, comprehensively with the explicit goal to entice farmers to breed milk cows producing milk with a high protein content. There were two ideas behind this endeavor. The first was that farmers take account of measurements and use them to pursue long term goals. Scientific methods to systematically estimate yield of cows and fat content of milk were developed and applied in the late nineteenth century (Atkins, 2010). Before long, these were used to develop new milk payment systems based on fat content but also to provide information about productivity of cows, which was used as an input (surely in Friesland) to successful breeding strategies (Knibbe & Molema, 2018). Fat content of Frisian milk increased considerably between 1910 and 1950 – contrary to many other parts of the Netherlands. This success was well known to farmers and agronomists. Less well known is that, around the middle of the twentieth century, cheap methods to estimate the protein content of milk were developed which in a comparable way would be used to introduce the MPC systems which are, by now, used worldwide. Even lesser known is that Friesland acted as the global innovator. The system consisted of payment based on fat as well as protein, based upon per cow information of the protein content of its milk. These methods were, against the background of the successful increase of the fat content of milk, explicitly introduced to enable and entice farmers to breed 'protein cows'. The second idea behind the introduction of MCP was that dairy products with a high protein content, such as cheese and milk powder, would become more important, as already had happened after 1950. It was supposed that price incentives (MCP) plus information about the productivity of individual cows would lead farmers to breed profitable 'protein cows' with milk with a higher protein content. However. Contrary to the situation about 50 years before, when, after the introduction of payment based on fat, sustained increases in the fat content of milk were witnessed, after the introduction of MCP no noticeable improvement in the average protein content of milk can be witnessed, even when the knowledge network was dense enough to disseminate the new techniques. Regional farmer organizations, dairy cooperatives, and milk control societies had large memberships and effective information services, protein was measured and MCP was introduced. But it did not work out. Despite this failure, MCP was continued and even spread globally. How come?

The remarkable absence of an increase in the protein content of milk raises two questions. First, why did the measurement and pricing of protein *not* result in the breeding of cows that produced milk with a higher protein content? And, second, despite these thwarted expectations, why was the system of protein pricing continued, maintained and eventually even extended? In this chapter, we mainly focus on answering the first question, to generate hypotheses on the impact of pricing systems compared to other factors on processes of biological innovation in dairy farming. We investigate how opposing rationales affected different forms of post-war bio-based productivity growth in dairy production. Although our analysis is empirical and we do not aim to contribute to theoretical debates, our perspective touches upon sociological accounts of how knowledge is produced (Latour, 1987). The literature informs about the creation of knowledge in scientific environments, and how these environments are entangled with but fundamentally different from the setting in which knowledge is used. Creation and utilization of bodies of practical knowledge follow different logics, and it is the difference between these logics that offers a fertile ground for historical analysis – certainly when different actors such as farmers, agricultural experts, dairy factories, and herd book societies are involved (Karel, 2005).

These different logics and complexities will become apparent from our analysis of the process of exploration and hesitating application of protein knowledge in the Dutch-Frisian dairy system during the decades between 1950 and 1980. The first date is chosen because it more or less marks the end of post-war reconstruction and the start of a new era of innovation and growth. The last date is chosen because the introduction of American genetic material into the Dutch population of dairy cattle (Holsteinization) took hold after around 1980. This was connected to a restriction of milk production (European milk quota were introduced in 1984), which provided farmers with a strong incentive to rapidly increase production per cow (see Table 7.2 on p. 179, indicating the regime change in the early 1980s).

The chapter is structured as follows. In the next section, we provide a characterization of the Frisian dairy sector and its major institutions and organizations. This is followed by a discussion of the introduction of protein as an element of MCP. Section 4 analyses the incapacity of regional organizations, especially breeding societies, to improve the protein content of milk between the late 1950s and early 1980s. In Section 5, we show that bio-based productivity growth focused more on improving land productivity, which was supported by government-subsidized land consolidation and water management improvements.

We argue that these bio-based innovations, at least from the perspective of Frisian dairy farmers, were a rational choice in the search for higher productivity. The chapter ends with a concluding section.

II. The dairy sector in Friesland: farms, factories, and organizations

Friesland is and was characterized by a large area of fertile clay- and peat soil based pastureland with highly specialized dairy farms. Up to the late nineteenth century, butter and low-fat cheese were produced. Arable production was surely after 1600 almost totally absent in this region. From about 1530 onwards, dairy products were increasingly shipped to cities such as Amsterdam, and later, after 1815, to the rapidly growing London market. After the introduction of dairy factories, which took off in the 1880s, the region dominated the Dutch production of butter and, after WW I, condensed milk and after World War II of factory cheese and condensed milk. This did not mean, however, that Frisian dairy production was an uninterrupted success story. In the mid- and late nineteenth century, there were serious setbacks to the export position of Dutch, and especially Frisian, butter. These setbacks were mainly due to the rise of Danish (after 1860) and New Zealand (after 1895) butter production and export. Countering this competition required modernization and an increase in scale of dairy production by moving it to factories. Dairy farmers became milk producers instead of butter and cheese producers. This transition from dairy production on farms to production in factories represented a challenge in terms of developing and disseminating new knowledge and the innovations and institutions enabling this dissemination. Moreover, the post-war period yielded other challenges, including coping with increases in labour costs and environmental impacts. During the successful transition to factory-based production, new farmer organizations and knowledge networks were created, to disseminate knowledge and to support farmers and factories in producing, selling and marketing their products. The Frisian Animal Health Department (*Gezondheidsdienst voor Dieren*), to cite an example, was established in 1919 and preceded comparable organizations in other Dutch provinces by decades. Guided by this organization and in cooperation with dairy factories, Friesland was by a large margin the first Dutch province to eradicate bovine tuberculosis (Abbo-Tilstra, 2002). The Association of Frisian Cooperative Dairy Factories (Bond van Cooperatieve Zuivel Fabrieken, BCZF) was founded in 1897 and increasingly functioned as a central organization in the Frisian knowledge network – even when private non cooperative factories continued to exist (the factories of the Lijempf being the most important ones). As the BCZF united all cooperative dairy cooperatives (around 80 factories in 1950), it was able to support small cooperative dairy factories that were unable to invest in R&D, lobbying activities, quality control of milk and dairy products and marketing. In addition, a dairy school was founded to train the management of dairy factories and a dairy consultant to support the industry with R&D was employed (Plantinga & Molema, 2017). By the early twentieth century, a dense, interconnected, and region-based sector consisting of dairy farms, processing industries, and supporting organizations had emerged. The significance of this interconnected network – resembling Michael Porter's notion of 'cluster' (Porter, 1998) – was highly efficient knowledge dissemination and production compared to a (counterfactual) situation in which companies had to rely on their own resources.

The transition from a traditional sector based on individual farms and farm production of dairy to factory production and a network of organizations was not entirely autonomous. After 1860, the Frisian farmers would learn the meaning of the phrase 'export quality' the hard way (Tjepkema & Wiersma, 1959). As stated, international competition on the all-important London market increased while producers of low-quality butter and margarine factories from the south of the Netherlands became important competitors at the low end of the market. Competition from Danish farmers was disturbing both in an economic and a cultural sense. After 1860, the Danish not only managed to increase the number of milk cows but also to triple milk yields per milk cow from about 1,000 kg a year to over 3,000 kg a year, increase the fat content of milk, and establish a very extensive system of almost 2,000 small butter factories. Danish exports rapidly increased, especially to the UK. Dutch and especially Frisian producers had dominated the London market for imported butter for a long time but Danish exports soon dwarfed Dutch exports. In 1890, 41% of English butter imports originated from Denmark, compared with 8% from the Netherlands (Croesen, 1932: 192; 195). This was possible because almost all Danish exports went to the UK: before 1914, almost 90% of Danish export production ended up on the London market, while Dutch butter was exported to many other countries too (Frietema, 1937: 4; 67; 69; 85; 92). But after 1895 total Danish butter exports also surpassed total Dutch butter exports, while to add insult to injury high-grade Danish butter received a higher price than high-grade Frisian butter. It was not just an economic loss but also a shock to the self-esteem of the Dutch dairy sector. These developments caused anxiety in Friesland, and members of the Frisian Society for Agriculture and Cattle searched for solutions. German dairy schools were visited from 1889 onwards, and Danish production practices of milk research inspired the Dutch-Frisian farmers to think about scientific innovations as well (Molema, 2017). Individual efforts to modernize dairy production on farms were abandoned, while efforts to shift production to factories were increased: within a limited number of years, age-old farm production of butter shifted to, for the time, rather large factories (Knibbe and Molema, 2018).

Moving dairy production to factories enabled a host of supply chain related developments. The increasing production of factory butter and cheese also led to the increased centralized availability of decreamed milk and whey. Traditionally, these products were returned to farms and used as feed. The centralized availability and the cost of returning these products to farms, combined with the risk of

Figure 7.1. Production of butter, cheese, powdered milk, and condensed milk, 1895-1985 (volume indexes (kg), 1933-100).
Source: Statistics Netherlands (CBS).

spreading animal diseases, in 1913 led 33 cooperative factories to establish the *Coöperatieve Condens Fabriek Friesland* (Cooperative Condensed Milk Factory, CCF), which, especially after World War I, managed to become a global player on the market for condensed skimmed milk (its 'Friesche Vlag' brand was one of the early, and successful, endeavours to apply branding to dairy). Also, increasing amounts of skimmed milk were used for the production of milk powder. This caused a shift from the use of decreamed milk and whey as feed and an increase in the amount of milk used for commercial purposes and, by implication, a shift to more use of purchased feed on farms. Market dependence therefore increased, on the output as well as on the input side, even when this process would take several decades. The dairy product mix for the Netherlands as a whole is shown in Figure 7.1. It is important to note the increased production of powder and condensed milk in the 1930s, a time of overabundance of dairy, and the decline in the production of these products during World War II, a time of scarcity in which the government (largely successfully) tried to use available milk more fully for domestic human consumption. After 1947, an impressive increase in the amount of cheese produced was seen. This meant that, in the long run, products with a relatively high protein content became increasingly important, which also meant that less and less milk protein was used as feed.

After World War II, the Frisian dairy sector was confronted with new economic and organizational challenges, originating at the national and international levels. European and global markets became increasingly important, especially after the introduction of the European Community. Rising costs were another challenge, as wages began to rise quickly in the early 1960s after the Dutch government relaxed stringent wage policies (Van Zanden, 1997: 117-19). Wages in the dairy industry increased by 40% to 50% between 1961 and 1965. As wage costs constituted a large part of non-milk production costs, the major challenge was to improve labour productivity (Bles, 1961). Throughout the 1960s and 1970s, a combination of mechanization and increase in scale of farms and factories took place. Larger production units resulted in a reduction in the ratio between labour and resources (i.e. milk), improving profitability. Also, an increase in the amount of milk produced allowed for larger investments in mechanization. The BCZF was once more crucial for this adaptation of the cooperative dairy factories. After World War II, the BCZF expanded into areas such as management consultancy, technology, and even engineering and factory construction. It also increasingly disseminated and translated knowledge from the international and national levels to the regional level while its function as a social and informational node in the network of cooperative factories should not be underestimated. In Friesland, the decentralized production of milk and dairy was increasingly combined with a centralized system of research and consultancy, which enabled a rapid spread of new ideas and new knowledge - like MCP.

III. The introduction of multi-component pricing in Friesland

The efficiency of the system of regional dairy organizations is underscored by the rapid implementation of protein pricing. In 1957, Frisian cooperative dairy factories started to pay farmers based not just on fat content, but also on protein content which, as far as we know, is the first example of multi-component pricing (MCP) in the world. This happened almost overnight, and many parts of the Netherlands quickly followed suit, as did other countries (Sneddona et al., 2013). Before 1957, cooperative factories based milk prices on the fat content of milk and bacteriological quality, but suddenly protein was added to this mix. The idea was not new. Jesse quotes professors Froker and Hardin who, in 1942, stated that 'an equitable method of paying producers' using some version of MCP was long overdue in dairy production (Jesse, 1994). Note the world 'equitable'. Also, California introduced an MCP system in 1962, but this was five years after the introduction of MCP in Friesland, while the Frisian system clearly spread to other parts of the Netherlands before 1962 (Cropp, 1978; Whitaker, 1978; Smith & Snyder, 1978; Novakovic, 1994). Outside the Netherlands and California, MCP developments were subdued for decades to come, and it was only after around 1980 that MCP started to spread in the USA (Cropp & Wasserman, 1993). MCP only became compulsory in the USA in 2000 – lagging the Netherlands by around 30 years (Covington, 1993).

The reason for the sudden implementation of protein content as an element of MCP in Frisian cooperatives was the increasing importance of protein content in the immediate post-war years, combined with scientific breakthroughs. Up to the 1950s, it was generally assumed that the protein content of milk correlated with the fat content. It therefore made no sense to reward farmers for a higher protein content, also because the measurement of protein content was more cumbersome and expensive than the measurement of fat content. Two factors changed this. First, the dissertation of Rommert Politiek (published in 1957, but the results were known to the Frisian cooperatives in 1956) convinced the Frisian dairy industry of the importance of protein pricing. Politiek showed that the protein content of milk was hereditary while the correlation of protein content with fat content was only limited and argued, based on experiences with fat, that it was possible to use price as an incentive to increase the protein content of milk (Politiek, 1957; Dijkstra, 1991). Second, the development of cheaper techniques to measure protein content brought mass implementation within reach.

Politiek was to become an important figure in Dutch dairy breeding. After finishing his dissertation, he soon became a leading figure in Frisian and Dutch agriculture. He was the oldest son of a farmer in the Frisian village of Wons, but chose to continue his studies at the Wageningen Agricultural School (nowadays Wageningen University and Research). After completing his studies, he studied the heredity of protein content, which resulted in a PhD from Wageningen in 1957 (Dijkstra, 1991). The research was based on measurements of protein content conducted by the FRS (Fries Rundvee Stamboek, Frisian cattle herd book) and the BCZF. In 1950 and 1952, financed by the FRS, the protein content of a small number of individual cows was already measured (Janse et al., 1953). These experiments functioned as a pilot study, inspired by research conducted in Sweden, in which Bonnier and Hansson investigated relationships between the heredity of the lactose, fat, and protein content of milk. In 1946 they concluded that 'the interdependencies between the percentages of fat, protein, and lactose are, very largely, genetically determined' (Bonnier & Hanson, 1946). This conclusion was confirmed by further investigations using larger groups of dairy cows (Hanson & Bonnier, 1949). Aside from this scientific inspiration, a further motivation was the increasing importance of protein compared to fat for dairy products. It was decided to expand the pilot study, resulting in more extensive data collection between 1953 and 1955, when 726 individual Friesian cows were investigated. The data were analysed in Politiek's dissertation, and his conclusions were clear. First, he stated that protein was becoming increasingly important (see also Figure 7.1), even more so for Friesland because in Friesland the shift from butter to cheese production was more profound. As the yields of cheese production were strongly affected by protein content, the shift from butter to cheese production meant that, according to Politiek, protein content should be valued more and therefore incorporated into the pricing system. Second, protein content was hereditary and, contrary to previous insights, largely independent of fat content. Therefore, the selected breeding of cows could improve the protein content of milk and thereby the efficiency of cheese production (Politiek, 1957).

Several months before the public defense of Politiek's dissertation, the key organizations of the Frisian dairy industry acted on his results. Politiek presented his findings to the general board of the FRS on 27 October 1956. In a reaction, the board stated that selective breeding for protein content was crucial, as long as sample testing was not too expensive.[1] While Politiek was working on his dissertation, techniques enabling the cheap testing of huge numbers of samples for protein content were developed. For example, Kofranyi published a method in 1950 that was subsequently adapted by the Laboratory for Dairy Research at Wageningen University (Lolkema & Van der Have, 1958). In coordination with the FRS, the BCZF developed plans to establish a 'protein laboratory' based on this technique that would enable the measurement of large numbers of samples and hence also the introduction of MCP (which required lots of testing).[2] A joint 'technical commission' of the BCZF and the FRS was set up and, in three separate reports, the financial and technical aspects of protein knowledge and the plans to shift to a new pricing system involving the measurement of protein content in the near future were summarized.[3]

The dairy cooperatives – the BCZF had 71 members in 1956 – had to approve these plans. This meant that the introduction of protein pricing was discussed by farmers, who were voting members of these cooperatives, in meetings between February and May 1957. An investigation of the minutes of the meetings suggests that the BCZF's plan was presented by the board of the cooperative, that the importance of protein content was stressed, and that the rationale behind the plan was explained. For the farmers, the new payment method promised a better income, especially as selective breeding would create cows that produced milk with a higher protein content. MCP may also have appealed to farmers' desire for 'fair' (*billijk*, meaning 'fair' as well as 'reasonable' and 'equitable') payment,[4] which may also explain the continued support for protein pricing despite the additional costs for farmers. More or less 50 years before, the same process took place but, at that time, fat was the component embedded in the price that farmers received for their milk (Knibbe & Molema, 2018). It seems that protein, as another component of a 'fair' price for milk, followed the same footsteps as the accounting and payout system for fat.

After general agreement by all 71 cooperatives, from may 1957 on milk prices were to quite an extent based on, in addition to fat content, protein content of milk.[5] To make this possible, a laboratory for the large-scale testing of protein content was established in just six weeks. Friesland took the lead, but its example was swiftly followed. In the same year, a protein laboratory was founded in the neighbouring province of Drenthe and, two years after the foundation of the

---

1 Friesch Rundvee-Stamboek board meeting minutes, 27 October 1956 (Tresoar, Leeuwarden (T), Het Fries Rundvee-Stamboek, 1879-1972, access number 93-49, inventory number 6).

2 Friesch Rundvee-Stamboek board meeting minutes, 18 January 1957 (T, access 93-49, inv. 6).

3 Friesch Rundvee-Stamboek board meeting minutes, 20 November 1956 (T, access 93-49, inv. 6); Friesch Rundvee-Stamboek board meeting minutes, 8 February 1957 (T, access 93-49, inv. 6).

4 In the economic Chapter 8 of Politiek (1957) he uses the phrase 'billijk' eight times to denote one of the fundamental variables of the design of an MCP scheme.

5 BCZF board meeting minutes, 13 February 1957 (T, Bond van Coöperatieve Zuivelfabrieken in Friesland, access 91, inv. 63); BCZF board meeting agenda, 13 March 1957 (T, access 91, inv. 63).

Table 7.1. Share of milk cows included in the protein control system (%)

Province	**1958**	**1959**	**1960**	**1965**	**1970**	**1975**
Groningen	5	17	19	13	52	63
Friesland	46	45	45	27	73	72
Drenthe	12	15	26	30	68	72
Overijssel	6	16	19	23	49	65
Gelderland	4	14	16	19	41	54
Utrecht	-	6	6	7	49	51
Noord-Holland	-	6	16	17	66	66
Zuid-Holland	-	4	15	23	43	49
Zeeland	-	3	4	5	42	50
Noord-Brabant	-	3	3	7	50	56
Limburg	-	1	1	4	51	59

Source: *Landbouwcijfers*, various issues.

Frisian protein lab, 10 similar institutions were up and running in other parts of the Netherlands.[6] Especially in the northern and eastern provinces of the Netherlands, MCP rapidly took hold and the milk of a large proportion of milk cows was tested (see Table 7.1). By around 1960, only three years after its introduction, measuring protein in combination with MCP was a widely applied practice, even though provinces such as Zeeland and Limburg, where dairy was less important, did not follow suit.

The BCZF protein laboratory generated huge amounts of data, and therefore had the potential to generate useful knowledge that could be used for breeding purposes. In the first full year of the laboratory's operations (1958), over one million samples were analysed. The information was disseminated to the farmers through local dairy factories. The annual reports of the 'De Lege Gaeeën' cooperative dairy factory in the village of Sijbrandaburen contain, for instance, information on milk yield, fat content, and protein content per cow and per farm.[7] The analysis of millions of data points proved, however, to be a major challenge, also because of the lack of computing capacity (Rooijendijk, 2010). Although large amounts of data held the promise of novel insights, they did not generate new breeding knowledge even when research again and again verified the conclusion of Politiek of the absence of correlation between fat and protein

6 Yearly report of the Stichting Centrale Melkcontrole Dienst 1959, p. 8.

7 Yearly report De Lege Gaeeën 1961/1962 (T, Friese coöperatieve zuivelfabrieken, access 94-1, inv. 1222).

content. A high fat content did not mean that the same cows produced milk with a high protein content. This supported using protein content as an element in the pricing system and underscored the possibility to breed 'protein cows'. The analysis conducted by the BCZF, moreover, also focused on differences in protein content between farms and factories and different subregions (with different soils) and at different times of the year. These factors did not seem to be systematically related to the protein content of milk, and much of the research pointed towards weather conditions, which did not, of course, offer opportunities to improve protein content.

Even then, the data did give farmers the opportunity to stop breeding cows producing milk with an exceptionally low protein content, as opposed to breeding protein champions. Getting rid of cows with milk with an extremely low fat contentlow was a strategy directly used when data on fat content per cow became available (Knibbe & Molema, 2018), leading to a fast initial increase of average fat content. But investigating milk of first-calf heifers from the Sijbrandaburen factory shows that neither average protein content nor the standard deviation of the protein content changed in any serious way between 1961/1962 and 1971.[8] This lack of progress, again, begs the question of why the availability of 'big protein data' did not translate into useful protein knowledge and improved protein content. Information was available, but it was not acted upon. Also, after 1960 the increase in protein testing stalled. In the provinces of Groningen and Friesland testing even declined between 1960 and 1968. Private factories such as Lijempf had been hesitant to test for protein and, according to the yearly reports of Lijempf, farmers simply did not ask for it.[9] Moreover, quite a few farmers at the large cooperative factory De Goede Verwachting did not take part in testing either.[10] MCP therefore seemed to share the fate of most innovations, fading away. In other words, while technical and economic and organizational innovation was swift and decisive, biological innovation did not follow suit and MPC started to wither away. However… after the decrease in sampling in Friesland and Groningen between 1960 and 1965 (partly due to a shift in the frequency of sample collection), the number of samples suddenly jumped to over three million in 1968. Protein testing became mandatory for ultimately all farmers.[11] This was a

8 1961/1962: n = 27, average = 3.40, and SD = 0.17; 1971: n = 82, average = 3.38, SD = 0.15. For fat: 1961/1962: n = 32, average = 4.26, and SD = 0.26; 1971: n = 82; average = 4.24, SD = 0.26 (T, access 94-1, inv. 1222, jaarverslagen 1961/1962 and 1971 (between these years the accounting system changed from agricultural years (May–April) to calendar years)).

9 Yearly Report Lijempf 1964 (T, Lijempf, access 111, inv. 116).

10 Yearly Report 1969 De Goede Verwachting (T access 94-1, inv. 492). Data on individual cows show that, for many farmers, information on the protein content of milk was lacking for the Goede Verwachting, even when such data were available for all cows of the members of the De Lege Gaeeën cooperative.

11 Yearly Report of the Provinciale Melkcontroledienst 1964; Yearly Report BCZF 1964. The Yearly Report of the De Goede Verwachting cooperative mentions that, from 1 October 1969 onwards, all samples tested for fat content were also tested for protein content.

response to changes in the minimum prices for butter and milk powder set by the European Community, which in 1969 resulted in a 38% decline in the implicit price of fat and a 73% increase in the implicit price of protein while in 1972 Frisian cheese factories introduced internal accounting systems based on MCP instead of just fat.[12] But this renewed interest was not led by farmers but by organizations and the EC. The questions, therefore, are why farmers were hesitant and why, despite this, the costly system continued to be used by cooperatives where in the end the farmers could decide.

IV. Organizational (in)capacity in promoting protein content by selective breeding

Selective breeding of dairy cattle holds the promise of improving productivity by changing the breed. In recent years, various historians have focused on similar processes of biotechnological adaptation, looking at animals and plants as 'biological artefacts' that can be changed. According to this 'envirotech approach', the distinction between nature and technology is blurred. For instance, the American historian Edmund Russell states that 'cattle, sheep, and horses [...] were not machines, but they were biological artefacts shaped by humans to serve human ends' (Russell, 2004). Following the envirotech approach, the milk can be considered an artifact consciously adapted by humans. Studies focusing on cattle breeding show that dairy cattle have been profoundly adapted to human demands, for one reason to produce a different kind and increasing amounts of milk. We should not just think of genetics but consistent with the envirotech approach also of health. Mention has already been made of the successful eradication of bovine tuberculosis in Friesland but after 1950 other diseases like Abortus Bang and Foot and Mouth disease would follow. Importantly, purposeful breeding became much more feasible after World War II, when artificial insemination (AI) became available and quantitative methods for assessing the potential hereditary value of the genetic material of sires were developed (Theunissen, 2008). This enabled rational breeding as propagated by scientists such as Rommert Politiek, who hoped that – with the right economic incentives, information, and data – farmers would change their behaviour. The experts knew very well that farmers were embedded in communities and cultures that, up to a point, were incompatible with their idea of rational farming, which required rational individuals. They tried to 'modernize' the countryside, a process that focused on rational farming as well as more individualistic farmers (Karel, 2005). This point of view is, like the ideas of Politiek, based on an ideal of the farmer as a rational individual and not as a member of a community.

---

12 'Verslag eerste jaarvergadering DOMO-BEDUM te Assen', *Officieel Orgaan FNZ* 15 January 1969; T access 91-871.

In Friesland, however, it was not only, or even mainly, agricultural experts who called the shots when it came to breeding. The widely-known and powerful FRS ultimately set the (many) goals for selective breeding and was in a position to select bulls that produced offspring with the required characteristics, while private owners of prize bulls with high 'nobility' (the actual phrase used) were held in high esteem (not least by owners). The question is whether these organizations, would react to the price incentives described in the previous section in a supposedly 'rational' way, focusing on 'high productivity cows' with a high yield of milk with a high fat and protein content, instead of focusing, as they traditionally did, on an large array of characteristics, and whether farmers followed suit.

A first view of the data suggests they did not change course, as the protein content of milk produced by Friesian cows did not rise significantly between the late 1950s and early 1980s (see Figure 7.2). In 1968, 11 years after the publication of his dissertation, Politiek was appointed professor at Wageningen University. In his inaugural lecture, he stated (mind the word 'our'):

> Considering the growing importance of milk protein as well as the difficulties which are to be expected with regard to selling milk fat, our organizations have as the first and hitherto only organizations taken the initiative to pay milk also based upon protein content and to estimate the protein content of the milk of individual cows. Analysing this data, however, shows that very little purposeful breeding to improve protein content has been realized to-date. This is also a consequence of the fact that milk fat has until now yielded more income for farmers than milk protein. (Politiek, 1968a)

Politiek therefore highlights the price of protein relative to the price of fat. Groot mentions that EEC stipulations were based on milk with 3.7% fat while protein was not mentioned, making it difficult to tie Dutch intervention prices to protein (Groot, 1968). At the same time, Posthumus stated that there had been some institutional inertia. He shows that herd books societies only started to introduce information on the protein content of milk in 1960 – but as this is only three years after the introduction of MCP, this lag seems in fact rather short.[13] The implicit price of milk protein relative to fat was, according to him, 1: 6, which also made it by far the cheapest source of animal protein available; a curious situation that was highly detrimental to progress with regard to breeding high protein cows (Posthumus, 1968). This situation, however, changed after 1970, without any noticeable changes in the protein content of milk – even when the increased

13 'Verslag omtrent de toestand en de verrichtingen van het Fries Rundvee-Stamboek van 1 mei 1959 tot en met 30 april 1960' *Mededelingen maandblad van het Friesch Rundvee-Stamboek en van de Bond van K.I.-verenigingen in Friesland* no. 150, July 1960. 'Voor het eerst werden, wanneer het aantal voldoende was, ook cijfers gegeven over het gemiddelde eiwitgehalte van de dochters van elke stier. Deze cijfers varieerden van 3,12% tot 3,56%. Het blijkt dus, dat het overervingsvermogen van de stieren voor het eiwitgehalte niet onbelangrijk uiteenloopt'. ('For the first time, data were provided on the average protein content of the daughters of bulls. This varied between 3.12% and 3.56%, showing that the inherited protein content varies considerably').

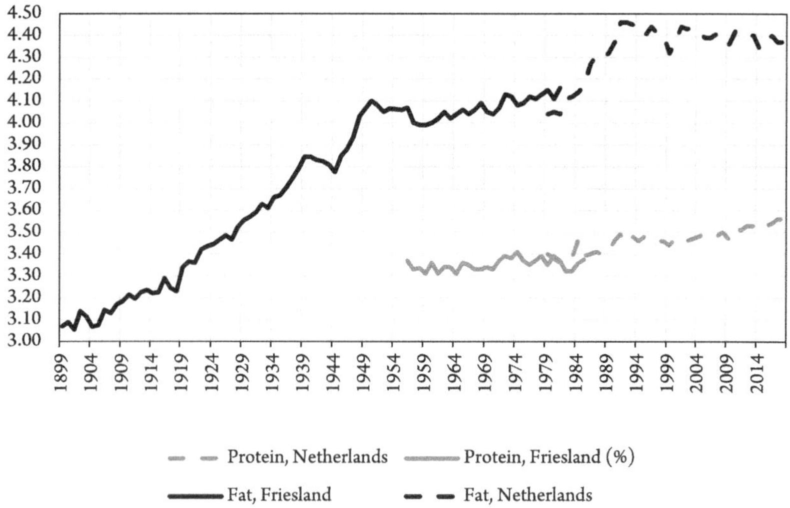

Figure 7.2. Fat and protein content of milk in Friesland and the Netherlands, 1899-2016
Sources: Fat: Merijn Knibbe and Marijn Molema, 'Institutionalization of knowledge-based growth: The case of the Dutch-Frisian dairy sector (1895-1950)', *Rural History* 29 2 (2018); Landbouwcijfers (several issues). The Frisian fat average is based on all cooperative factories from 1957 onwards, and on the factories of Wirdum and Achlum prior to 1957. The Frisian protein average is based on *Landbouwcijfers* from 1957 onwards and on Politiek (1957) for 1956. Politiek states that, according to the scarce data available, protein content had not increased before 1956.

protein content after around 1985, enabled by the use of Holstein-Friesian bulls, showed it could be achieved.

But ws it really about prices? A second look at the technology and culture of cow breeding may, at least partly, explain the lack of increase in the protein content of milk between 1957 and 1980, despite price incentives and the available data for individual cows. First, the internal affairs of the FRS partially explain the lack of progress. Theunissen showed in his research on the transformation of the Friesian cow during the post-war years that the FRS was a highly conservative organization. The international reputation of the Friesian breed made members of the FRS hold on to the features that developed in the late nineteenth century: a relatively small, strong cow, bred for both milk and meat production (Theunissen, 2012). In the 1960s, just after the introduction of protein pricing, there was considerable debate about future breeding goals.[14] The initial outcome of this was that a whole array of features of Friesian dairy cattle remained important – productivity being only one of these. Much attention was devoted to characteristics such as

14 FRS board meeting minutes, 28 November 1962 (T, access 93-49, inv. 9).

'milkability' and the importance of 'blood' and 'nobility' (Keestra, 1963). Protein content, by contrast, remained largely ignored by bull breeders throughout the 1960s. This lack of interest in protein content compared to other characteristics by 'organised breeding' may explain the lack of progress. Politiek mentions in his inaugural lecture that the mechanization of milking required a slightly different type of cow, which meant that breeding had multiple purposes, making it more difficult to increase milk yields and the fat and protein content of milk (Politiek, 1968b: 36-51).

Another explanatory factor for the lack of improvement in protein content is the availability of breeding knowledge. During the 1960s and 1970s, Artificial Information (AI) societies started to provide better information on the 'breeding value' of sires or the extent to which a sire positively influenced partly hereditary traits such as udder health, milk yield, and milk fat and protein content. These societies originally intended to prevent sexually transmitted diseases in the bovine population, but it soon became apparent that, as AI enabled a thousandfold increase in the number of offspring of the best sires, the system could also be used to improve the breed faster and more purposefully. In the course of time, ever more sophisticated methods to gauge the breeding value of bulls were introduced; methods that tried as much as possible to obtain an objective measurement of the extent of the hereditariness of traits as well as the amount of change.

The results were also mainly based on 'big data' analysis, as milk yield or udder health are, for instance, hereditary to an extent – but are also dependent on the interplay between many different genes. At the time, these genes were not known, let alone their interplay, something which even today is far from clear. For a long time and even today, breeding aimed at changing complex traits of animals or plants is largely or even mainly trial and error, and information is fuzzy. However, it was clear even then that, according to the statistics, some sires increased milk yield or the fat or protein content in the milk of their daughters while others did not. The AI societies – which became ever larger due to mergers and growth in membership – provided this information to farmers. According to Hamming, a number of systems were used to indicate the value of genetic material. Over the years, these systems provided more hereditary and phenotypical characteristics and more information on non-milk traits, such as udder health or ease of calving (Hamming, 2013: 12-15). Again, Rommert Politiek was instrumental, as he introduced the first two of these systems (Van Maanen & Wubbenhorst, 2019). However, even though these systems did contain information on protein, this was only one of multiple breeding criteria. Also, it took time to generate meaningful information on the quality of the genetic material of dairy cows and bulls. Gathering statistically meaningful results on the characteristics of the offspring of a bull took at least five to six years. Then, even if a bull turned out to be a 'protein bull' and could be used to increase the protein content of milk, it would take another two to three years before his daughters became productive. This meant that it would take about eight to nine years for an effect to be seen in the average yields of a specific farm, and even longer for it to be seen in the average of the total number of cows. So, although the entire period of stagnation lasted about 25 years, this explains at least in part the stagnation in protein content. Information was available. But farmers did not choose protein bulls, seemingly.

The stagnation only ended in the 1980s, when both the fat and protein content of milk and milk yields started to increase. The cause was the use of sperm from American Holstein-Friesian bulls. The AI societies were crucial in effecting this sea change, whereas the FRS was more reluctant. The management of the FRS continued to be convinced of the value of the traditional pure-bred Friesian cow, stressing the importance of the robustness of the traditional animals, even though they produced less milk with a lower fat content than Holstein-Friesians. The AI breeding societies, however, selected Holstein bulls, resulting in offspring producing much more milk with a higher fat and protein content. The result was that both the milk yield and the protein and fat content increased, and the stagnation in bio-based growth in dairy cows was overcome (Theunissen, 2008; 303-04).

V. More cows means more feed: improving land productivity

So far, this chapter has been a story of missed opportunities and an ever more enigmatic situation. Despite the high hopes of improving protein content through selective breeding, pricing and measurement systems and ever better information on sires Frisian dairy farmers did not achieve significant results until the 1980s. This is especially remarkable given the large investments in the extensive collection of data on protein content and the existence of strong regional organizations such as the Dairy Association and the FRS. In theory, it should have been possible, as with fat content, to improve the protein content of milk. We have discussed the factors of lock-in, organizational incapacity, and the complexity of cattle breeding. Together, these 'inertia' factors might explain part of the riddle. However, if we consider other forms of productivity, especially land productivity, an additional more fundamental explanation can be discerned. Table 7.2 provides an overview of the different components of bio-based productivity growth and shows that the most significant productivity growth was achieved in land productivity.

Table 7.2. Components of productivity growth in Friesland

	1950	1955	1957	1960	1965	1970	1975	1980	1990	% CHANGE
										1950-1990
Milk yield (kg)	3,800		4,200	4,400	4,400	4,500	4,900	5,500	7,200	93%
Fat content (%)	4.1	4.0		4	4.1	4	4.1	4	4.4	7%
Protein content	n/a	n/a	3.4	3.4	3.4	3.3	3.4	3.4	3.5	3%
Number of dairy cows (x 1,000)	231	247		267	266	315	348	387	298	29%
Pasture area (x 1,000 ha)	224	225		230	224	225	212	205	200	-11%
Pasture production, kg of starch equivalents per ha, the Netherlands	2,340	2,770		3,200	3,330		4,060	4,750	5,530	136%
Milk cows per 100 ha of grass land	105*	110		116	119	140	164	188	148	79%

*Figure for 1949

Sources: Number of cows, pasture area, fat content, and protein content: *Jaarcijfers voor de landbouw*, several issues. Milk yield per cow: 1950 calculated by multiplying milk delivered to factories by 1.04 and dividing this by the number of cows. Pasture starch value production: *Verslag over de Landbouw in het jaar [...]*, several issues. 'Starch value' is a somewhat outdated and limited metric to calculate the productivity of pastures. It has, however, the invaluable characteristic that it is, other than more sophisticated metrics such as 'dry matter', easier to calculate using historical data (Dijkstra, 1952; De Boer & Hoving, 2005). Starch value data in this table are for the Netherlands as a whole; Dijkstra (1952) states that the starch production value was 3,340 kg per hectare in Friesland in around 1950.

The number of dairy cows increased spectacularly between the late 1940s and the 1980s (when milk quota were implemented), while the available acreage of pasture declined slightly, mainly due to the increased use of land for roads and, especially, houses and other buildings. As can be seen in Table 7.2, the density of dairy cows increased up to 1980, even as the milk yield per cow also increased. In 1949, 105 dairy cows were held on 100 hectares, and this had almost doubled by 1980, when 188 cows were held on 100 hectares of pasture land. After a peak of 198 cows per 100 hectares in 1983, a combination of a rapid increase in production per cow and a government that curbed milk production meant that the number of cows per hectare declined rapidly, although milk production per hectare did not. The result of this growth in cattle density was that more feed was needed. In theory, there are several options for increasing feed production, but not all options were available in the agricultural environment of Friesland.

First of all, the production factor of land increasingly became a limiting factor in the post-war period. Although extensive meadows and wasteland were reclaimed in the pre-war years and up to the 1950s, the agricultural area declined after the 1960s due to housing and road construction and the creation of nature reserves.[15] Purchasing feed, as in pig and poultry farming, was not widely applied in Frisian dairy farming. Furthermore, large dairy farms without pasturage, such as in the United States (especially California), were virtually non-existent in the Netherlands (Gilbert & Wehr, 2003). Increased grassland productivity was therefore most commonly applied as a successful strategy to achieve the required improvements in land productivity to feed the increasing number of dairy cows (Knibbe, 1999; De Boer & Hoving, 2005: 37-60). This process caused profound changes in the agricultural landscape. By applying a combination of, among other things, improved water management, mechanization, enlargement of plots, and fertilizer use, land productivity was improved. While prior to 1965 extensive areas of peat and clay land were flooded during winter months, after 1975 better drainage (or at least more extensive and earlier drainage) caused this to be a thing of the past, enabling the earlier use of land, but more importantly the increased use of chemical fertilizers. This meant that more calories could be produced per hectare of cultivated land – pasture land and increasingly also forage maize cultivation. As a result of cultivation works (often in the context of land consolidation), drainage was improved, which meant that grass growth started earlier in the year. This enabled the increased and wider application of chemical fertilizers which, in combination with the use of new grass types, resulted in higher yields (Jagtenberg, 1975).

In addition, new forms of feed extraction were developed. At the end of the 1930s, the Frisian Society of Agriculture investigated various methods of storing grass for the winter months. In the first instance, attention was paid to grass drying factories which – after a few trials – were established on a large scale,

---

15 The agricultural acreage increased slightly in Friesland between 1951 and 1959. *Landbouwcijfers 1960* (Den Haag: LEI), p. 40.

especially in Friesland in the late 1930s (Priester, 2000).[16] After World War II, the less capital-intensive method of drying silage was introduced, with a Frisian farmer using the method for the first time in 1947. The grass was dried on the land after mowing and then ensiled in a day, whereby the silage was sealed airtight. In the 1970s, the planting of fodder maize also increased, initially on the sandy soils in the south of the Netherlands, but followed by other agricultural areas. The consequence of these changes in forage production was that the dairy cattle could be supplied with feed with a higher nutritional value during the winter months. These new methods were accompanied by the mechanization of feed extraction, which also resulted in an increase in efficiency (Priester, 2000).

Land productivity improvements were promoted by the national government, which heavily invested in land consolidation (*ruilverkaveling*). Initially, the goal of this was to enlarge agricultural plots and improve water management. Up to the 1950s, such land consolidation was relatively modest. However, at the end of the 1950s and during the 1960s, it was broadened to a general improvement of agricultural areas. The main goal was to improve land productivity, although goals such as the preservation of biodiversity and the landscape became important later on (Van den Bergh, 2004: 37-40; 59; 125). Government investments were motivated by the broader 'structural policy' to improve productivity in the agricultural sector by means of, among other things, scale expansion and mechanization.[17] From the perspective of the farmers, land consolidation offered an opportunity to improve productivity and thereby to maintain their income (which was generally a challenge).

The larger number of dairy cows that could be fed shows that land productivity increased between 1950 and 1990. This improvement in land productivity is in striking contrast to the more modest improvement in the milk yield per cow, and the lack of improvement in fat and protein content between 1950 and 1980. This gives rise to an alternative explanation, which is that farmers were probably less motivated to invest their time and energy in hard-won gains of improving protein content, whereas government policies enabled improvements in land productivity combined with an increase in the number of cows. Interestingly, this also contrasts with the conservative stance of the FRS. As such, Frisian farmers were not conservative, but rather rational in their choice of the most efficient path of bio-based productivity growth - even when this rationality was not price but production and income based: extensive instead of intensive growth. More feed, fertilizer and cows instead of more productive cows.

---

16 *Jaarverslag Vereniging tot behartiging van de belangen van de coöperatieve grasdrogerijen in Friesland (V.C.G.) 1940* (T, Vereniging van coöperatieve grasdrogerijen in Nederland, access 93-31, inv. 8).

17 Centrale Cultuurtechnische Commissie, *Meerjarenplan voor ruilverkaveling en andere cultuurtechnische werken in Nederland* (1958) pp. 25-27.

VI. Conclusion and discussion

By focusing on the attempts of Frisian dairy farmers and the dairy industry to improve the protein content of milk, this contribution adds to the literature on the knowledge-driven productivity growth of dairy farming in the post-war period. Improving protein content offers a lens for understanding the mechanisms behind bio-based productivity growth. The case study is especially interesting as it constitutes the case of an early adopter – Friesland was the first region in the world to implement protein pricing – where, despite all good intentions, the desired results did not materialize. In this chapter, we have offered answers to the questions of why protein content became a relevant theme within the Dutch-Frisian dairy cluster. We have assumed that the knowledge networks were dense enough to reach farmers with the new techniques and information. Therefore, their decision *not* to focus on improving protein content through selective breeding needs to be explained by other reasons. Why, therefore, did protein content remain stable for such a remarkably long period? In this respect, our analysis relates to sociologist accounts of the difference between the creation and utilization of knowledge.

The explanation of the stagnation in productivity growth in terms of protein content points to the inability of experts to deal with the complexities of cattle breeding. Breeding was time-consuming, and moreover there was a struggle to prioritize the different elements in the dairy cattle. Contrary to the ideas of experts and dairy companies, farmers did not just base their breeding decisions on a narrow, price-based rationality. To an extent, this inability can be excused. As Theunissen argues, twentieth-century breeding had never been based on narrow economic rationality. Knowledge on the fat content of milk was not used for a rigorous scheme of breeding high productivity cows. It was, however, used to exclude exceptionally 'bad' cows from breeding, which did lead to a rapid increase in average fat content, even when this was not the sole purpose of breeding. This, however, was misunderstood by the experts. Moreover, a degree of conservatism resulted in a lock-in of dairy breeding. It turns out that, due to these factors, the translation of theoretical knowledge into more practical 'useful knowledge' was highly complex. When it came to the implementation of protein knowledge, farmers decided to prioritize other ways of improving production. The easiest way was mainly realized by improving land productivity, as great gains were made in the scale enlargements and mechanization of grass production. It appears that farmers preferred these subsidized practices over the uncertain and time- and energy-consuming strategy of cattle breeding. Raising the protein percentages in the milk of their cows was therefore low on their list of priorities, if it was on it at all.

Despite the lack of progress in protein content, the payment systems based on protein content were maintained. The main reason might be that MCP appealed to farmers' desires for 'fairer' (*billijk*) payments – as, we have seen, was already stated in the PhD dissertation of expert Rommert Politiek. Of course, it has been known for some time that 'fairness' is fundamental to real-life and effective incentive structures (Kahneman et al., 1986). This is not only underscored by the very quick and overwhelming support for protein pricing in the spring of 1957, but also by the continued support for protein pricing when it turned out that it did not lead to more 'rational' breeding, and despite the additional costs to farmers. Another explanation is highly technocratic. From the beginning of the 1970s onwards, cooperative Frisian dairy factories (which, again, moved in tandem) introduced detailed accounting systems, guided by the BCZF. In these systems, production costs of different kinds of cheese and other dairy products were not just based on fat content but on protein content as well. This, of course, required knowledge of the protein content of milk and was hence consistent with extensive measurement as well as with MCP.[18] These hypotheses, however, require more extensive research.

This chapter represents a case study of roughly 30 years of exploration and hesitant implementation of protein know-how. Friesland was the first region to implement milk payment based on protein content, enabling a detailed assessment of the contribution of a novel form of knowledge. The results underscore the importance of economic, social, and cultural contexts in the process of agricultural innovation. Because of the risks and opportunities faced by farmers, they had to prioritize which knowledge and techniques they implemented first. Raising protein percentages was, in the 1960s and 1970s, overshadowed by other strategies to improve productivity, foremost the strategy of improving feed production. Intensive growth based on evidence-based knowledge is testament to daily practices in which preferences correlate with the social, economic, and cultural environment. Only after changes took place in this environment, from the 1980s onwards, did the protein percentage increase with the help of knowledge and techniques that had already been available for over 20 years.

---

18 Again, as in 1957, it is surprising how quickly these systems, once agreed upon, were introduced. The first factories changed their accounting systems, using the BCZF protein prices, within months (see T access 91, inv. 871).

Bibliography

Abbo-Tilstra, B. (2002) *Om de sûnens van het Fryske folk. Tuberculose en haar bestrijding bij bevolking en veestapel in Frylsân, 1890-1940*, Leeuwarden

Atkins, P.W. (2010) *Liquid Materialities. A history of milk, science and the law*, Farnham

Auderset, J. and P. Moser. (2018) *Die Agrarfrage in der Industriegesellschaft. Wissenskulturen, Machtverhältnisse und natürliche Ressourcen in der agrarisch-industriellen Wissensgesellschaft (1850-1950)*, Vienna

Bergh, S. van den. (2004) *Verdeeld land. De geschiedenis van de ruilverkaveling in Nederland vanuit een lokaal perspectief, 1890-1985*, Wageningen

Bles, J. (1961) 'Aanpassing in het zuivelbedrijf', *Bolwerk*, January, pp. 8-9

Boer, H.C. de and I.E. Hoving. (2005) 'Literatuurstudie van landbouwkundige aspecten van herinzaai van grasland', Wageningen University and Research Praktijkrapport Rundvee 70

Bonnier, G. and A. Hanson. (1946) 'Studies on Monozygous Cattle Twins VII. On the genetical determination of the interdependency between the percentages of fat, protein, and lactose in the milk' *Acta Agriculturae Suacana* 2, pp. 171-84

Brassley P., C. Martiin, and J. Pan-Montojo. (2016) 'Similar means to secure food supplies across Western Europe. A conclusion', in Paul Brassley, Carin Martiin, and Juan Pan-Montojo (eds.), *Agriculture in Capitalist Europe, 1945-1960*, London, pp. 267-74

Covington, C. (2015) '14 years of multiple component pricing: What has changed?', Progressive Dairy 15 April 2015 available at: https://www.agproud.com/articles/23188-14-years-of-multiple-component-pricing-what-has-changed

Croesen, V.R.IJ. (1932) Geschiedenis van de ontwikkeling van de Nederlandsche zuivelbereiding in het laatst van de negentiende en het begin van de twintigste eeuw, The Hague

Cropp, R. and W. Wasserman. (1993) 'Issues and options for using Multiple Component Pricing to set pricing in federal milk marketing orders', *Dairy Markets and Policy, Issues and Options* 0-8 January

——— (1996) 'Milk pricing and pooling in California', *Dairy Markets and Policy, Issues and Options* O-10

Dijkstra, H. (1952) 'De zetmeelwaarde-productie van het grasland. Opbrengsten en kosten op een aantal weidebedrijven over de boekjaren 1947/48, 1948/49 en 1949/50', Landbouw Economisch Instituut, Bedrijfseconomische mededelingen No. 4

Dijkstra, W. (1991) 'De veeveredelaar. Rommert Douwe Politiek', in P.C.M Hoppenbrouwers (ed.), *Een loopbaan in de Landbouw: twaalf portretten van markante figuren in agrarisch Nederland*, Groningen, pp. 178-81

Frietema, H.J. (1937) *Productie en prijsvorming op de Engelsche markt van Nederlandsche, Deense en Koloniale boter*, Haarlem

Gilbert, J. and K. Wehr. (2003) 'Dairy industrialization in the first place: urbanization, immigration, and political economy in Los Angeles', *Rural Sociology* 68, no. 4, pp. 467-90

Gorman, Hugh. S. and Betsy Mendelsohn. (2010) 'Where does nature end and culture begin? Converging themes in the history of technology and environmental history', in M. Reuss and S.H. Cutcliffe (ed.), *The Illusory Boundary: Environment and Technology in History*, Charlottesville, pp. 265-90

Groot, C. (1968) 'Marktaspecten van eiwit mede in verband met de EEG' in *Eiwit*, The Hague, pp. 52-62

Hamming, I. (2013) 'Historie fokwaardeschatting. In veertig jaar tijd van melkindex via plussysteem naar totaalindex', *Veeteelt*, 20, no. 1/2, pp. 12-15

Hansson, A. and G. Bonnier. (1949) 'Further studies on the genetical determination of the composition of cows' milk with regard to fat, protein and lactose', *Acta Agriculturae Suacana* 3, 179-88

Hernández-Adell, I. and J. Pujol. (2016) 'Economic growth and biological innovation: the development of the European dairy sector, 1865-1940', *Rural History*, 27, no. 2, pp. 187-212

Jagtenberg, W.D. (1975) 'Verhoging van de graslandproduktie noodzakelijk en mogelijk', Bedrijfsontwikkeling 6, pp. 413-17

Janse, H.C., H.G.A. Leignes Bakhoven, and E.T. Roelofs. (1953) 'Het eiwitgehalte en de verhouding tot het vetgehalte in de melk van Friese koeien', *Mededelingen maandblad van het Friesch Rundvee-Stamboek en van de Bond van K.I.-verenigingen in Friesland* no. 61, November 1953

Jesse, E.V. (1994) 'Proposals for Multiple Component Pricing in midwest federal milk marketing orders', University of Wisconsin-Madison, Department of Agricultural Economics, Marketing and Policy Briefing Paper Series Paper No. 45

Kahneman, D., J.L. Knetsch, and R.H. Thaler. (1986) 'Fairness and the assumptions of economics', *The Journal of Business*, 59, No. 4, pp. S285-S300

Karel, E. (2005) *De maakbare boer. Streekverbetering als instrument van het Nederlandse landbouwbeleid 1953-1970*, Groningen

Keestra, J. (1963) *Melkbaarheidsonderzoek bij het zwartbonte vee in Friesland*, Wageningen

Knibbe, M. and M. Molema. (2018) 'Institutionalisation of knowledge-based growth: the case of the Dutch-Frisian dairy sector (1895-1950)', *Rural History* 29, no. 2, pp. 217-35

——— (1993) 'Landbouwproductie en -productiviteit, 1807-1997' in R. van der Bie and P. Dehing (eds.) *Nationaal goed. Feiten en cijfers over onze samenleving (ca.) 1800-1899*, Voorburg/Heerlen

——— (1993) *Agriculture in the Netherlands 1851-1950. Production and institutional change*, Amsterdam

Latour, B. (1987) *Science in action. How to follow scientists and engineers through society*, Cambridge MA

Lolkema, H. and A.J. van der Have (1958) 'Ervaringen met de bepaling van het eiwitgehalte in de melk I', *Landbouwvoorlichting* 15, pp. 559-64

Maanen, G. van and L. Wubbenhorst. (2019) 'Nominatie grootste Wageninger. Rommert Politiek'

Mokyr, J. (2002) *The gifts of Athena. Historical origins of the knowledge economy*, Princeton

Molema, M. (2017) 'Collective organization of knowledge in the early phase of the Dutch-Frisian dairy industry (c. 1885-1914)', *Historia Agraria* 73 (2017): 91-118

Novakovic, A.M. (1994) 'Lessons and questions from fifty years of milk marketing orders', Session on marketing orders https://www.semanticscholar.org/paper/Lessons-and-Questions-from-Fifty-Years-of-Milk-Novakovic/7baac33641d536cf5d901a528f8bb7baebd12c7f

Olmstead, A.L. and P.W. Rhode. (2002) 'The Red Queen and the Hard Reds: productivity growth in American wheat, 1800-1940', *The Journal of Economic History*, 62, no. 4, pp. 929-66

Orland, B. (2016) 'Turbo-cows. Producing a competitive animal in the nineteenth and early twentieth centuries', in S. Schrepfer and P. Scranton (eds.) *Industrializing Organisms. Introducing Evolutionary History*, London, pp. 167-89

Plantinga, R. and M. Molema (2017) *'Weten is kunnen, kennis is macht'. Een geschiedenis van de Bolswarder Zuivelschool (1880-1995)*, Leeuwarden

Politiek, R. (1957) *De invloed van erfelijkheid en milieu op de samenstelling van de melk bij Friese koeien en de praktische mogelijkheid van selectie op het eiwitgehalte*, Drachten

Politiek, R.D. (1968a) *Problematiek en perspectief in de veeteelt*, Wageningen

——— (1968b) 'Aspecten van de opvoering van de produktie van melkeiwit', in *Eiwit*, The Hague, pp. 36-51

Porter, M. (1998) 'Clusters and the new economics of competition', Harvard Business Review 76, no. 6, pp. 77-90

Posthumus, G. (1968) 'Het eiwitgehalte van melk en de zuivelindustrie', in Eiwit, The Hague, pp. 63-81

Priester, P.R. (2000) 'Boeren met machines. Het melkveehouderijbedrijf', in W. Schot and Albert de la Bruhèze (eds.) *Techniek in Nederland in de twintigste eeuw. Vol.* 3, Zutphen, pp. 99-125

Rooijendijk, C. (2010) *Alles moest nog worden uitgevonden. De geschiedenis van de computer in Nederland*, Amsterdam

Russell, Edmund. (2004) 'Introduction. The garden in the machine: toward an evolutionary history of technology', in Susan R. Schrepfer and Philip Scranton (eds.), *Industrializing organisms: introducing evolutionary history*, New York/London, pp. 1-16

Smith, B.J. and Stephen D. Snyder, 'Effects of protein and fat pricing on farm milk prices for the five major U.S. dairy breeds', American Journal of Agricultural Economics, 60-1 (1978), pp. 126-31

Sneddona, N.W., N. Lopez-Villabosa, R.E. Hicksona, and L. Shalloo. (2013) 'Review of milk payment systems to identify the component value of lactose', *Proceedings of the New Zealand Society of Animal Production* 73, pp. 33-36

Stine, J.K. and J.A. Tarr. (1998) 'At the intersection of histories: technology and the environment', *Technology and Culture*, 39, no.4, pp. 601-40

Theunissen, B. (2013) 'Breeding for nobility or for production? Cultures of dairy cattle breeding in the Netherlands, 1945-1995', *Isis*, 103, pp. 278-309

——— (2008) 'Breeding without Mendelism: theory and practice of dairy cattle breeding in the Netherlands 1900-1950', *Journal of the History of Biology*, 41, 637-76

Thoen, E., T. Soens, and P. Brassley (eds.). (2015) *Struggling with the environment: land use and productivity*, Turnhout

Tjepkema, T. and J. Wiersma. (1959) *Erf en Wereld*, Drachten

Uekötter, F. (2012) *Die Wahrheit ist auf dem Feld. Eine Wissensgeschichte der deutschen Landwirtschaft*, Göttingen

Whitaker, M.D. (1978) 'Component pricing of milk to farmers: a more equitable way', *Western Journal of Agricultural Economics* 2, pp. 121-30

Zanden, J. L. van. (1997) *Een klein land in de 20e eeuw. Economische geschiedenis van Nederland 1914-1995*, Utrecht, pp. 117-19

Zwarts, H. (2021) Knowledge, networks, and niches. Dutch agricultural innovation in an international perspective, *c.* 1880-1970, Wageningen

FERNANDO COLLANTES

8. From massification to upgrading*

A long run view of recent trends in dairy consumption in Spain

I. Introduction

Recent changes in food consumption are an important element in ongoing public debates about consumer society and its implications. While some celebrate the abundance and diversity that characterize today's foodways (Pinker, 2018), in contrast to the widespread hunger that prevailed during most of our history as a species, others warn about the public health problems (among them, an emergent obesity epidemic) and environmental impacts associated to contemporary diets (Pollan, 2008). Moreover, for food system researchers changes in consumer behaviour are also relevant because of their influence upon the strategies of other actors in the production chain, such as farmers, processing industries and supermarket chains. Over the last half century, said strategies have been strongly conditioned by the end of a historical era, sometimes referred to as 'nutritional transition'. This is a term coined by public health scholar Barry Popkin (1993) to describe a historical pattern of Western diet change involving sustained increases in food intake and a reorientation from plant-based foods (cereals, legumes, potatoes) to animal-based foods (meat, dairy, eggs, fish). The end of sustained increases in food intake has forced producers to face what they sometimes call 'the problem of the fixed stomach': the possibility that, since consumers are essentially satisfied with their diet, they will not make any further changes to it, which prevents processors' sales from growing as usual (Pollan, 2008: ch. 6).

Both dimensions can be perceived in the case of dairy products, which is the focus of this chapter. For optimists, few commodity groups can illustrate so tellingly the transition from scarcity to abundance and variety, as a walk through

* I am very grateful to Laurent Herment and the other editors of this book, as well as the external reviewers, for fantastic feedback on previous versions of this chapter. I have also benefited from comments made by participants at the Uppsala workshop on "Animal husbandry in 20th century Europe", the Rural History 2019 conference at Paris (especially Jonas Albrecht), the NISALDes seminar organized by José Miguel Martínez Carrión (especially Roser Nicolau), and the III International Congress of Agricultural History (SEHA) at Salamanca (especially Xavier Cussó and Pablo Delgado).

From Breeding & Feeding to Medicalization, Rural History in Europe, 17 (Turnhout: Brepols, 2023), pp. 183–208
BREPOLS PUBLISHERS 10.1484/M.RURHE-EB.5.135026

the dairy aisle of any contemporary supermarket will prove. For pessimists, milk and dairy products are, together with meat, one of the most powerful carriers of our diet-related problems: they have made a major contribution to the excess fat intakes that are allegedly making degenerative diseases more prevalent, as well as to the increasing greenhouse gas emissions and other environmental impacts associated with intensive, industrialized farming. Beyond these assessments, it is a fact that changes in dairy demand have strongly conditioned the strategies of supply-side actors. Once the decades after the Second World War were left behind, consumer demand began to lose momentum, and this had a profound impact on the shaping of the two main lines of change in the production chain. On the one hand, the strategies of companies have reoriented towards product innovation. Substantial investments in R&D and marketing are expected to bring about new, ever more sophisticated products that make consumers interested once again in expanding their expenditure on food. On the other hand, there is a parallel process of structural adjustment going on. Most farms and many processing firms have disappeared, while power becomes increasingly concentrated in just a few actors, most remarkably supermarket chains. This has led to significant waves of social unrest, especially among dairy farmers (Valenze, 2011; Velten, 2011; Smith-Howard, 2014; DuPuis, 2002; Vorley, 2007).

This chapter poses a deceivingly simple question: to use economist Louis Malassis' (1997) term, what is the new 'model of food consumption' about? After this introduction, a review of previous research presents the issue, more complex than it may perhaps seem at first sight. Throughout the rest of the chapter, we will focus on the specific case of Spain, which is in many ways representative of similar transformations taking place in dairy consumption across Europe over the last few decades. A section is devoted to presenting the case of dairy consumption in Spain, inserting recent trends in a longer-run socioeconomic perspective. After that, a novel methodology is proposed and applied to our case in order to identify the drivers of the new model of dairy consumption. A comment and a contextualization of the results comes next. The concluding section suggests some implications for ongoing debates on contemporary diets, food systems, and consumer societies.

II. Great acceleration or successive models of food consumption?

Narratives of a 'great acceleration' (McNeill and Engelke, 2014) after the Second World War are far from uncommon in historical analyses of diet change. After all, and as historian Peter Scholliers (2014: 341) points out in an authoritative state of the art, 'post-1945 developments in food evolve faster, have a broader reach, are more international, and cause irrevocable transformations for more people than ever before'. Scholliers (2014: 347) reviews diet change from a perspective that considers the whole of the period between 1945 and the present day, and stresses that consumers in the global North have had to allocate an ever-smaller share of their budget to food, as an extraordinary increase in the productivity of the food system implied that 'more diverse food became available at affordable prices'.

Scholliers also suggests that within this general pattern there was not a single, pre-determined path of diet change. This suggestion has been explored at length by other historians recently. Josep Pujol and Xavier Cussó (2014), for instance, compare diet change in several European countries from the mid-nineteenth century to the early twenty-first century and conclude that there were several different paths according not only to usual suspects such as international differences in economic growth, but also to environmental conditions and food system dynamics in each country and region. Another historian, Ernst Langthaler (2018), also underlines the diversity of trajectories by comparing diets in 1961 to diets in 2011 in different countries and world regions. Finally, the team of environmental historians and ecological economists led by Manuel González de Molina presents diet change in Spain in terms of a great acceleration that, from the 1960s onwards, deteriorated both consumers' health and the environment, as southern Europe's traditional, organic food system (and the Mediterranean diet that was associated to it) was displaced by an industrial food system (González de Molina et al., 2020).

These views of recent trends in diets share a preference for long-run analysis and tend to portray the present as the result of one single historical movement that would have started in the mid-twentieth century. (A partial exception is Langthaler's engagement with 'food regime analysis', a research programme that is reviewed later in this chapter). Other researchers, however, have found a break in the trajectory of diet change in the late twentieth century. Much of this research is organized around concepts and arguments that were already proposed in the 1990s and early 2000s.

Economist Louis Malassis (1997), for instance, found in the 1990s that a break had taken place around 1980, when the 'mass, agri-industrial model of food consumption' came to an end. Around 1980, and after several decades of increasing food intakes, most consumers would have become satiated from a biological point of view. Food intake then ceases to grow, and diet change starts to move in a different direction: now it mostly consists of substitutions of cheap foods with more expensive varieties of those same foods. For instance, consumers become interested in higher-quality varieties of food products. Also, they become regular consumers of new, highly processed foods that companies have developed by combining an ever-larger number of ingredients in the framework of ambitious R&D programmes. Complementary lines of change in consumer behaviour would include the rise of so-called functional foods, intimately linked to new health anxieties, and service foods, as in the case of consumption that takes place in restaurants, cafeterias, or bars. Occasionally Malassis even seems to favour 'agri-industrial, mass restoration' as a tentative label for this new model of food consumption, but this does not capture well the variety of possible trajectories of change that he is describing.

Sociologists and geographers also developed a historical sequence of this kind. David Goodman and Michael Redclift (2005) coined the term 'quality turn' to refer to the post-Fordist pattern of food consumption. According to them, the late twentieth century was witness to the emergence of a new vanguard of well-off consumers who would use their remarkable level of income to move away from Fordist food and into alternative foodways comprising organic foods, foods with protected designations of origin, or locally produced foods, to name a few (see also Goodman et al., 2012). In a related fashion, for Philip McMichael (2013) the food regime of the decades after the Second World War, in which the United States played a hegemonic role, would have been replaced in the late twentieth century by an emerging neoliberal, 'corporate' regime, in which the World Trade Organization coordinates the pursuit of a pro-globalization agenda. From the viewpoint of diets, the essential characteristic of this new regime is the contrast between 'food from somewhere' and 'food from nowhere': a vanguard of affluent consumers transitions towards a diet rich in high-quality products, while lower-class consumers remain trapped in a Fordist diet that eventually becomes detrimental to their health (see also Magnan, 2012).

Barry Popkin (1993) also proposed a late-twentieth century rupture in his influential theory of the nutritional transition. In the latter's original version, formulated in the early 1990s, the health problems associated to widespread consumption of large quantities of food, especially an increasing prevalence of cardiovascular and other degenerative diseases, were motivating some 'behavioral change' among consumers. The resulting new diet would feature less fats and less processed foods. Contrary to previous trends towards higher intakes of animal foods, there would be a reorientation towards carbohydrates, fruits and vegetables. In an article written only a few years after Popkin's seminal piece, geographer David Grigg (1999: 11) quantified the fall in the consumption of sugar, animal fats and dairy products, to reach the conclusion that, beginning as early as in the 1960s in the most developed countries, a 'fundamental break with the past' was taking place.

Finally, there were also historians among those who stressed that sort of 'fundamental break'. In the 1990s again, historian Massimo Montanari (1993) published a wide-ranging monography on food history, significantly entitled *Hunger and abundance*. According to Montanari, in the decades after the Second World War there was an increase in intakes, meat being the paradigmatic example (notice the parallelism with Malassis' mass, agro-industrial diet), but later on an alternative, more moderate style of consumption began to take shape. In his social history of eating in the United States, Harvey Levenstein (1993) placed this shift within a parallel shift in nutritional science: the rise of 'negative nutrition', that stresses the need to reduce the consumption of some specific foods (see also Biltekoff, 2012). Another major work of synthesis, Felipe Fernández-Armesto's (2002) global history of food, also found a break in the last decades of the twentieth century, as consumers begin to move away from cheap, mass-produced food and into a new pattern of food consumption based on identity, quality, exclusiveness, artisanal methods, and organic productions (this is basically Goodman and Redclift's quality turn and McMichael's food from somewhere). The actual magnitude of these shifts, however, is not evident, as an increasing number of historical studies on the rise of obesity during the twentieth century seems to suggest (Oddy et al. eds., 2009).

To be fair, the proponents of a late-twentieth century rupture have been more successful at identifying elements of breakaway with the mass, agro-industrial diet and the 'classic' nutritional transition than at combining those elements in an integrated description of what the new model of food consumption is actually about. It seems necessary to clarify the respective roles of the two major forces pointed out by different research programmes. There is, on the one hand, the 'fixed stomach': consumers are no longer increasing their food intake, and some may even reduce it. On the other hand, there is the possibility of what we will call 'upgrading', following the term used by economist Nick von Tunzelmann (2010) to refer to demand differentiation in the food market. Even if their material intake is stagnant or declining, consumers may still increase their monetary expenditure on food if they want their diets to be more oriented towards quality, variety or (put in general terms) any qualitative characteristics of food that they have reasons to value as superior to those of the foods that they were consuming until then. An integrated quantitative assessment of both forces is needed.

III. Dairy consumption and socio-economic change in Spain

The case of dairy products may provide, as has been argued in the introduction, a good lens to examine recent changes in food consumption in the global North. The specific case of Spain, on its part, has the additional value that it presents a very sharp version of the rupture that takes place in the late twentieth century (Collantes, 2015a). Contrary to what was happening in the United Kingdom, France and other north-western European countries, the consumption of dairy products in Spain remained strikingly low until well into the twentieth century (figure 8.1). In the nineteenth century, milk played a minor role in the diet of most Spaniards and, unlike the cases of cheese in Italy or yoghurt in Greece, this was not compensated with a remarkable level of consumption of other kinds of dairy products. Even though the consumption of milk began to spread more widely during the early decades of the twentieth century, the outbreak of the Spanish Civil War in 1936 halted that trend. Milk consumption would only rise again after a long postwar period that lasted until the 1950s. Under this new wave of growth, however, milk consumption became massive, which led to a correction of the calcium deficiency problem that most of the population had been suffering from (Cussó, 2005; Cussó et al., 2018) as well as to significant improvements in biological welfare indicators such as height (Martínez-Carrión, 2016). Especially during the period from 1965 to 1980, milk consumption grew

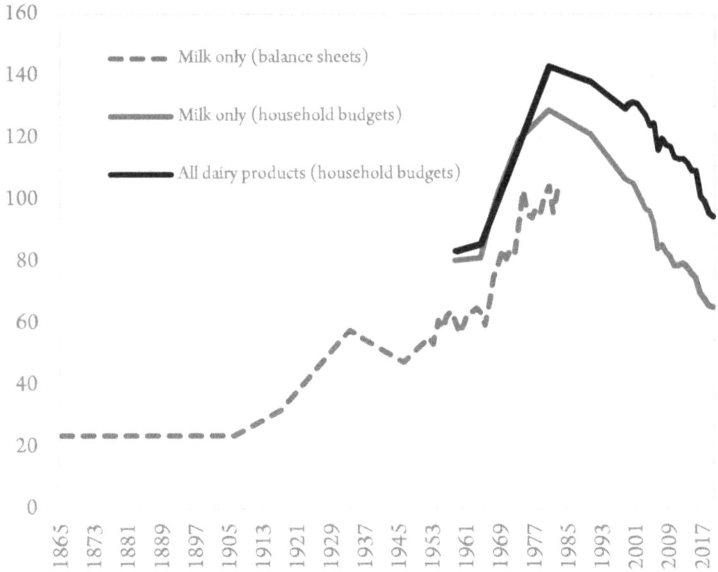

Figure 8.1. Consumption of milk and dairy products in Spain, 1865-2019 (kilograms per person and year)
Sources: 1865-2007: Collantes (2020: 287); 2007-2019: Instituto Nacional de Estadística (www.ine.es, *Encuesta de presupuestos familiares*).

at an unprecedented rate, not only for Spanish but also for European standards. This means that the rupture that took place across Europe in the late twentieth century, when the consumption of milk started to decline, was sharper in Spain. Milk consumption decreased and this led aggregate dairy consumption to decline as well.

It is true that talking about aggregates and averages can hide noteworthy variations according to different types of product and heterogeneous consumer profiles. Yet, it is striking that these patterns of composition and segmentation also underwent a transformation in the late twentieth century. Prior to that moment, milk was the most consumed dairy product. Still in 1990, more than 80 per cent of the calories that consumers took by means of dairy products came from milk. In the last few decades, in turn, consumers have reoriented towards other dairy products, in particular yoghurts and refrigerated desserts (table 8.1). In the early years of the twenty-first century, the calorie intake associated to dairy products other than milk came to be larger than that associated to milk itself (Collantes, 2015a: 256).

This reorientation from milk to other dairy products hides even more diversity within each of the product groups. Around 1980, the world of dairy consumption gravitated around just one type of milk (whole milk), one type of yoghurt (natural

Table 8.1. Household consumption of dairy products (kilograms per person and year, except otherwise noted)

	1958	1964/5	1980/1	1990/1	2007	2019
Liquid milk (litres)	77.9	78.7	125.1	117.6	85.3	63.3
Preserved milk[a]	1.2	2.8	3.7	1.8	0.4	0.3
Cheese	1.5	1.5	4.3	5.7	6.7	6.5
Butter	0.5	0.4	0.4	0.3	0.3	0.4
Yoghurt[b]		0.2	5.7	7.6	19.3	15.8
Other[c]				1.6	7.8	6.4
Total	**83.4**	**85.7**	**143.0**	**138.1**	**119.8**	**94.6**

Notes: [a] Evaporated, concentrated, condensed and powdered milk (except in 1958: only condensed milk); [b] Includes all sorts of fermented milk; [c] Mostly milk shakes, yoghurt shakes, ice creams and refrigerated desserts other than fermented milk.

Source: Instituto Nacional de Estadística (1959; 1965-95; www.ine.es, *Encuesta de presupuestos familiares*).

yoghurt) and a narrow range of cheeses (such as the *manchego*, based on ewe's milk, and melted cheese in portions). There has been an explosion in the number of available products ever since. Milk has ceased to be only whole: it can be skimmed and semi-skimmed. Whatever the fat content, milk may also be fortified with calcium or vitamins, and it may or may not have lactose. Yoghurts can be natural yoghurts, but they can also have some added flavour or contain fruit bits. They can be solid, but they can be liquid as well. Their milk can be fermented with natural ferments, but also with probiotics. Fermented milks may or may not be oriented towards regulating intestinal circulation or the accumulation of cholesterol in the blood. Possibilities have multiplied in the world of cheese too: consumers can choose from a wide range of regional varieties and an ever-expanding selection of cheese coming from other countries. Product presentation may vary from pieces of hard cheese and tubs of creamy cheese to portions of melted cheese and grated and sliced versions of a great variety of cheeses. In other words, the aggregate trend of dairy consumption conceals widely different trajectories for different products, ranging from the unquestionable decline of whole milk to the rise of some of its most sophisticated by-products. The diversity of the last few decades breaks clearly away from the previous model of consumption, which steadily gravitated around whole milk.

Contrasts between different types of consumers also present a rupture in the late twentieth century (table 8.2). The phase of great expansion prior to 1980 had been a phase of massification, based on the spread of frequent milk drinking among consumers that until then had only been making sporadic consumption. There had been a remarkable reduction in regional disparities, as milk drinking had become widespread in regions where, as in Andalusia and the Mediterranean

Table 8.2. Relative levels of household dairy consumption

	1964/5	1980/1	1990/1	2006	2019
Upper-class households (lower-class households=100)[a]	168	101	85	127	126
North region (Mediterranean region=100)[b]	237	161	148	120	123

Notes: [a] Fourth quartile vs. first quartile according to socio-economic status; [b] Among four macro-regions, the North is the region with the highest level of consumption, while the Mediterranean is together with Andalusia the region with the lowest level.

Sources: 1964/5-2006: Collantes (2015b: 108); 2019: Ministerio de Agricultura, Pesca y Alimentación (www.mapa.gob.es, *Panel de Consumo Alimentario*) and Instituto Nacional de Estadística (www.ine.es, *Cifras de población*).

regions, average levels of consumption had traditionally remained low. There had been a noteworthy reduction in inequality according to social status too. Around 1980, the traditional gap between upper- and lower-class households had vanished: regular milk drinking had ceased to mark status. During a short interlude, dairy consumption became even higher in lower-class households. However, the transition towards the new model of dairy consumption has been led by upper-class households again. From the late 1990s onwards, a social gap re-appeared. This was particularly apparent for dairy products other than milk, as well as for the subperiod leading up to the Great Recession that started in 2008. The social gap is not at all as large as it had been before the massification of milk drinking (still around 1965, for instance), but it is significant and has not been blurred by the Great Recession and its aftermath. The rebound of the social gap stands out even more if we position it in relation to the evolution of the territorial gap, which (especially up to the Great Recession) has continued to become smaller.

These late-twentieth century breaks in the trend, composition and segmentation of dairy consumption were in parallel to others in the trajectory of the dairy chain (Langreo, 1995, 2005, 2006; Fernández Prieto and Lanero eds., 2020; Collantes, 2016, 2019), as well as of Spain's economy and society more broadly (Collantes, 2017). During the period 1950-1980, major changes took place in Spain's dairy system. Farmers were joined by an ever more influential processing industry. Both farmers and processors introduced process innovations that triggered large increases in production and productivity. Dairy farming became more intensive and more specialized, as farmers substituted autochthonous breeds with high-yield, Friesian-type cows and became increasingly involved in commercial and technological interaction with processors. Processors incorporated machinery and industrial methods of business organization for the large-scale production of processed milk, while at the same time providing incentives for farmers to install refrigerating tanks on their farms.

The resulting growth wave was also made possible by the close connection that developed between processors, the State, and scientists. Franco's regime implemented policies for the promotion of dairy industrialization, among them price regulations and monopolistic concessions for the supply of the country's main urban markets. Scientists, especially physicians, were unanimously positive about the impacts that frequent milk drinking would have on consumers' health. Business, politics, and science reinforced each other. In the meantime, major changes were taking place in the broader Spanish economy and society. Long-run transformations such as industrialization, de-agrarianization and urbanization culminated during these years. A middle-class consumer society, involving increasing diversification of household expenditures as well as decreasing inequality, was finally taking shape.

However, in the late twentieth century the dairy chain and Spain's economy and society at large shifted onto a different trajectory. In the dairy chain, the wave of expansion and transformation of the previous period was sharply followed by an era of rationalization and severe structural adjustment. Most dairy farms eventually disappeared or reoriented towards other productions. Employment in the processing industries stopped growing and came to be concentrated in an ever-smaller number of relatively successful firms. A key actor when it came to the management of structural adjustment trends were supermarket and hypermarket chains, which emerged as coordinators of the other nodes in the chain in a way like that in which processors had previously emerged as coordinators of farmers. Innovation remained important, but now less oriented towards process innovation and more towards product innovation. Leading processors were allocating increasing resources to R&D activities that were expected to allow them to launch new products that would dispel the ghost of the 'fixed stomach' and forge a new mode of profitmaking based on upgrading.

The political and scientific context was also radically altered. Following the accession of post-Franco's Spain into the European Economic Community in 1986, the State was forced to discontinue its major policies for the promotion of dairy industrialization, while farmers would eventually have to adapt to the EEC's system of dairy quotas and the bleak prospects of growth associated to it. Meanwhile, unconditional enthusiasm about dairy products was gradually replaced among physicians by a more nuanced view and, in time, by widespread acceptance of the 'lipid hypothesis', according to which excess dairy intakes were detrimental to consumer health through a higher risk of cardiovascular disease. While all these changes were taking place, Spain was going through a reorientation of its economy from manufacturing to construction and services. Said reorientation allowed for a new wave of economic growth to unfold between approximately 1985 and 2007. This was also a period of consolidation, diversification, and renewal in consumer society beyond food. The Great Recession, however, drastically deteriorated the country's macroeconomic results and (even more so) its social results.

IV. Characterising recent trends: a new empirical strategy

Considering the evidence presented in the previous section, the case of dairy products in Spain provides a promising perspective for the study of recent trends in diet change in the global North. Here we find the pieces alluded to by the literature clusters reviewed above, in particular the end of the expansion that had taken place under the mass, agro-industrial model of consumption and the emergence of a large number of upgrading options for consumers. What has been the result of the tension between them? What is the structuring of these pieces within the new model of food consumption?

A simple methodology is presented now to give an answer to these questions. Our starting point is framed in ecological economics, from which we take the idea that one must study the co-evolution of material and monetary flows in economic systems (Fischer-Kowalski et al., 2011). Upgrading in food consumption can be seen as an increasing disconnection between consumer expenditure on food (the monetary flow) and consumer intake of food in terms of kilograms or calories (the material flow). As a matter of fact, this is what Malassis (1997) has in mind when he proposes that the average price of the calorie consumed is taken as an indicator of the degree up to which consumers are making qualitative substitutions (see also Clark, 2007: 196). The underlying logic is that the transition towards more expensive, more sophisticated products would be reflected in increasing prices per calorie consumed. Under the agro-industrial model of mass production and consumption, the price of foodstuffs would tend to decline, and so would the price per calorie consumed. However, at a later stage, the shift to new product varieties that have a similar caloric content but are more sophisticated than the older ones would exert an upward pressure on the price per calorie consumed. In other words, the point is to measure the extent up to which consumer expenditure on food is increasing faster than their material intake of food.

This empirical strategy allows us to identify upgrading as a possible driver of the new food consumption model, but it relies on two implicit assumptions that cannot be taken for granted. The first of them is that consumers' material intake remains constant. In formal terms:

$$r(E) = r(K) + r(E/K)$$

where r is annual compound rate of growth, E is consumer expenditure on food, and K is consumption in terms of calories or kilograms. Change in food expenditure can be decomposed into two elements: change in the number of calories consumed and change in the price of the calorie. If the number of calories consumed is stable, then change in the price of the calorie equals change in consumer expenditure. Upgrading drives the new consumption model. This is the logic implicit in Malassis, but we obviously need to consider alternative scenarios of consumers who are still increasing their intake and consumers who have started to reduce it. It is within this bigger picture that we can assess whether upgrading is strong enough to drive the new consumption regime.

A second implicit assumption is that the relative price of the different foods remains stable through time. This is not very realistic for a long-run analysis. Upgrading is not at all the only factor that may cause observable changes in the price of the calorie. Let us imagine, for instance, a preindustrial famine in which scarcity makes the price of food skyrocket and the consumption of food plummet. In a situation like that, the monetary value of the calorie is increasing, but consumers are not upgrading – they are just suffering the consequences of rising relative prices. This example may not seem too relevant for the context of the food system of the last few decades, when an extraordinary increase in actual and potential production has taken place. But, precisely because of said increase, it does not seem that we are in a world of stable relative prices either. It seems rather more plausible to expect decreasing relative prices, which is what actually triggers the severe structural adjustment that we have described for Spain's dairy chain in the previous section (Langreo, 2005, 2006). In that case, it is important to notice that, in case consumers were satisfied with their initial diet and showed no interest for upgrading, their expenditure on food would not be stable, but declining: lower food prices would make it possible for them to buy the same food with less money.

In other words, upgrading can be more important than it seems. We must not contrast it with an alternative scenario of stable expenditure, but with an alternative scenario of decreasing expenditure that represents the 'fixed stomach' situation more realistically. To that end, we will imagine what would have happened in case consumers had felt fully satisfied with their dairy consumption of 1990. This date suits our purpose because it is still during a time of transition between the pre-1980 mass consumption model and the new model that, as we have seen above, takes shape after the mid-1990s. It is also suitable because it is only from 1987 onwards that we have sufficiently disaggregated data on dairy consumption: the Food Consumption Panel commissioned by the Spanish Ministry of Agriculture, Fisheries and Food gives us data for as many as 17 dairy foods.[1] Finally, around 1990 we find a pattern of dairy consumption that qualifies as good enough according to nutritional advice, especially when it comes to having a sufficient intake of calcium. Some nutritionists may object that there was an excess of whole milk consumption, which would be contributing to the emerging problem of excess intake of saturated fat. Regardless of this argument, it must be noted that the price of semi-skimmed and skimmed milks was very similar to that of whole milk, so that a hypothetical substitution of the latter with

1 This is the list of 17 items: whole, semi-skimmed and skimmed milk; concentrated, evaporated, condensed and powder milk; natural, flavoured and fruit yoghurts; fresh, melted, and hard cheese; yoghurt and milk shakes; butter; and one last residual category of 'other dairy products' (most likely, cream, ice creams, ice cream cakes and other refrigerated desserts).

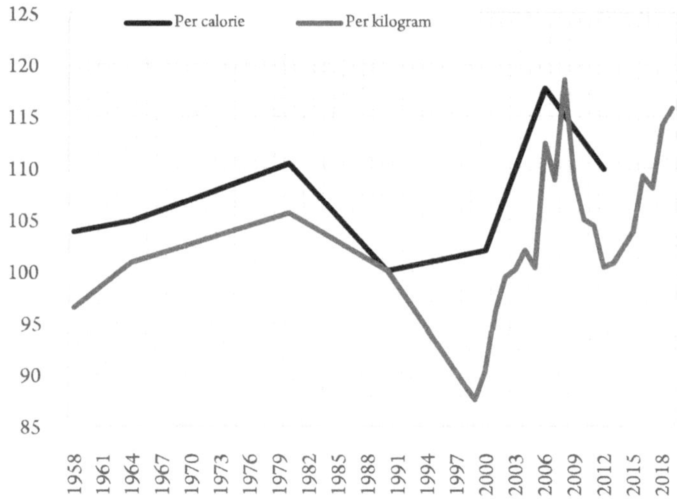

Figure 8.2. Unit value of dairy consumption (constant 2013 euros per calorie or kilogram), 1990=100
Sources: expenditure and consumption (kilograms): Instituto Nacional de Estadística (1959; 1965-95; www.ine.es, *Encuesta continua de presupuestos familiares*, *Encuesta de presupuestos familiares* and *Índice de precios al consumo*); consumption (calories): estimated with Spanish food composition tables, mainly Moreiras et al. (2001). The estimation of calorie intake and expenditure from 1990 onwards includes consumption outside the home; see Collantes (2015a: Appendix) for details.

the former would not have important consequences upon our results. The major problem of Spanish dairy consumption around 1990 was that the substitution of raw milk with processed milk, which was more reliable from the standpoint of consumer health, was still incomplete. In our estimation we allow for a notion of the 'fixed stomach' in terms of drinking processed milk only.

We then estimate how much it would have cost consumers to buy their 1990 diet with the price levels of 1991, 1992 and so on until 2020. It can be objected that this counterfactual assumes that the relative prices observed for the years after 1990 would have evolved similarly in case consumers had not changed their diet at all, but it is doubtful that this can have a major impact on our results. Some of the short-run deviations would cancel one another (for instance, not-so-declining milk prices and not-so-high prices for other dairy products) and, perhaps more importantly, we must consider that by the start of our period Spain's dairy chain already featured excess capacity. The room for reallocating resources and competing via prices was remarkable during the whole of our period and, in consequence, the long-run picture provided by the results seems reliable.

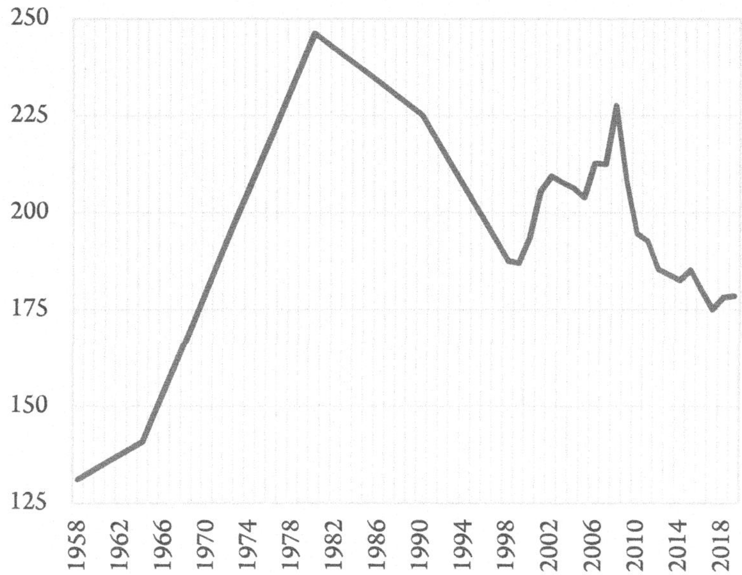

Figure 8.3. Consumer expenditure in dairy products (constant 2013 euros per person and year)
Source: Instituto Nacional de Estadística (1959; 1965-95; www.ine.es, *Encuesta continua de presupuestos familiares*, *Encuesta de presupuestos familiares*). The general price index (Instituto Nacional de Estadística, www.ine.es, *Índice precios al consumo*) was used as deflator.

What are the results? Figure 8.2 gives us a first image of the importance of upgrading in dairy consumption from the 1990s onwards. After having grown moderately during the massification period that lasted until 1980, the price of the dairy calorie (or the dairy kilogram) began to decline rapidly. In the ten years prior to the Great Recession, however, it started to grow again and it did so much more vigorously than in the past. The Great Recession interrupted this trend, but, after the most critical years were left behind, from 2013 onwards the unit value of dairy consumption has been growing rapidly again.

Was upgrading powerful enough to start a new cycle of increasing consumer expenditure on dairy products? As we can see in figure 8.3, consumer expenditure had expanded much during the massification of the 1960s and 1970s, but after that started to contract. In the ten years prior to the Great Recession, upgrading was powerful enough to start a new phase of expenditure growth. This was halted by the Great Recession and, interestingly, has not resumed in the last few years. That is, in the most recent years the unit value of dairy consumption has been on the rise again, but it has not been able to make expenditure grow again. These results must be put in the context of declining material intakes of dairy products.

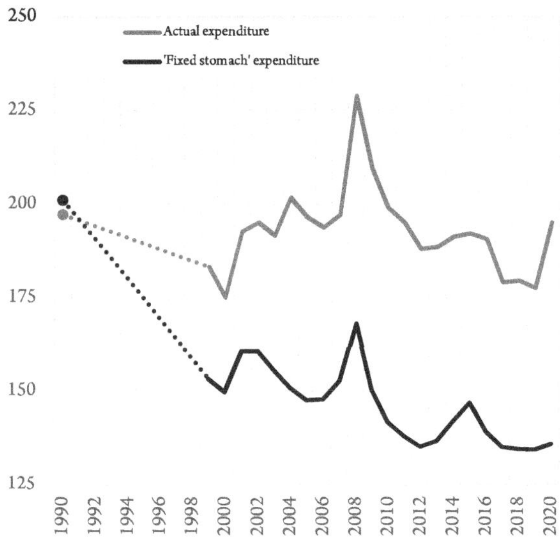

Figure 8.4. Actual expenditure on dairy products and 'fixed stomach' expenditure (constant 2019 euros per person and year), 1990-2020
Source: Ministerio de Agricultura, Pesca y Alimentación (1988-91; www.mapa.gob.es, *Panel de Consumo Alimentario*, "Datos de consumo"); 'fixed stomach': consumption (kilograms) in 1990 valued according to prices in the following years (constant 2019 euros), in both cases disaggregated for each of the 17 dairy products listed in the *Panel*. The general price index (Instituto Nacional de Estadística, www.ine.es, *Índice precios al consumo*) was used as deflator.

Upgrading was powerful enough to compensate the effect of declining material consumption during the years before the Great Recession, but not afterwards.

All in all, as has been argued, the nature of the new model of food consumption is grasped in a better way if, instead of comparing actual expenditure to a flat line of zero growth, we compare it to a counterfactual estimation of the expenditure that consumers would have made in case they had considered that their stomach was already 'fixed' at the start of the period. Even though the flat line is the relevant one for the profitmaking strategies of producers, the 'fixed stomach' estimation presented in figure 8.4 is the one that is relevant from the viewpoint of consumer agency and behaviour. As we can see, had consumers had enough with their 1990 pattern of dairy consumption, they would have had the chance of reducing their expenditure on a steady basis for the following thirty years.

The distance between both series, actual expenditure and (hypothetical) 'fixed stomach' expenditure, becomes remarkable. During the ten years prior to the Great Recession, the gap widened rapidly, reflecting the extent up to which

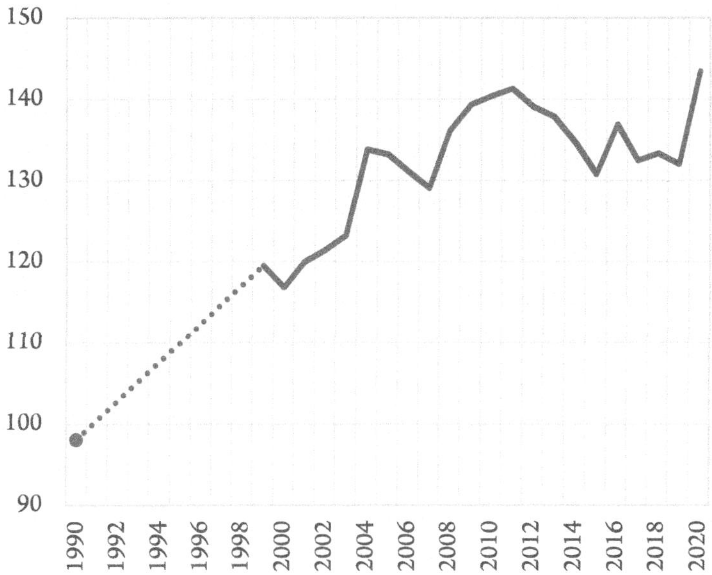

Figure 8.5. 'Over-expenditure' on dairy products (actual / 'fixed-stomach' expenditure, %)
Source: calculated from figure 4.

consumers did not think of their stomach as 'fixed'. Upgrading led this gap to an order of magnitude of 30-40 per cent, which is where it has remained ever since in spite of the Great Recession (figure 8.5). The Great Recession and its aftermath have not reverted the dynamics of the new consumption model. They have simply stabilized them, as consumers devote a remarkable share of their dairy expenditure to upgrading the pattern of consumption that they originally had in the 1980s. Under the extremely anomalous circumstances of 2020, including lockdowns and restrictions to food consumption in restaurants and cafeterias, the gap between actual and 'fixed stomach' expenditure even became higher than ever before.

It must be clarified that these results consider dairy consumption in households only. Unfortunately, the sources for the study of consumption outside the home are of lower quality and do not lend themselves easily to calculating precise measures of quantities, prices, and food product varieties. Consumption outside the home could be incorporated to our calculations at a later stage, but it would require adopting many working assumptions. In any event, household consumption was clearly dominant, ranging between 80 and 90 per cent of total consumption for the whole of our period (Collantes, 2015a: 256).

V. Upgrading and the making of a new model of dairy consumption

As the 1990s progressed, a new model of dairy consumption emerged in Spain. Its two major elements were the ones that have been pointed out by previous research on the breakdown of the mass, agro-industrial model of consumption. On the one hand, consumers began to moderate their intake of dairy products, leaving behind the rapid expansion that had been taking place in the decades prior to 1980. On the other, consumers became interested in upgrading their dairy consumption by incorporating a wider array of more complex, more sophisticated foods.

The first element, the moderation of intake, can be illustrated by the decline in the consumption of whole milk, which was only partially offset by growth in skimmed and (more importantly) semi-skimmed milk. A major difference between the times of massive milk consumption and today is that scientific messages about milk have become much less compact and insistent. During the last thirty years, the previously existing consensus among nutritional scientists about the amount of calcium that is necessary for a healthy diet has been eroded. When it comes to calcium (the major nutritional reason for drinking milk), a surprisingly wide range of scientific recommendations has emerged on a global scale (Kiple and Ornelas eds., 2000). This has probably contributed to weakening what until then had been a highly cohesive network of physicians, policymakers and agribusiness leaders that had insistently (and successfully) propagated pro-milk messages in Spain during the period 1950-1980 (Collantes, 2020).

Another element that must be kept in mind has to do with lactose intolerance. Spain is not a country where rates of lactose intolerance are particularly high from a global perspective, but it does have rates that are somewhat higher to those in northern Europe (Patterson, 2000). Medical and social attitudes towards intolerance have changed substantially. The *Revista Española de Lechería* (Spanish Dairy Journal), the main professional journal related to the dairy chain between around 1950 and 1980, contains abundant qualitative information showing that during that period physicians had insistently recommended intolerant consumers that they should overcome their 'problem' and drink milk anyway. Physicians had also been prone to linking any complaints about indispositions to a lack of the right habits and orientations among consumers, rather than to a biological condition. This pressure on the lactose-intolerant population, which must have been particularly effective when applied on mildly intolerant consumers, has been relaxed as the scientific understanding of lactose intolerance has been strengthened. The availability of lactose-free milks, such as those made from almond or soy, took place later than this change in attitudes towards intolerance, but of course reinforced its implications.

We must also consider milk's poor performance in terms of consumers' sensorial experience. For a long time, the advertising of milk had focused on health and reliability (Collantes, 2020). In no small measure, it still does. However, as

Table 8.3. Product shares in dairy expenditure (%)

	1958	1964/5	1980/1	1990/1	2007	2019
Liquid milk	70	71	57	54	35	28
Preserved milk[a]	9	9	8	6	1	1
Cheese	15	14	22	26	29	37
Butter	6	4	2	1	1	2
Yoghurt[b]	0	0	9	11	24	23
Other[c]	0	2	2	2	10	10
Total	**100**	**100**	**100**	**100**	**100**	**100**

Notes: [a] Evaporated, concentrated, condensed and powdered milk (except in 1958: only condensed milk); [b] Includes all sorts of fermented milks; [c] Mostly milk shakes, yoghurt shakes, ice creams and refrigerated desserts other than fermented milk.

Source: Instituto Nacional de Estadística (1959; 1965-95; www.ine.es, *Encuesta de presupuestos familiares*).

some heterodox publicists warned during the wave of massive growth already, little has been done to position milk as a tasty, enjoyable product. Processors have done little in order to produce milk that consumers will find pleasurable. In the 1960s and 1970s they consistently made technological choices that were openly harmful for milk's taste, such as favouring sterilized, UHT milk over pasteurized milk and opting for homogenizing the fats in whole and semi-skimmed milks (Langreo, 1995). Through a path-dependent process, these technological choices were crucial at shaping innovation at a later stage. When a few years ago one of the country's most reputed gastronomic journalists, Mikel López Iturriaga (2011), wrote in the digital version of the country's leading newspaper an article that was highly critical of the quality of Spanish milk (which he referred to as "insipid water-like liquid"), readers quickly replied with a few hundreds of mostly approbatory remarks.

While they were becoming uninterested in milk drinking, consumers reoriented towards products with a higher degree of processing, such as cheese, yoghurts and other refrigerated desserts (table 8.3). These were products with a price per calorie (or per unit of calcium) much higher than that of milk, too (table 8.4). As a matter of fact, these new dairy products started to occupy centre stage in the profit-making strategies of the country's main processors, while a process of severe restructuring set in for dairy farmers and for those processors specialized in liquid milk only (Collantes, 2019).

Table 8.4. Price of the calories and calcium contained in a selection of dairy products (2008)

	Price (euros per litre/kg)	Composition (per 100 grams)		Price per...	
		Kilo-calories	Calcium (mg)	Calorie (1,000)	Kg of calcium
Whole milk	0.83	65	121	1.28	0.69
Semi-skimmed milk	0.84	43	132	1.95	0.64
Semi-skimmed and fortified milk	1.01	43	160	2.35	0.63
Fresh cheese (standard)	5.10	174	186	2.93	2.74
Fresh cheese (low-fat)	4.94	66	186	7.47	2.66
Hard cheese	9.37	396	835	2.37	1.12
Yoghurt with added flavour	1.59	88	133	1.81	1.20
Fat-free yoghurt	2.12	38	143	5.55	1.48
Fat-free and fortified yoghurt	3.44	38	320	9.01	1.08
Milk shake	1.40	62	62	2.27	1.27
Ice cream	4.01	177	177	2.27	2.86

Sources: prices: Ministerio de Agricultura, Pesca y Alimentación (www.mapa.gob.es, *Panel de Consumo Alimentario*, 'Datos de consumo'); calories and calcium: Moreiras et al. (2001), Ortega et al. (2005) and Mans (2013).

Table 8.5. Two types of growth in consumer expenditure on dairy products

	1964-1980	2000-2006[c]
Growth in per capita expenditure[a] (%)	3.6	1.7
– Decomposition of growth		
– Calorie intake[a]	3.2	−0,7
Expenditure per calorie[a]	0.3	2.4
Main source of growth in consumer expenditure	Calorie intake (91%[b])	Expenditure per calorie (143%[d])

Notes: [a] Annual compound rate of change; [b] $r(K) / r(E)$, where r is annual compound rate of change, K is calorie intake and E is expenditure; [c] Includes consumption and expenditure outside the home; [d] $r(E/K) / r(E)$.

Sources: see figures 2 and 3.

The reorientation of consumer behaviour was vigorous in the ten years prior to the Great Recession, which are the ones that fit best the narrative of upgrading as the driver of the new food consumption model. During those years, consumers expanded their dairy expenditure back again. The previous expansion had taken place during the era of massification, in the 1960s and 1970s. Frequent milk drinking was then spreading at a very fast pace among different social strata and regions, and it was that spread that had driven expenditure growth. After massification had been completed, in the 1980s and much of the 1990s consumer expenditure on dairy had been decreasing, as consumers were beginning to moderate material intakes and take advantage of the declining relative price of their traditional basket of dairy consumption. Why, if these two forces remained in operation, did expenditure rise again around the turn of the century? Basically because, even though consumers were taking less calories from dairy products and were facing lower dairy prices, they were willing to reorient their pattern of consumption towards those products featuring a higher degree of processing and a higher price. Therefore, this was a mode of expenditure growth that was essentially different to that of the massification years. The driver was not material intake but upgrading (table 8.5).

Why did consumers decide to upgrade? Partly, a quality turn of the kind described by Goodman, Redclift and others was going on, as in the cases of cheese with a protected designation of origin, yoghurts produced by means of semi-artisanal techniques, and organic dairy products. However, when involved in blind sampling experiments most consumers were actually uncapable of telling higher-quality dairy products from their lower-quality alternatives (Barreiro et al. 2020). Most of the upgrading had to do instead with products that, far from contesting the conventional Fordist logic, reinforced it by spreading it to new niches. We find, for instance, new products with a high degree of processing that promised consumers additional gains in health: milk fermented with probiotics, milk-based drinks designed to tackle cholesterol problems, dairy products fortified with extra calcium or vitamins. All these products were substantially more expensive than their traditional counterparts, but a new generation of consumers came to be interested in them for alleged health reasons. We also find lines of upgrading that seem to have been motivated by consumers' wish to enjoy a more pleasurable sensorial experience. It is interesting to note that, in the case of most yoghurts and desserts, especially in their more conventional, more popular varieties, this experience resulted in no small measure from higher sugar intakes and contributed to deteriorate, rather than improve, the health balance of consumer diets.

It is true that the Great Recession interrupted these trends towards upgrading before they had become widespread. Prior to the recession, upgrading had been performed mostly by upper-status groups, who were in a better position to substitute conventional milk with fortified milk, ordinary cheese with cheese that had a protected designation of origin, or natural yoghurts with probiotic yoghurts. Lower-status groups lagged and had a more traditional pattern of dairy consumption. When the Great Recession, leading to a fall in household incomes and an increase in social inequality, severely damaged the economic perspectives of many consumers, the result was an abrupt end to the short era of renewed growth in dairy expenditure. Consumers stopped buying cheese and refrigerated desserts as frequently as prior to the recession, while at the same time remained resolutely uninterested in going back to massive milk drinking.

All in all, the legacy of the vigorous upgrading of the years prior to the recession is still among us. The recession halted its spread but did not really motivate consumers to return to the kind of dairy consumption pattern with which the model of mass consumption had been closed in the 1980s. With all its limitations, in the aftermath of the Great Recession consumers have gone on making an expenditure in dairy products that is substantially larger than if they had felt satisfied with the pattern of consumption that they had reached in the 1980s and that qualified as good enough from a nutritional point of view. In the 1990s, consumers started a journey towards new objectives. They performed some 'over-expenditure' to transition towards a diet that they felt was more satisfactory. The storm of the Great Recession interrupted many of these journeys but did not lead consumers to think that maybe their stomach was already fixed after all and that they should get back.

VI. Conclusion

The case of dairy products in Spain provides a good illustration of the new path taken by diet change in the global North over the last few decades. Contrary to great acceleration narratives, the mass, agro-industrial model of consumption and the longer-run nutritional transition to which the latter belonged have been left behind. The wave of expansion in milk drinking that made massification possible among different social groups and regions in the 1960s and 1970s began to fade away in the 1980s. By then, the contribution of dairy products to the Spaniards' diet could be considered satisfactory in nutritional terms. Calcium deficiency, a problem that had affected most people until then, had been overcome. However, this did not lead consumers to think of their stomach as 'fixed'. In time, and especially from the late 1990s onwards, consumers ended up defining a new round of dairy-related targets. Since said targets were more expensive than the previous ones, pursuing them led consumers to increase their expenditure on dairy products way beyond what would have been necessary if they had felt satisfied with their 1980s pattern. This is the sense in which we can talk of an 'over-expenditure' of 30-40 per cent.

Upgrading was central to the new model of food consumption. This had been already suggested by a number of research programmes, but in a way that did not make it easy to position upgrading within a bigger picture that would also include the other major force shaping the new model: the way in which in an affluent society some consumers moderate their food intake. The Spanish figures of 'over-expenditure' on dairy presented in this chapter are even more striking if we consider that the consumption of what under the previous consumption model had been the main product, liquid milk, has declined sharply, leaving cheese, yoghurt and refrigerated desserts as the only possible drivers of demand growth. Our study case has also allowed us to assess in a more explicit way the role played by a critical event such as the Great Recession on the unfolding of the new model of food consumption. It has been concluded that, even though the recession interrupted the deepening and spread of upgrading, the latter's centrality was not threatened. Although consumer 'over-expenditure' on new dairy targets fell from its 2011 peak of 41 per cent, it comfortably remained within the 30-40 per cent band. Under the anomalous circumstances of 2020, it even reached a new peak of 43 per cent. There have not been any signs of a return to the consumption pattern of (say) the 1980s.

The new model of food consumption has co-evolved with the production chain and its transformations. The end of the massification phase moved processing industries (and, indirectly, supermarket chains) to diversify their outputs in the hope that consumers would continue to increase their expenditure on highly processed dairy products that for some reason they would think of as superior to those that they were consuming already. The rapid decline taking place in milk consumption has increased the pressure for companies to reorient their business model to other dairy products. This reoriented strategy has had its share of success, especially in yoghurt and other refrigerated desserts, but offsetting trends in the milk market has proved difficult.

The result has been an abrupt process of structural adjustment by means of which the production and commercialization of dairy products has become increasingly concentrated on an ever-smaller number of farms, processing companies and large retailers. The new demand conditions have been particularly adverse for most farmers and many processors. Farmers have found that any expansion in consumers' dairy expenditure is to be driven by product sophistication (as in the years prior to the Great Recession), not by increasing intakes. Under such conditions, the growth prospects of the demand for the commodity that they supply, raw milk, are as bleak as if farmers were actually facing the ghost of the fixed stomach. Many processors, on the other hand, lack the capabilities and resources required to capture substantial shares of the most dynamic niches in the dairy market. These niches are largely dominated by a small number of leading processors that are strong on R&D and marketing, as well as by supermarket chains by means of their increasingly popular own brands.

The analysis undertaken here, which could be replicated for other products and countries, poses two questions for further research. The first has to do with the causes of upgrading and 'over-expenditure'. The fact that these trends have been more powerful during upswings in economic growth and among those social groups with a higher socio-economic status suggests that the interplay of household incomes and relative dairy prices has played a role that would deserve a systematic analysis. It can also be speculated that this view of causation should be complemented with a study of the social, cultural, and business dynamics that shaped consumer preferences in that direction instead of others.

This leads us to a second question related to the evaluation of the new model of food consumption. Much has been written about recent trends in food consumption from the viewpoints of public health and environmental sustainability, but not so much in the framework of the debate on our consumer societies. This is a debate that has been brought back to life recently by several researchers, ranging from those who celebrate the abundance of the present (Pinker, 2018) to those who argue that consumer capitalism has moved us to abandon a notion that is essential for the good life: the notion of sufficiency, that is, the self-assessment that one has enough and feels well enough (Skidelsky and Skidelsky, 2012). In the meantime, others, such as economist Mariana Mazzucato (2021), highlight that the issue is not growth as a rate, but growth as a direction: the degree up to which growth contributes to achieving relevant social targets. It may be fruitful to position consumer reorientation towards food upgrading within this broader debate. Up to what extent does upgrading entail genuine gains in social well-being? Could this 'over-expenditure' on food have been directed along more promising lines? What have been the strengths and weaknesses of upgrading from the viewpoint of the promotion of the good society, the good life, and human flourishing?

Bibliography

Barreiro, J. M., Losada, F. and Ruzo, E. (2001): 'Valor de marca, calidad percibida y calidad real: un análisis comparativo del mercado de la leche', *Revista Española de Estudios Agrosociales y Pesqueros*, no 190, 195-221

Biltekoff, C. (2012): 'Critical nutrition studies', in J. M. Pilcher (ed.), *The Oxford handbook of food history*, New York

Clark, G. (2007): *A farewell to alms: a brief economic history of the world*, Princeton

Collantes, F. (2015a): 'Dairy products and shifts in Western models of food consumption since 1950: a Spanish perspective', *Rural History*, 26 (2), 249-68. doi:10.1017/S0956793315000060

——— (2015b): 'Más allá de los promedios: patrones de segmentación del consumo de productos lácteos en España, 1964-2006', *Investigaciones de Historia Económica*, 11 (2)

——— (2016): 'Food chains and the retailing revolution: supermarkets, dairy processors and consumers in Spain, 1960 to the present', *Business History*, 58 (7), pp. 1055-76

——— (2017): *La economía española en 3D: oferta, demanda y largo plazo*, Madrid

——— (2019): 'From organized to disorganized capitalism? Market versus nonmarket coordination in Spain's dairy chain', *Journal of Agrarian Change*, 19 (2), p. 295-318

——— (2020): 'Calidad alimentaria, transición nutricional y capitalismo lácteo en España (1965-1990)', *Ayer*, 118

Cussó, X. (2005): 'El estado nutritivo de la población española 1900-1970. Análisis de las necesidades y disponibilidades de nutrientes', *Historia Agraria*, 36, p. 329-58

———, Gamboa, G. y Pujol, J. (2018): 'El estado nutritivo de la población española, 1860-2010: una aproximación a las diferencias de género y generacionales', *Nutrición Hospitalaria*, 35 (supplement 5)

DuPuis, E. M. (2002): *Nature's perfect food: how milk became America's drink*, New York

Fernández-Armesto, F. (2002): *Near a thousand tables: a history of food*, New York

Fernández Prieto, L. and Lanero, D. (eds.) (2020): *Leche y lecheras en el Siglo XX. De la fusión innovadora orgánica a la Revolución Verde*, Zaragoza

Fischer-Kowalski, M., Kraussman, F., Giljum, S., Lutter, S., Mayer, A., Bringezu, S., Moriguchi, Y., Schütz, H., Schandl, H. and Weisz, H. (2011): 'Methodology and indicators of economy-wide material flow accounting', *Journal of Industrial Ecology*, 15 (6), p. 855-76

González de Molina, M., Soto, D., Guzmán, G., Infante-Amate, J., Aguilera, E., Vila, J. y García, R. (2020): *The social metabolism of Spanish agriculture, 1900-2008: the Mediterranean way towards*, Cham

Goodman, D., DuPuis, E. M. y Goodman, M. K. (2012): *Alternative food networks: knowledge, practice, and politics*, London

——— and Redclift, M. (2005): 'Modernisation and the international food system: re-articulation or resistance', in C. Sarasúa, P. Scholliers and L. van Molle (eds.), *Land, shops and kitchens: technology and the food chain in twentieth-century Europe*, Turnhout

Grigg, D. (1999): 'The changing geography of world food consumption in the second half of the twentieth century', *Geographical Journal*, 165 (1), p. 1-11

Instituto Nacional de Estadística (1959): *Encuesta sobre cuentas familiares, marzo 1958*, Madrid

Instituto Nacional de Estadística (1965-95): *Encuesta de presupuestos familiares*, Madrid

Kiple, K. F. and Ornelas, K. C. (eds.) (2000): *The Cambridge world history of food*, Cambridge

Langreo, A. (1995): *Historia de la industria láctea en España: una aplicación a Asturias*, Madrid

——— (2005): 'El sistema alimentario español desde la perspectiva de los sectores productivos (I)', *Distribución y Consumo*, 81

——— (2006): 'Entre los nuevos productos, la calidad estándar y los productos de calidad diferencial. El ejemplo del sector lácteo: leche de consumo, refrigerados y quesos', *Distribución y Consumo*, 85

Langthaler, E. (2018): 'The soy paradox: the Western nutrition transition revisitied, 1950-2010', *Global Environment*, 11 (1), pp. 79-104

Levenstein, H. (1993): *Paradox of plenty: a social history of eating in modern America*, New York

López Iturriaga, M. (2011): 'La triste realidad de los lácteos españoles', *El País*, 22 June

McMichael, P. (2013): *Food regimes and agrarian questions*, Winnipeg

McNeill, J. R. and Engelke, P. (2014): *The great acceleration: an environmental history of the Anthropocene since 1945*, Cambridge

Magnan, A. (2012): 'Food regimes', in J. M. Pilcher (ed.), *The Oxford handbook of food history*, New York

Malassis, L. (1997): *Les trois âges de l'alimentaire: essai sur une histoire sociale de l'alimentation et de l'agriculture*, Paris

Mans, C. (2013): 'Densia forte y la osteoporosis', *Investigación y Ciencia*, <www.investigacionyciencia.es>

Martínez-Carrión, J. M. (2016): 'Living standards, nutrition and inequality in the Spanish industrialization. An anthropometric view', *Revista de Historia Industrial*, 64, pp. 11-50

Mazzucato, M. (2021): *Mission economy: a moonshot guide to changing capitalism*, London

Ministerio de Agricultura, Pesca y Alimentación (1988-91): *Consumo alimentario en España*, Madrid

Montanari, M. (1993): *El hambre y la abundancia: historia y cultura de la alimentación en Europa*, Barcelona

Moreiras, O., Carbajal, Á., Cabrera, L. and Cuadrado, C. (2001): *Tablas de composición de alimentos*, Madrid

Oddy, D. J., Atkins, P. J. and Amilien, V. (eds.) (2009): *The rise of obesity in Europe: a twentieth century food history*, Farnham

Ortega, R. M., Mena, M. C. and López, A. M. (2005): 'Leche y lácteos: valor nutricional', in J. Aranceta and Ll. Serra (eds.), *Leche, lácteos y salud*, Madrid

Patterson, K. D. (2000): 'Lactose intolerance', in K. F. Kiple and K. C. Ornelas, *The Cambridge world history of food*, Cambridge

Pinker, S. (2018): *Enlightenment now: the case for reason, science, humanism, and progress*, New York

Pollan, M. (2008): *In defense of food: an eater's manifesto*, New York

Popkin, B. (1993): 'Nutritional patterns and transitions', *Population and Development Review*, 19 (1), pp. 138-57

Pujol, J. and Cussó, X. (2014): 'La transición nutricional en Europa occidental, 1865-2000: una nueva aproximación', *Historia Social*, 80, pp. 133-55

Scholliers, P. (2014): 'Post-1945 global food developments', in P. Freedman, J. Chaplin and K. Albala (eds.), *Food in time and place: the American Historical Association companion to food history*, Oakland

Skidelsky, E. and Skidelsky, R. (2012): *How much is enough? Money and the good life*, New York

Smith-Howard, K. (2014): *Pure and modern milk: an environmental history since 1900*, New York

Valenze, D. (2011): *Milk: a local and global history*, New Haven

Velten, H. (2010): *Milk: a global history*, London

von Tunzelmann, N. (2010): 'Competencies, capabilities and time: towards a dynamic knowledge-based theory of value', presentation at the DIME Network meeting (Pisa)

Vorley, B. (2007): 'Supermarkets and agri-food supply chains in Europe: partnership and protest', in D. Burch and G. Lawrence (eds.), *Supermarkets and agri-food supply chains: transformations in the production and consumption of foods*, Cheltenham

Part 2

Veterinarians and antibiotics

KARL BRUNO

9. Reproduction work

Veterinarians, artificial insemination, and the production of reproductive health in Swedish dairy cattle, 1910-1960

I. Introduction

In 1958, professor Sigurd Tovborg Jensen of the Danish Veterinary and Agricultural University presented a striking image of late 1950s dairy farming when he discussed how 'modern methods of cattle breeding' were giving cows a 'mechanized sexual life [...] controlled by learned men in white coats' (quoted in Fritzbøger, 2015: 85).[1] He referred to artificial insemination (AI), a way to achieve pregnancy without mating by inserting pre-collected semen into cows' reproductive organs. AI was increasingly common across dairy-producing countries in 1958. Fewer and fewer farmers brought their cows to a bull. Instead, semen was gathered from bulls kept at central locations and then driven to farms for inseminations. Cattle reproduction was being disaggregated, with responsibilities for its management passing from the cowmen of earlier days to scientists and technicians.

In Sweden, the setting of this chapter, these 'men in white coats' were veterinarians. In the late 1950s, Swedish AI was carried out by vets or under their close supervision, and included considerable veterinary attention to cows' reproductive health. This differed from conditions in Denmark as well as in countries like Britain and the United States, where independent non-veterinary technicians played much bigger roles. In this chapter, I investigate how early Swedish AI became centred on veterinary expertise and why vets sought to manage this new technology. Hitherto unexplored, these issues help us understand more of the scientific and medical underpinnings of Swedish dairy farming's gradual development towards intensification and specialization, a process placing new demands on cattle and humans alike.

Earlier research on the twentieth-century history of Swedish farming affirms the importance of AI for the modernization of cattle breeding and cattle husbandry: it saved labour and increased occupational safety, but also enabled a breeding revolution by allowing individual bulls to inseminate many more

1 All translations in the chapter are by the author.

From Breeding & Feeding to Medicalization, Rural History in Europe, 17 (Turnhout: Brepols, 2023), pp. 211–236
BREPOLS PUBLISHERS 10.1484/M.RURHE-EB.5.135027

cows, especially after the 1960s introduction of deep-frozen semen (Bruno, 2020; Bruno, 2022; Flygare and Isacson, 2011; Martiin, 2011; Martiin, 2017; Rendel, 2003). Existing studies do not, however, consider the associated problem of managing cows' reproductive health, or the role of veterinary expertise and other forms of labour in dairy farming, in detail (but see Israelsson, 2005: chapter 12, about nineteenth century barn work). Abigail Woods has addressed such issues for the case of Britain, showing how British veterinary leaders translated 'the demand for more milk into a question of reproductive control' as they sought to redefine the veterinary role in dairy farming (Woods, 2007; quote from Gaudillière, 2007: 526). She has also put this example into a longer historical context of vets recurrently having to 'reconsider their roles, identities, priorities and relationships' in response to changing circumstances (Woods and Matthews, 2010: 29; Woods, 2012). Shifting veterinary roles in other countries have also been explored by, for example, Susan Jones (2003) and Kristin Asdal and Tone Druglitrø (2014). My chapter explores related concerns for the Swedish veterinary profession. I argue that Swedish vets worked hard to make the 'question of reproductive control' central to AI's organization, a process resulting in a far-reaching professional reconfiguration (the role played specifically by vets is not prominent in histories of AI, but is touched upon in Wilmot (2007); other significant studies that address AI in cattle include Derry (2015); Risan (2003); Smith-Howard (2014); and Theunissen (2012)). In turn, the veterinary interest in AI also became part of a longer and broader re-negotiation of social and professional roles in cattle husbandry.

This process started in the late nineteenth century, when Swedish agriculture gradually reoriented itself towards animal production, particularly dairy. Veterinarians, seeking to stabilize their social and economic position and establish themselves as the sole legitimate providers of animal healthcare, accordingly became interested in addressing cattle husbandry's problems. A multi-faceted ruminant medicine emerged from around 1910. A crucial aspect of it, in focus here, was what I call *reproduction work*: interventions into cows' reproductive functions. The notion of reproduction work plays on but should not be confused with reproductive work, a feminist term referring to unpaid, unrecognized caregiving and domestic labour. Reproduction work, in contrast, was a service to farming that afforded both monetary compensation and professional recognition. Swedish vets saw it first as a way to establish and then, particularly in relation to AI, to maintain professional relevance in a changing agricultural system. Yet their position was tenuous. Reproduction work's veterinary status was never definitive. Arguments abounded over how and by whom cows' reproductive health should be produced. Agricultural workers, particularly, saw in reproduction work a path to future relevance in cattle husbandry. Rather than cowmen and farmhands, the new, rationalizing dairy farming needed modern husbandry specialists, and reproduction work promised to remake its practitioners in this image.

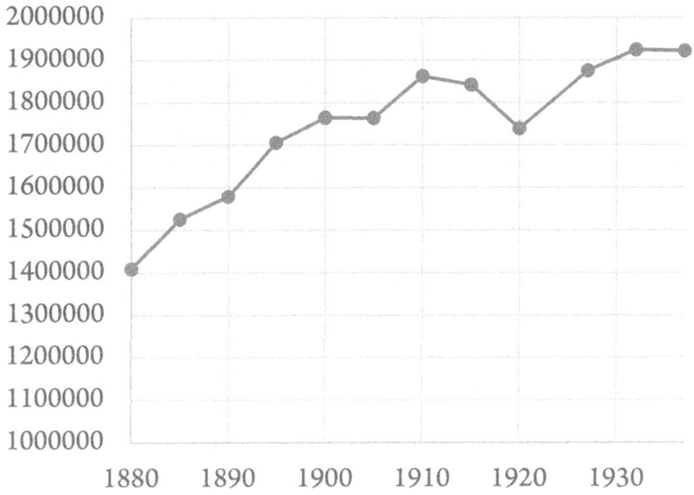

Figure 9.1. Number of cows in Sweden from 1880 to 1937.
Source: The Swedish Central Bureau of Statistics, *Statistical Abstract of Sweden*, https://www.scb.se/hitta-statistik/aldre-statistik/innehall/statistisk-arsbok-1914-2014/ Note that some of the data points are given as estimates in the source.

II. Reproduction Work and the Modernization of Cattle Husbandry

II.1. Reproduction Work Comes to Sweden

Reproduction work came to Sweden early in the twentieth century, and would over subsequent decades reshape both the structure of the veterinary profession and vets' relationship to farming. Its introduction was part of a broader, market-driven transformation of agriculture. From the late nineteenth century, cattle husbandry grew increasingly focused on the commercial production of dairy. Dairy farming eventually became important both at the large farms and estates that used hired labour to manage herds with up to hundreds of milk cows, and at most of the many small family farms with only a few cows that made up the vast majority of agricultural units. The number of cows in the country grew by almost 40% between 1880 and 1930 (with the national herd peaking at just short of two million cows in the early 1930s), while the amount of milk delivered to dairies grew more than fivefold between 1890 and the mid-1930s (figures 9.1 and 9.2). As Sweden remained a highly agrarian society – more people still worked in farming than in industry in the early 1930s – milk production also became an important branch of the national economy (Morell, 2011).

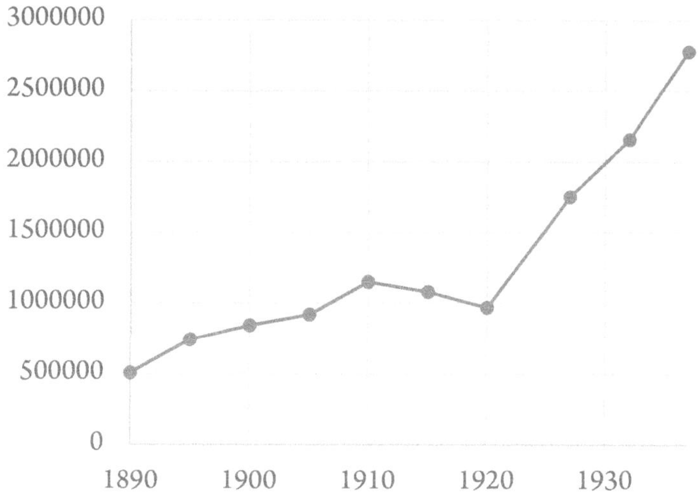

Figure 9.2. Amount of milk delivered to Swedish dairy plants from 1890 to 1937, in tonnes
Source: The Swedish Central Bureau of Statistics, *Statistical Abstract of Sweden*, https://www.scb.se/hitta-statistik/aldre-statistik/innehall/statistisk-arsbok-1914-2014/

The turn of the century also saw changes in husbandry practices that reflected cattle's increasing economic importance. Housing, feeding, and breeding cows became matters of new concern, both for individual farmers and for the public and semi-public institutions that promoted agriculture. Cattle breeding was sponsored by the state from 1891. It was led primarily by large landowners, who were able to keep large breeding herds. These leading breeders formed pedigree-breeding associations, seeking to establish delineated breeds with specific exterior and production characteristics. Two main dairy breeds were eventually created: the black-and-white Swedish Friesian and Swedish Red-and-White Cattle (Korkman, 1988; Rendel, 2003: 46-49).

The socio-economic transformation of farming also affected veterinarians. They increasingly engaged with animal production, particularly in the good farming areas where cattle herds were larger and active breeders were concentrated. Earlier, veterinary work with cattle had mostly been obstetrical: removing retained placentas, facilitating problematic calvings, or performing embryotomies: cutting up dead or dying foetuses in uteri to save the mother. From around 1900, however, a broader ruminant medicine emerged, eventually becoming its own teaching subject at the Veterinary College of Sweden in 1913.[2] Its primary clients were the relatively small number of estates and better-off farmers with larger

2 Government Bill 1913:268, pt. V.

herds, and unless otherwise specified, the chapter focuses on veterinary work in relation to these groups.

Reproduction occupied a significant part of the new ruminant medicine, since achieving successful and timely conceptions was a perennial objective in dairy farming. Milk flows depend on cows completing successive full-term pregnancies, so cows that struggled to conceive were economic liabilities, and widespread sterility in a herd could ruin the farmer. Furthermore, infertility interfered with systematic breeding. Earlier, it had been seen as an 'unavoidable evil' to which farmers could only respond by 'handing over [the cow] to the butcher'. From the early 1900s, however, drawing on developments in Denmark and Switzerland, vets argued that other options existed.[3] In Denmark, generally an important inspiration for Swedish farming, vets debated two causes of cow infertility. Martin Poulsen blamed ovarian pathologies, either cysts or persistent corpus lutea (yellow bodies in the ovaries, normally present during pregnancies but seen to 'persist' in some infertile cows). These could be treated by rupturing them through the rectum. Poulsen's rival Jacob Albrechtsen instead saw the primary cause as endometritis – inflammation of the uterine lining – and proposed uterine flushings as the proper treatment. Both approaches were extensively discussed in *Svensk veterinärtidskrift*, the principal Swedish veterinary journal, after Albrechtsen published his findings in 1906.[4]

Equally important was that the likewise-new technique of rectal pregnancy diagnosis allowed vets to determine pregnancies as early as six weeks after mating. This enabled a structured follow-up of fertility conditions. In 1908, veterinarian Anders Bergstrand argued for its systematic combination with sterility treatments in entire herds:

> It is of primary importance to *systematize* the matter. It is not suitable to occasionally examine and treat one or a couple of cows in a herd that do not become pregnant, and then leave it at that. If any real result is to be achieved, one must approach the problem systematically and include the whole herd in the pregnancy control. [...] If a veterinarian is requested to examine and treat a few cows in a herd for infertility, he should [...] arrange to perform a rectal examination of all cows whose pregnancies cannot be diagnosed externally. [...] The cows that are not pregnant but that the owner nevertheless wish to keep [...] should be treated as indicated. The veterinarian should generally not engage in diagnosing pregnancies on other terms.[5]

This systematization – also expressed in Bergstrand's recommendation that vets should consistently record the dates of calving, mating, repeated oestrus, treatments, and diagnosed pregnancy – was seen to have significant economic

---

3 Quotes from Harry Stålfors (1907), 'Äggstockscystor och permanenta corpora lutea såsom orsaker till sterilitet hos ko', *Svensk veterinärtidskrift*, 12, 1, pp. 13-30 (p. 14).

4 See particularly vols. 12 (1907) and 13 (1908); on the new methods, see also Woods (2007): 469.

5 And. Bergstrand (1909), 'Ofruktsamhet hos kor', *Svensk veterinärtidskrift*, 14, 1, pp. 8-19 (pp. 16-17).

potential, and not only for farmers. It promised to extend veterinary responsibilities from treating acute obstetrical cases to monitoring herds' reproduction status, making both healthy and sick animals potential patients.[6]

Vets' interest in providing reproduction services was not purely economic. They were trying to establish themselves as a respected profession and the sole legitimate providers of animal healthcare. Being able to provide a recurring, recognizably valuable service to the central branch of animal production was therefore important in their struggle for professional recognition. This was particularly so as vets were mostly trained in equine medicine, and now entered a domain with more competition from unlicensed practitioners (Palm, 1981: 50). This made it problematic for them that both treating sterility and diagnosing pregnancies built on manipulations that were not self-evidently medical and could be learnt with relative ease. When non-veterinary workers on large farms began to attempt them, vets accordingly insisted that the procedures *were* medical interventions, part of the obstetrical sphere they claimed for themselves. A 1910 meeting of the Swedish District Veterinary Officers' Association (district vets were public employees and the primary providers of veterinary services to farms) expressed that it would be 'unsuitable' and 'risky' for non-vets to perform 'rupturing and flushing'. Without veterinary education and broad clinical experience, they would not be able to judge when the procedures were appropriate nor perform them safely under varying conditions. The most invasive technique was the Albrechtsen treatment, which involved grabbing the cervix with a pair of special pliers and then passing a catheter into the uterus to flush it. Allowing farm workers to do this might even, the meeting argued, constitute 'animal cruelty'.[7]

Demarcating against unlicensed practitioners was a constant concern, and subsequent attempts by non-vets to practice reproduction work was generally bundled up with the broader issue of 'quackery' and condemned as such.[8] In Swedish discourse, this did not necessarily imply fraudulent behaviour but simply referred to unlicensed medical practice (Eklöf, 2001: 112-13). In a veterinary context it nevertheless encompassed both animal healthcare based on folk understandings and traditional remedies, and activities similar to veterinary interventions, such as lay castrations or non-veterinary calving assistance. Veterinary leaders were increasingly concerned with the latter group, which also helps explains their strong reaction against lay practice of sterility treatments.

Leading estate owners, who had called for 'forceful measures' against sterility on account of the 'considerable economic loss' it entailed, ultimately did not

6 Cf. also the British developments discussed in Woods (2007).

7 Minutes, Swedish Society of District Veterinary Officers special meeting, 28 March 1910, § 11, *Svensk veterinärtidskrift*, 15 (1910), 6, pp. 211-19 (p. 218).

8 Minutes, SSVS general meeting, 27-28 November 1911, § 11, *Svensk veterinärtidskrift*, 17 (1912), 1, pp. 21-31 (p. 25); SSVS' Board of Directors Annual Report 1911-1912, SSVS' archives (SSVSA), SA 72:1, Swedish Veterinary Museum (SVM). On vets and quackery until the early twentieth century, see Lundgren (2001): 24-27; Palm (1981): 42-51.

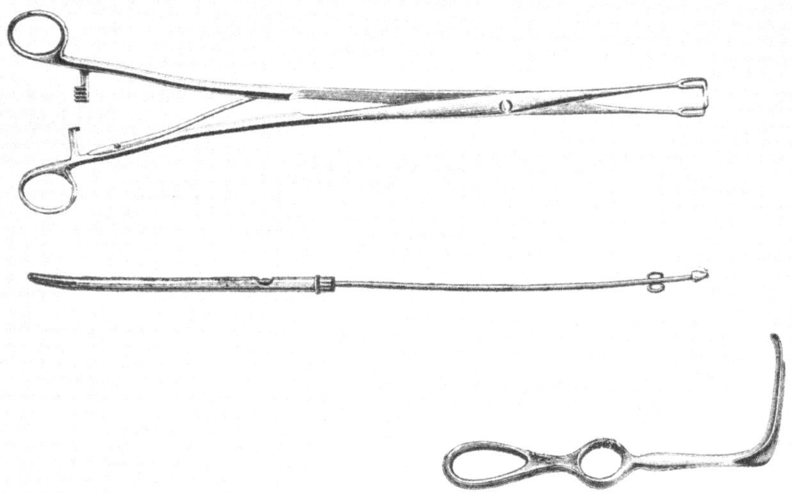

Figure 9.3. Instruments for the Albrechtsen treatment of infertility, as modified by the Veterinary College's lecturer in obstetrics, Harry Stålfors.
Source: *Svensk veterinärtidskrift* 12, no. 12 (1907), 96. Reproduction: National Library of Sweden.

press the issue of non-veterinary services. Seeing vets as allies in their efforts to modernize husbandry, they were unwilling to offend veterinary sensibilities. Instead, the estate owners made a joint proposal with the Swedish Society of Veterinary Surgeons (Svenska veterinärläkareföreningen, SSVS – the chief professional association of Swedish vets) to the government, which then funded courses in pregnancy diagnosis and sterility treatments for interested district vets. Fifteen years later, most of them offered these services.[9]

II-2. Reproduction Work and Artificial Insemination

Artificial insemination was introduced in Sweden in the mid-1930s and quickly attracted veterinary interest. This was bound up with a scientification both of veterinary reproduction expertise and of cattle breeding. Before 1930, veterinary reproduction work had built on learning and applying manual techniques for diagnosing pregnancies, flushing uteruses, rupturing corpora lutea, facilitating deliveries, and removing retained placentas. Few vets had a scientific understanding

9 'Huru skall frågan om sterilitetsbehandling och dräktighetskontroll bäst lösas?', *Svensk veterinärtidskrift*, 15 (1910), 8, pp. 305-10 (p. 305); 'Kurser i sterilitetsbehandling hos kor', *Svensk veterinärtidskrift*, 15 (1910), 9, p. 431; 'Frågor och svar', *Landtmannen*, 8 (1925), 14, p. 263.

of reproductive physiology and pathology. This gradually changed from the late 1920s, when research in reproduction began at the Veterinary College and also shaped veterinary education in the field, bolstering vets' claims to expertise. The leading proponent of this development was veterinarian Nils Lagerlöf, who had made a scientific name for himself through researching bull infertility. He was appointed professor of obstetrics and ruminant medicine in 1934, becoming the principal Swedish authority on cattle health and disease (Bruno, 2020; Bruno, 2022; Einarsson, 2010; Rendel, 2003: 125-27).

Lagerlöf's appointment coincided with the start of AI experiments in Sweden, at the semi-private Institute for Animal Breeding at Wiad outside Stockholm. Sweden had an established tradition of successful plant breeding research since the late nineteenth century, particularly at the Swedish Seed Association's experimental farm at Svalöv, and the Wiad institute had been created in 1928 in part to extend the application of genetic knowledge into animal breeding as well. Directed by geneticist Gert Bonnier, it performed research geared to improving farm animals from an economic point of view. From 1935, one of its projects sought to use AI to develop methods for progeny testing. This meant identifying good breeding bulls based on their daughters' production, a radical innovation in cattle breeding since the prevailing system of judging bulls built only on pedigrees and exterior evaluations.[10] Bonnier's interest thus lay squarely in AI's genetic potential. But the trials attracted veterinary interest. Some saw them as 'interference by laymen' in the 'purely veterinary work' of manipulating cattle reproduction. Lagerlöf accordingly emphasized to Bonnier that for AI 'to be risk-free' it had to be 'performed by an expert, i.e., veterinarian, who can judge possible pathologies of the genitalia etc.' Thus echoing the arguments earlier invoked for veterinary control over sterility treatments, he insisted on veterinary participation at Wiad, which Bonnier agreed to.[11] Lagerlöf and his assistant Stig Settergren also introduced AI to a wider audience of Swedish vets in a 1936 article in *Svensk veterinärtidskrift*.[12]

Settergren then studied AI in Denmark – the AI pioneer of Europe – and published another article discussing its relation to other reproduction work. In Denmark, he noted, 'all AI work is completely in veterinary hands', and the authorities discouraged non-veterinary involvement. The AI vet performed pregnancy diagnoses and sterility treatments in conjunction with the inseminations. 'From a purely veterinary point of view', Settergren argued,

---

10 'PM angående upprättandet av en praktisk-vetenskaplig forskningsanstalt på husdjursförädlingens område', Department of Animal Breeding and Genetics' archives, Ö2:2, Swedish University of Agricultural Sciences' archives (SLU); Rendel (2003): 69-76; 80-84.

11 Hans Östberg to Nils Lagerlöf, [date unreadable] September 1936; Lagerlöf to Östberg, 2 October 1936, Department of Obstetrics-Gynecology's archives (OG), Ö1:2, SLU.

12 Nils Lagerlöf and Stig Settergren (1936), 'Om konstgjord sädesöverföring hos husdjuren', *Svensk veterinärtidskrift*, 41, 7, pp. 267-85.

> this offers the advantage that one gets the opportunity to *treat infertile animals* at a much earlier stage than is otherwise the case and this treatment can – if needed – be *repeated more often* than otherwise. [The vet] is thus able to exercise *constant control* over the sexual health of the animals.[13]

At Lagerlöf's behest, in late 1940 SSVS adopted the similar position that 'artificial insemination shall be considered veterinary work and be performed by a veterinarian', and acknowledged its interrelation with pregnancy diagnoses and sterility treatments.[14] These positions fulfilled the systematizing ambitions that Anders Bergstrand had outlined in 1908. With AI organized in line with the Danish model, veterinary fields of work would significantly expand since every enrolled cow would be examined for pregnancy, and, as needed, treated for sterility. While not all vets were positive to AI, most would have welcomed new potential sources of work in the uncertain times of war and with bleak memories of the 1930s agricultural crisis still vivid.

In its 1940 decision, SSVS not only affirmed AI as veterinary work, but also framed it as a veterinary specialization in its own right, envisioning future vets whose sole line of work would be reproduction work.[15] This reflected experiences from the Institute for Animal Breeding, where vets performed experimental inseminations on farms around Wiad. Pregnancy diagnoses and sterility treatments remained, however, the remit of the ordinary district veterinarian, whereas the institute's vets only did the actual AI work. This setup led to delays in treating infertile cows, further exacerbated by the troubling transport conditions prevailing in wartime. 'When the insemination veterinarian discovers a problem, for example a [ovarian] cyst, it can take months before the regular veterinarian visits the farm and treats the cow', noted institute vet Torsten Modig in 1941. This was a public relations problem, since animal owners were dissatisfied and had 'expressed their surprise that the inseminating veterinarian [...] was not allowed to perform sterility treatments'. Modig argued that 'ideally', the proper treatment should be administered as soon as possible, by the regularly visiting AI vet.[16] This was in fact how AI then became organized at the institute. A small number of vets were trained along those lines there during the war by Lagerlöf's student Allan Bane, who had adapted many of the AI methods used to Swedish conditions.[17]

Settergren's emphasis on AI as an opportunity to follow up sexual health in herds was also connected to a new etiological understanding of infertility being developed by Lagerlöf. Since the early 1930s, he had argued that hygienic

---

13 Stig Settergren (1940), 'Om den artificiella inseminationens praktiska tillämpning inom dansk nötkreatursavel', *Svensk veterinärtidskrift*, 45, 9, pp. 275-87 (p. 278, p. 285).

14 Minutes, SSVS annual meeting, 20 October 1940, § 25, *Svensk veterinärtidskrift*, 46 (1941), 4, pp. 116-17 (p. 116).

15 Minutes, SSVS annual meeting 1940, § 25.

16 Torsten Modig, 'Motivering för att vederbörande inseminationsveterinär skall få utföra dräktighetsundersökningar och sterilitetsbehandlingar', 5 February 1941, OG, F8A:1, SLU.

17 The methods themselves were imported from Britain and the Soviet Union. See Einarsson (2011a).

conditions were dismal in many barns and that infertility could be caused by many common diseases – including the major infectious diseases of brucellosis and bovine TB but also various other infections – as well as by inborn predispositions. Improving hygiene and distinguishing between inherited and environmental causes were therefore, he reasoned, crucial to combatting infertility. In 1944, Lagerlöf summarized his views in *Svensk veterinärtidskrift* and introduced the term 'sexual health control' as a shorthand for a system of herd-level veterinary services to that effect, including comprehensive examinations of cows' general reproductive health, pregnancy diagnoses, and sterility treatments. He argued that farmers increasingly 'realized that [sexual health control] is profitable, yes, even a necessary precondition for a rational cattle husbandry. That it would also be of considerable importance for smaller herds is now apparent'. After the war, he suggested, it might be time to attempt a state-led program establishing a more extensive sexual health control.[18]

Lagerlöf's 1944 piece did not specifically link the argument for broader sexual health control to AI. The connection was nevertheless obvious to his readers, since mandatory sexual health control was de facto being implemented in that context. When AI moved into practical farming in 1943, it maintained the Wiad model. The new cooperative AI associations hired vets to perform the work, and their remit always included sexual health control as well as inseminations. Animal owners could not opt out: signing up for AI also meant signing up for the system of veterinary examinations and treatments that constituted sexual health control.

II-3. Artificial Insemination and the Veterinary Profession

The expansion of AI triggered a demand for more personnel but also debates about their training. The ad hoc activities at Wiad would not suffice. Twelve vets had been trained there by 1943, but thirty AI associations were expected to exist within a few years, each requiring one or two vets. Lagerlöf therefore told the board of the Veterinary College that it was important to 'as soon as possible resolve how to best train the veterinarians who are to work in this field', and the board subsequently petitioned the government for funds for regular AI courses.[19]

This initiative was opposed by Bonnier, who argued that the training could remain at Wiad and pointedly stated that AI primarily had a 'breeding purpose', implying that means (veterinary participation) should not be confused with ends (improved cattle). The National Board of Agriculture incorporated Bonnier's views in its comments to the government, arguing for training at Wiad for

18 Nils Lagerlöf (1933), 'Några synpunkter på sterilitetsproblemet inom nötboskapsaveln', *Avelsföreningens för SRB tidskrift*, 6 (1933), 1, pp. 2-21; Nils Lagerlöf (1944), 'Några synpunkter på sterilitetsproblemet inom nötkreatursaveln: Profylaktiska åtgärder och behovet av en utökad sexuell hälsokontroll', *Svensk veterinärtidskrift*, 49, 5, 6, 7, pp. 133-44, pp. 163-73, pp. 205-19 (p. 217).

19 Lagerlöf to the Board of Directors, Royal Veterinary College, 8 November 1943, Swedish Ministry of Agriculture's archives (MoA), E1:2882, Swedish National Archives (SNA).

'veterinarians and others suited for the purpose, who intend to engage in [AI]'. Lagerlöf's counterargument was that as a reproductive technology, AI had to be taught in a context where 'experts and specialists in [animal reproduction]' were available, meaning at the Veterinary College. Doing it elsewhere would mean 'breaking out a part of the veterinary education from its organic context'. If it became necessary to train non-vets for AI in the future, they too had to be trained at the college.[20] Lagerlöf became highly involved with the issue, going as far as to personally make his case to the Parliament's Standing Committee on Agriculture, to the desired effect. Funding was provided and he could hire Allan Bane to teach AI at the college.[21]

The output of these courses remained constrained by the number of interested vets available, however, and though AI – as an observer later put it – 'employed two years' worth of Veterinary College graduates' around the war's end, that still was not enough to meet the demand.[22] In late 1944, Lagerlöf then recognized that a significant number of veterinarians were among the Baltic refugees who had fled to Sweden from the advancing Red Army, and sought to train them for AI work. This effort has been accounted for by Stig Einarsson (2011b) and will not be discussed here, beyond pointing out that training the Baltic vets were also parts of Lagerlöf's attempts at maintaining AI as veterinary work. He saw it as the only way to 'manage the question' in the face of what was then becoming increasingly loud demands to open up AI positions for other groups.[23]

Around 1945, then, AI represented both a radically modern promise to breeders and a growing field of work for a veterinary profession that otherwise considered future agricultural development, and therefore its employment prospects, 'highly difficult' to predict.[24] It was also splitting vets' practice. Before AI, practically all farm practice had been handled by the district veterinarians. AI was, however, creating a cadre of reproduction specialists that were not public employees (working instead for the cooperative AI associations), and therefore also a new career path. While there were only 13 AI veterinarians in Sweden by the end of 1945 (as compared with 248 district vets), AI was expanding rapidly and was by the war's end the easiest path to permanent employment for a recent graduate.[25] This new career path had been created through the integration of AI with existing veterinary routines of reproduction work, a move strongly

20 Bonnier to the Board of Agriculture, 8 December 1943; Board of Agriculture to the Cabinet, 13 December 1943; Nils Lagerlöf, 'V. P. M. angående utbildning vid veterinärhögskolan i artificiell insemination avgiven med anledning av lantbruksstyrelsens remissyttrande i ovannämnda fråga', 26 December 1943. All in MoA, E1:2882, SNA.

21 Lagerlöf to Hj. Jeppsson, 22 June 1944, OG, Ö1:2, SLU; JoU 1944:1, item 56.

22 Ivan Christenson to Lagerlöf, 11 May 1948, OG, Ö1:6, SLU.

23 See Lagerlöf's notes for an SSVS board meeting, 17 January 1945, OG, F8A:5, SLU; and Einarsson (2011b).

24 F. Wilhelmson (1945), 'Något om befordringsmöjligheterna på veterinärbanan', *Svensk veterinärtidskrift*, 50, 5, pp. 127-39 (p. 139).

25 Medicinalstyrelsen (1947), *Det civila veterinärväsendet år 1945*, Stockholm, Medicinalstyrelsen, p. 46.

supported by Lagerlöf's argument that a systematically organized sexual health control was a precondition for rational breeding. In itself, that argument was largely uncontested in Swedish dairy farming. But the link between inseminations, sexual health control, and the veterinary profession was not as unanimously accepted, as we will see next.

III. Reproduction Work Contested

III-1. Barn Foremen between Labourers and Specialists

Although vets had succeeded in defining reproduction work as veterinary work when struggling for professionalization in the 1910s and 1920s, its status would become contentious in conjunction with AI. As discussed above, veterinary leaders had denounced non-veterinary reproduction work as 'quackery'. But one person's quack was someone else's legitimate specialist, and several groups associated with cattle husbandry's modernization disputed the veterinary claims on exclusive rights to reproduction work. One such group was the hired barn foremen who managed cattle herds on large farms and estates. As part of its general promotion of animal husbandry, the state sponsored year-long theoretical and practical training courses for this group, seeking – as Carin Israelsson (Martiin) has discussed – to 'redefine cattle husbandry from a female, hidden field of work with inadequate resources to a male, high-productive and high-status one' (Israelsson 2005: 249).[26] In practice, however, barn work often remained unprestigious, and graduated students were increasingly discontent with the mismatch between their formal knowledge and their position on the estate. They were keen to improve their stature and recognition as specialized employees, ambitions reflected in the creation of a national organization and a journal for barn foremen in 1930.[27]

The debates in their journal reveal that some foremen saw reproduction work as an important way to establish themselves as valuable husbandry specialists. In 1941, E. Rohmer, the national organization's chairman, argued that 'every interested and competent barn foreman should be able to perform' pregnancy diagnoses and sterility treatments, and that 'part of this competence also includes [...] the removal of retained placentas and the flushing of a cow'. Boldly moving onto what had until then been strictly veterinary territory, Rohmer asserted that the necessary skills could be acquired without 'particular difficulty'. Unlike the preoccupied district vet, barn workers had, after all, their cows 'in sight from service until calving'.[28] Ultimately, Rohmer sought to include reproduction work

---

26 Government Bill 1912:1, pp. 199-208.

27 'Anmälan', *Ladugårdsförmannen*, 1 (1930), 1, 1.

28 H. H. (1941), 'Praktiskt utförande av klämningar och dräktighetsundersökningar', *Ladugårdsförmannen*, 12, 2, p. 16; 'Diskussionsspalten: Dräktighetsundersökning hos kor', *Ladugårdsförmannen*, 13 (1942), 2, 3, pp. 32-35, pp. 46-47 (p. 32). See also 'Diskussionsspalten:

in the daily routine of the barn. There was no need to frame infertile animals as patients, he implied: reproduction work was not a clinical encounter necessitating the trappings of veterinary expertise, but part of the day-to-day care for dairy cattle. It could be handled by barn workers, just as identifying oestrus or connecting milking machines could. But appropriating reproduction work also helped elevate those workers from low-status labourers to 'interested and competent' specialists in their own right.

This was a matter of status and pay but also involved broader issues. Increasingly unionized, agricultural workers fought for better working conditions and to end the *stat* system of payment in kind for agricultural labour, struggles that certainly informed the barn foremen's attempts to redefine themselves (Morell 2011: 176-77). Even so, tensions never ran that high before the war. Local vets less in thrall to SVSS' perspectives (but perhaps worn out from overwork) sometimes even provided barn workers with informal training.[29] After the war, however, the matter of reproduction work as an avenue of professional advancement grew considerably more loaded, as it became intertwined with the question of AI's development and expansion.

III-2. Lay Inseminators or Veterinary Nurses?

From 1945, both large and small farmers' interest in AI grew rapidly, reflecting a need to conserve labour as increasing numbers of young people left for the cities, and often also a desire to get access to higher-quality genetic material.[30] As we saw above, the expansion of AI hinged on the number of available vets. There were, however, international precedents for handing over AI tasks to non-veterinary personnel: in the United Kingdom and the United States, AI fieldwork was handled by independent non-veterinary technicians, sometimes known as lay inseminators. There, AI was thus separated from the provision of veterinary services, not only making it cheaper but also decoupling its expansion from the availability of veterinary personnel.

This model was making inroads in Denmark, too, where AI's rapid expansion had left vets unable to stop the training of technicians to work alongside them. Swedish veterinarians found this deeply problematic. Though their position had stabilized – district vets had received the coveted status of state officials in 1934 – they nevertheless remained preoccupied with the issue of quacking. There were few legal restrictions on non-veterinary animal healthcare unless it fell under statutes of animal cruelty, which vets themselves conceded would rarely apply. Quacking seemed, furthermore, particularly troublesome in the context of AI, as

Steriliteten i våra besättningar', *Ladugårdsförmannen*, 7 (1936), 3, 4, 5, pp. 31-32, pp. 44-45, pp. 54-56; G. V., '"Djurkvacksalvare etc."', *Ladugårdsförmannen*, 8 (1937), pp. 105-06.

29 "Opponenten" (1942), 'Dräktighetsundersökningen hos kor', *Ladugårdsförmannen*, 13 (1942), 2, pp. 22-24 (p. 23).

30 SOU 1944:65; Morell (2011): 179-81.

vets both considered obstetrics and reproduction as prime quacking fields and (much as medical doctors did) saw 'half-educated' people as a particular threat.[31] This made them worried about AI interests effecting the training – perhaps in agricultural schools where vets had little insight or influence – of non-academic and non-medical groups of agricultural workers, like the barn foremen, to a level of independent technical skill in AI that they could then leverage in animal healthcare more generally, undercutting vets.

Acting to pre-empt this, Lagerlöf and the Veterinary College's vice-chancellor, Gerhard Forssell, proposed as a solution to AI's staffing problems the training of veterinary 'nurses'. These nurses – exclusively women – would be taught inseminations as part of yearlong course giving them a broad overview of animal healthcare.[32] The plan was intended as a compromise that satisfied farmers' and breeders' demands for AI technicians without threatening veterinary interests. It hinged on coding inseminations as female. While tending cows had historically been mostly female work, Lagerlöf and Forssell saw the issue primarily in relation to the gendering of medicine. As Motzi Eklöf (2000: 156) has discussed for human medicine, doctors perceived female unlicensed practitioners as less threatening than male ones, since women could not be mistaken for 'real' physicians. Similarly, the vets saw no risk that young women would be able – or want – to put themselves off as medical authorities vis-à-vis farmers. From the vets' point of view, it was therefore safe to give them access to reproduction work. Accordingly, Lagerlöf would later propose more far-reaching training for them in order to pre-empt any demand for Swedish male workers being trained in Denmark, where the (male) AI technicians were now doing sterility treatments as well:

> It would be a great misfortune for animal healthcare if barn foremen were to be trained in artificial insemination, pregnancy diagnosis, and sterility treatments. One has every reason to presume that this would give rise to a corps of legalized quacks, incurring great damage on the animal-owning public and surely also much annoyance to practicing veterinarians. […] The only solution at present is to give the female veterinary assistants […] such a thorough and complete training in artificial insemination and pregnancy diagnosis that no demand for barn foremen trained in Denmark arises. The advantage of thus trained female veterinary assistants [are] […] that an educated woman hardly will turn into a quack, and that it [will] be easier to keep their work under veterinary control.[33]

31 SSVS' Board of Directors Annual Report 1936-1937, attachment B, SSVSA, SA 72:1, SVM; Minutes, SSVS annual meeting, 17 October 1937, § 28, *Svensk veterinärtidskrift*, 43 (1938), 3, pp. 106-14.

32 Nils Lagerlöf, 'P.M. angående utbildning av examinerade veterinärbiträden', 28 April 1945, OG, F8A:1, SLU.

33 Nils Lagerlöf to the SSVS' Board of the Directors, 15 February 1946, OG, F8A:1, SLU.

SSVS decided accordingly in April 1946.[34] Anything was seen as preferable to male assistants being given the means to quack.

III-3. Milk Recording and the Politics of Artificial Insemination

By that time, a gulf was opening up between veterinary and non-veterinary framings of AI's staffing issues. Leading AI breeders, who had created the National Association of Swedish AI-Breeding Associations (Riksorganisationen Sveriges Seminföreningar, RSS) to advance their shared interests, were increasingly frustrated with the veterinary position. RSS insisted, in line with Bonnier's initial conception, that AI was a genetic technology that should be used to identify the best bulls and to use those bulls intensively. Its leaders were no longer willing to delay AI's expansion over what they saw as narrow and shortsighted veterinary interests. The war was over, and an RSS delegation had been able (in the words of a participating vet) to 'greedily' inform themselves of AI's organization in Denmark.[35] Afterwards, most non-veterinary representatives and proponents of AI shared the sentiments of animal husbandry advisor Erik L. Larsson, who forcefully argued that Swedish AI 'ought to be organized also according to other perspectives than purely veterinary'.[36]

This meant giving the AI associations not nurses but male inseminators, recruited among people already working in farming: 'farm-foremen [lantmästare] and useful milk-recording assistants', as agronomist Tore Lindström put it.[37] Lindström's argument highlights another labour issue that became important in connection with early AI: how to counteract the tendency of 'flight from the barn'.[38] Though the *stat* system had been abolished in 1945, working in animal farming remained unattractive. AI, however, was seen as a modern, technological reinvention of cattle husbandry that might entice barn foremen and, especially, milk-recording assistants to remain in the field.

Milk recording was a cooperatively organized effort to follow up quantitative and qualitative production levels. The fieldwork was carried out by milk-recording assistants (kontrollassistenter), who were agricultural school graduates subsequently trained in state-sponsored courses. They regularly visited participating farms to measure milk yields and butterfat levels and check fodder consumption. This data was used to calculate yearly production results both for herds and individual cows, a practice seen as integral to a rational organization of production

34 Minutes, SSVS extra meeting, 28 April 1946, § 11, *Svensk veterinärtidskrift*, 52 (1947), 3, pp. 66-72.

35 Ernst Pålsson to Lagerlöf, 1 November 1945, OG, Ö1:2, SLU; G. E—n (1945), 'Danmarks seminavel har väldig omfattning', *Lantmannen*, 29, 43, pp. 955-56.

36 Erik L. Larsson (1946), 'Seminverksamheten, tjururvalet och veterinärerna', *Lantmannen*, 30 (1946), 22, pp. 535-36 (p. 535).

37 Tore Lindström (1946), 'Omkostnaderna vid olika former för tjurhållning', *Avelsföreningens för SRB tidskrift*, 19, 1, pp. 24-32 (p. 31).

38 By. (1937), 'Flykten från ladugårdarna!', *Ladugårdsförmannen*, 8, 1, p. 6.

(Roos, 2004). But the job itself was low-paid and ambulatory, and few assistants – who all had a degree of schooling – envisioned it as a lifetime pursuit. Yet 1940s debates in the journal published by their association suggest that most saw possibilities of advancement to better positions in agriculture as increasingly doubtful.[39]

By war's end, milk recording only extended to a minority of farms, and in 1946, a group of Conservative, Liberal, and Agrarian politicians introduced motions in parliament on its reform. Emphasizing milk recording's 'importance for the nation's food supply' and noting that many milk-recording assistants lacked the 'necessary competence', they called for a study of how milk recording could be expanded and the milk-recording associations be given better possibilities to 'compete for the best manpower in the field'. These were perennial issues, and the points were in line with numerous other motions introduced in the preceding decades. In 1946, however, they coincided with the government's planning of a new, rationalization-focused agricultural policy – and with motions from another group of Agrarian parliamentarians seeking state funding to expand AI.[40] In response, the Minister of Agriculture appointed a committee with a dual remit: studying milk-recording reform *and* the organization of AI. Gösta Liedberg, a leading agricultural politician from the Conservative Party, was named chairman. Both the dual remit and the choice of Liedberg signalled the government's view: Liedberg was himself a board member of an AI association and had earlier motioned for AI training to other 'suitable persons' besides veterinarians.[41] This resounded in his instructions: one the matters to be considered was the

> lack of sufficiently qualified employees in [milk recording]. The salaries and conditions of employment are not seen as attractive enough, and the work does not completely fill the employee's working hours.

Here, the Minister saw a possible synergy with AI:

> There ought to be possibilities to transfer some artificial insemination tasks [from veterinarians] to suitably trained laymen. [...] It does not seem precluded that milk-recording employees, after proper training, could to some extent get tasks there, alongside those that they have in milk recording.[42]

The minister thus promoted a development that the veterinary profession feared but milk-recording assistants and AI breeders desired. The latter kept the pressure up as the committee sat for two years: RSS' executive secretary Börje Olsson

39 Folke Qvennerstedt (1940), 'Vad är att göra för att höja kontrollassistentens löneförmåner och hans anseende?', *Kontrollassistenternas arbetsfält*, 4, 6 & 7, p. 54; Gustaf Carlsson (1940), 'Kontrollassistenternas löneförhållande och anseende', *Kontrollassistenternas arbetsfält*, 4, 8, p. 61; 'Diskussionsspalten', *Kontrollassistenternas arbetsfält*, 7 (1943), 1, 2, pp. 6-7, p. 14.

40 Mot. AK 1946:375; mot. AK 1946:287.

41 Mot. AK 1944:234.

42 *Berättelse till 1947 års lagtima riksdag om vad i rikets styrelse sig tilldragit*, pp. 270-71.

suggested in a 1947 interview that if AI worked well abroad using lay technicians, it would work in Sweden too. 'We should', Olsson argued, 'be allowed to train milk-recording assistants, barn foremen, etc., for this purpose. That would also create better livelihoods in these professions and make it easier to recruit good people to them.'[43]

However, Nils Lagerlöf's expertise in this field could not be bypassed, so he had also been appointed to the committee and predictably considered the idea of transferring 'veterinary matters to the milk-recording assistants' to 'make their profession more desirable' to be 'a very dangerous tendency'. He strived to moderate the committee's stance, pointing out what he saw as the complexity of AI work and problematic experiences from countries using independently working AI technicians.[44] In the end, he was able to extract a compromise. The committee's 1948 final report opened for the training of male AI assistants, but confirmed that they could only be trained at the Veterinary College and only work under direct veterinary supervision. The report also proposed to continue training women as well, to provide redundancy in case of an armed conflict that would require military mobilization of the male population.[45]

III-4. Lay Inseminators and Sexual Health Control

The Liedberg committee report, which dealt with other aspects of AI as well, generated considerable comment and the government elected to postpone its decision on its proposals. Meanwhile, having resigned to the coming of male AI assistants, the veterinary priority was to deny them training in sterility treatments and pregnancy diagnosis, reserving the broader practice of sexual health control for veterinarians. Lagerlöf envisioned a future in which the district vets would be increasingly involved in AI, bringing the two career paths together again. In early 1949, he suggested to a colleague that things had gotten to a 'crossroads'. Unless district vets engaged more with AI, it would be difficult to stop the AI associations from hiring male assistants, after which things would 'degenerate to quackery', a 'considerable danger to the animal-owning public, the AI associations, and the veterinarians'.[46]

This was not only about the distribution of tasks in AI but about broader concerns about the future role of vets in farming. Around the same time, SSVS' encouraged all practicing veterinarians to take more interest in AI. This was important not in the least as

---

43 'Semineringen och veterinärerna', *Skaraborgs läns nyheter*, 14 April 1947.

44 Quote from Lagerlöf to Åke Tönne, 27 August 1946, OG, F8A:3, SLU. See also Lagerlöf to Nils Lannek, 10 September 1946, OG, F8A:3, SLU; Lagerlöf to Weddig Borg, 13 September 1946; Lagerlöf to Willy Hallgren, 14 April 1947, both in OG, Ö1:2, SLU; Lagerlöf to J. Flatla, 21 February 1948; Lagerlöf to G. Mortensen, 31 March 1948, both in OG, Ö1:6, SLU.

45 SOU 1948:36, pp. 76-85.

46 Lagerlöf to Lars Borglin, 4 January 1949, OG, F8A:1, SLU.

> [veterinary] work is starting to decrease because the number of horses has decreased, farms without animals have been introduced, [and] because the fight against tuberculosis and brucellosis has progressed far enough [to require less veterinary manpower]. It is important that every veterinarian in the country is kept fully occupied.[47]

The passage is revealing, since the prevalence of infectious disease had been an important argument for the expansion of veterinary sexual health control. But as brucellosis and TB were nearing eradication, vets saw their participation in AI (and thus their provision of such control) as more rather than less important, as the fight against two major diseases would no longer require as much manpower. The chief impulse behind this position was a growing pessimism among farm practitioners, who saw the increasing number of transitions to livestock-free farms up close. In a letter to Lagerlöf, district vet Willy Hallgren complained about the situation's 'sadness' and suggested that the district veterinarians had 'their best days behind them'.[48] Further threats to veterinary farm practice, including reproduction work, would thus have been particularly troubling at a point when the future of such practice was already seen as very uncertain.

In February 1950, as the government finally prepared to present its AI bill, the Swedish Veterinary Society (Sveriges veterinärförbund, SVS – which had replaced SSVS in 1948) doubled down. It informed RSS and all AI associations that AI technicians 'under no circumstances' should be allowed to work independently, and that vets should perform all sexual health control. RSS' leaders were outraged, and when the government bill turned out not to address work distribution in detail, they petitioned the Minister of Agriculture directly. They noted that the training of AI assistants was not addressing farmers' needs, as it had been designed with the 'economic interests of the [veterinary] profession' in mind. The Veterinary College refused to train technicians for independent work (that is, handle sexual health control), so as to make any expansion of AI contingent on the limited supply of vets. RSS thus did not question sexual health control's integration with AI, but attacked its status as an exclusively veterinary practice. It now demanded that the AI organizations be given formal influence over the Veterinary College's training of assistants.[49]

SVS rejected this proposal as 'absurd' and claimed that it demonstrated a simplified view of the veterinary role in AI. Their preferred work distribution was unrelated to veterinary economic interests, but reflected biological realities that made the treatments themselves less important than the question of who

47 Fritiof Wilhelmson et al. (1949), 'Upprop till de praktiserande veterinärerna från Svenska Veterinärläkareföreningens arbetsutskott för artificiell insemination', *Svenska veterinärläkareföreningens medlemsblad*, 1 (1949), 2, pp. 45-46 (p. 45).

48 Hallgren to Lagerlöf, 7 February 1949, OG, Ö1:6, SLU.

49 'Arbetsfördelningen mellan veterinär och icke veterinär personal inom AI-verksamheten', *Medlemsblad för Sveriges veterinärförbund*, 2 (1950), 3, pp. 29-30 (p. 30); 'RSS och veterinärbiträdenas utbildning', *Medlemsblad för Sveriges veterinärförbund*, 2 (1950), 23-24, pp. 177-84 (p. 178).

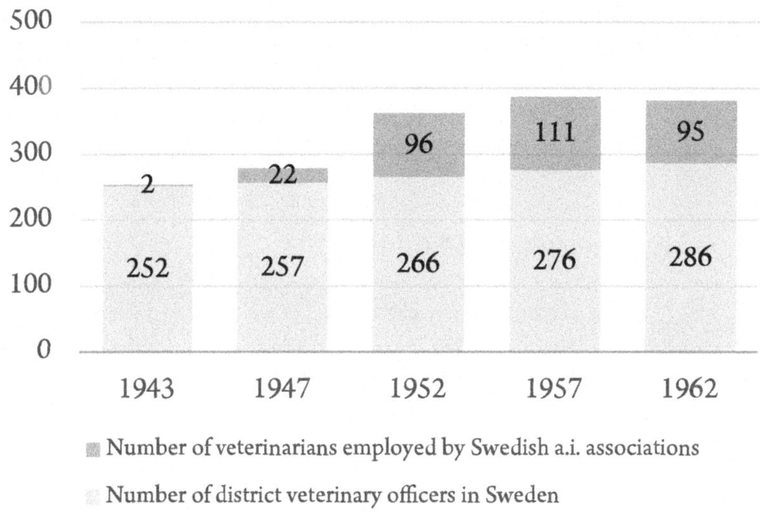

Figure 9.4. The number of AI vets and district veterinary officers in Sweden between 1943 and 1962
No sources for the number of district vets in 1952 and 1957; the bar sizes are based on my extrapolations. Sources: Medicinalstyrelsen/Veterinärstyrelsen, *Det civila veterinärväsendet*, https://www.scb.se/hitta-statistik/aldre-statistik/innehall/sveriges-officiella-statistik-sos/halso--och-sjukvard-19111996/detcivilaveterinarvasendet_sos_1916-1949; Nils Lagerlöf, 'Veterinära tjänster inom den artificiella inseminationen', 1 June 1952, OG, F8A:1, SLU; Sveriges veterinärförbund (1961): 269; Nordblom (1982): 37.

carried them out. Drawing on Lagerlöf's scientific understanding, SVS argued that a 'rational combatting of sterility is not carried out by performing sterility treatments. Of very significant importance is, however, the determination of the causes of sterility'. This focus on etiology ruled out too big a role for laymen: it required a mobilization of medical expertise.[50]

While available sources do not reveal if there was political mediation, the two sides did eventually reconcile, publishing a joint statement in December 1950 that blamed the conflict on 'misunderstandings' while reiterating the desire and need to work together.[51] For the time being, the veterinary profession retained its influence over AI's technical management and its de facto monopoly on reproduction work. Through the 1950s, this looked to vets as a rare positive aspect of a rationalizing agriculture where tractors replaced horses and increasing numbers of

50 'RSS och veterinärbiträdenas utbildning', p. 180, p. 184.
51 'Gemensamt uttalande av Riksorganisationen Sveriges Seminföreningar och Sveriges Veterinärförbund', *Medlemsblad för Sveriges veterinärförbund*, 3 (1951), 5, pp. 29-30 (p. 29).

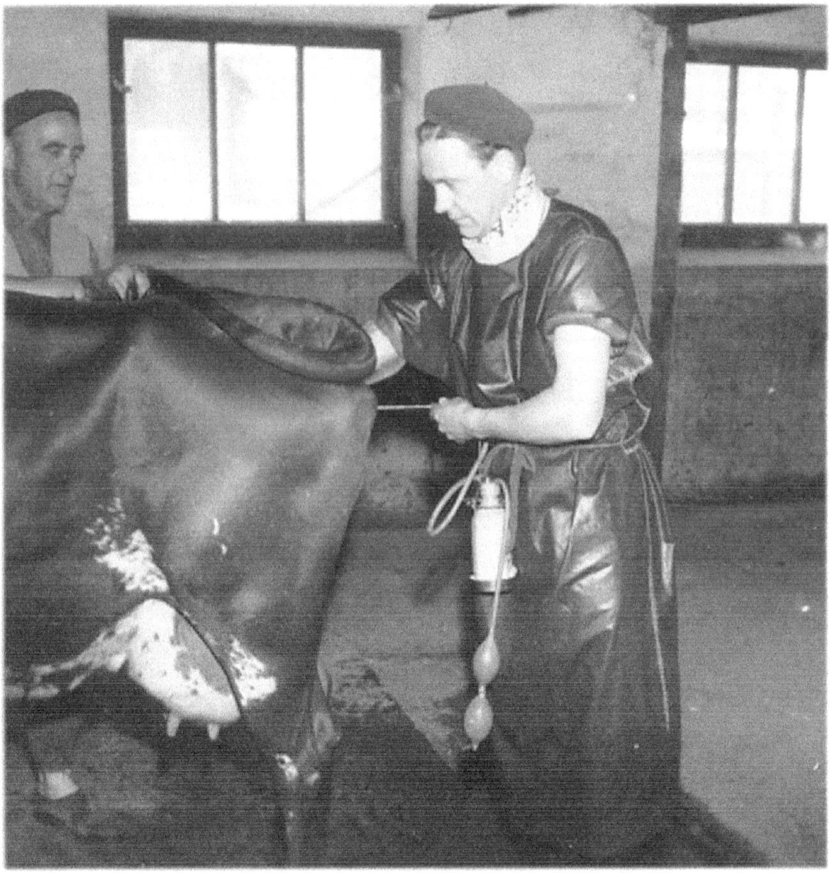

Figure 9.5. In 1959, veterinarian Gunnar Pettersson of the Skaraborg AI association flushes the uterus of a cow as part of his sexual health control of the herd. Photographer unknown.
Source: The Swedish Veterinary Museum's collections.

farmers took the previously unthinkable step of farming without cattle. Though veterinary representatives denied that their interest in AI was economically motivated, it was clearly of immense importance to them. In collegial correspondence in 1953, Lagerlöf characterized the veterinary engagement with AI as successfully having 'preserve[d] an essential branch of work' for the 'veterinary workforce'.[52] Without this 'preservation', far fewer vets could have remained in farm practice in the 1950s and 1960s. By the mid-1950s, there were around 110 AI veterinarians in

52 Lagerlöf to K. G. Wessman, 13 July 1953, OG, F8A:6, SLU.

Sweden, at a time when there were about 270 district vets (figure 9.4). It should be noted that a number of the AI vets were the Baltic refugees that Lagerlöf had trained during the immediate postwar period, who were not yet fully licensed as veterinarians in Sweden and so had few other professional options open to them. Nevertheless, they held positions designated as veterinary and for all intents and purposes functioned as veterinarians in the AI associations that employed them. It is therefore still valid to conclude that a significant new segment of the veterinary profession had been created: privately employed veterinarians doing nothing but reproduction work.

Ironically, the 1950s was nevertheless the peak of veterinary involvement in Swedish AI. Under continuing rationalization pressures, subsequent decades saw a gradual reduction in veterinary manpower, and the training of AI technicians eventually became coordinated with the training of milk-recording assistants after all (Rendel, 2003: 155-57; Roos, 2004: 132). This trend has continued, so that most AI today is being handled by the farmers themselves or by their direct employees. Yet Swedish AI remains in some ways a veterinary practice, as present-day regulations still mandate veterinary oversight. And though the agricultural landscape is wholly different today, reproduction work is still an important part of veterinary work on farms.

IV. Conclusion

The twentieth-century development of Swedish dairy farming towards a more specialized and higher-output production system depended on a range of human interventions into cows' physiological processes, linked to new developments in the animal sciences. Here, I have discussed one such development: a set of veterinary techniques to systematically combat infertility and monitor cows' reproductive health. Part of a new ruminant medicine that emerged as commercial dairy farming grew increasingly prominent on large farms in Sweden in the late nineteenth century, vets' established practice of reproduction work later allowed them to capture AI as a field of work. They did so not only out of an interest in reproductive medicine, but also in response to concerns over their profession's future role in a less animal-centric agriculture and on account of a need to articulate and re-articulate its position and its relevance in response to transformative technological change (cf. Kendra Smith-Howard's (2017) argument about vets and antibiotics). These have been perennial issues for vets, and the Swedish case does not diverge greatly from veterinary attempts elsewhere at claiming new technologies and at latching on to new production systems. Reproduction work proved a strikingly powerful lever in the early Swedish post-war context, however, where agricultural rationalization and a ballooning interest in AI very quickly restructured the veterinary profession and created the professional identity of the AI veterinarian. By 1955, one sixth of all vets in Sweden (if one includes the Baltic vets), and close to 30% of farm-oriented practitioners, worked exclusively in the context of AI.

Yet questions over reproduction work remained divisive. Other groups involved with cattle husbandry, such as barn foremen and milk-recording assistants, also saw in it a path to professionalization: it promised the possibility of pivoting from low-paying, low-status jobs to becoming more of modern specialists, better adapted to new production systems. Breeders and animal owners agreed, also seeing such a pivot as a way to counteract the exodus of working people from animal farming. Though they held diverging views, there was thus a broadly shared understanding – also highlighted by Karen Sayer (2019) in another context – of particular skills in handling animals as pathways to modernity in post-war farming. This vision of reproduction work as a key to modernity in the barn seemed of almost existential importance to those heavily invested in dairy farming – such as vets, breeders, barn foremen, milk recorders and agricultural policymakers – which helps explain why the matter became so contentious in the 1940s and early 1950s.

To conclude with an absence: this chapter has dealt with questions of organization and professional responsibility, but has not engaged in any detail with the intimate human-animal interactions in which the skills of reproduction work were applied. Yet reproduction work was ultimately about the actual practices of rupturing cysts, flushing uteri, and passing insemination catheters: forms of human-animal interaction that of necessity built on a deep understanding of cattle physiology and behavior. Indeed, the fact that they knew 'their' cows as individuals was a credible argument deployed by barn foremen when they sought to move onto the veterinarians' turf. This area warrants further research, for like Sayer I see interactions between humans and livestock as sites where both identities and knowledge were produced and where social and technical practices were shaped. Considering them in greater detail will give further insights into how reproduction work developed and how it changed Swedish farming.

Bibliography

Asdal, Kristin and Druglitrø, Tone (2014), 'Veterinærene: En profesjons arbeid på grensen: for jurisdiksjon og velferd', Slagstad, Rune and Messel, Jan (eds.), *Profesjonshistorier*, Oslo, Pax, pp. 700-34

Bruno, Karl (2020), 'Disciplining Cattle Reproduction: Veterinary Reproductive Science, Bull Infertility, and the Mid-Twentieth Century Transformation of Swedish Dairy Cattle Breeding', *Studies in History and Philosophy of Science*, 84, pp. 106-18

——— (2022), 'Use and Users of Artificial Insemination in Swedish Dairy Cattle Breeding, 1935–1955', *History and Technology*, vol. 38, no. 4, 317-43, published online 13 March 2023, https://doi.org/10.1080/07341512.2023.2174655

Derry, Margaret (2015), *Masterminding Nature: The Breeding of Animals, 1750-2010*, Toronto, University of Toronto Press

Einarsson, Stig (2010), 'Nils Lagerlöf – grundaren av specialiteten husdjursandrologi', *Svensk veterinärtidning* 62, 4, pp. 39-44

——— (2011a), 'Allan Bane – den första AI-läraren i Sverige', *Svensk veterinärtidning* 63, 4, pp. 37-40

——— (2011b), 'Baltiska veterinärers flykt till Sverige', parts 1 & 2, *Svensk veterinärtidning*, 63, 11, pp. 43-46, 55-60

Eklöf, Motzi (2000), *Läkarens ethos: Studier i den svenska läkarkårens identiteter, intressen och ideal 1890-1960*, Linköping: Linköping University

——— (2001), 'Doctor or Quack? Legal and Lexical Definitions in Twentieth-Century Sweden', Jütte, Robert, Eklöf, Motzi and Nelson, Marie C. (eds.), *Historical Aspects of Unconventional Medicine: Approaches, Concepts, Case Studies*, Sheffield, European Association for the History of Medicine and Health, pp. 103-17

Flygare, Iréne A. and Isacson, Maths (2011), 'The Tension Between Modernity and Reality, 1945-2010', Myrdal, Janken and Morell, Mats (eds.), *The Agrarian History of Sweden: From 4000 BC to AD 2000*, Lund, Nordic Academic Press, pp. 214-56

Fritzbøger, Bo (2015), *Mellem land og by: Landbohøjskolens historie*, Copenhagen, University of Copenhagen

Gaudillière, Jean-Paul (2007), 'Afterword: The Farm and the Clinic: An Inquiry into the Making of our Biotechnological Modernity', *Studies in History and Philosophy of Biological and Biomedical Sciences*, 38, 2, pp. 521-29

Israelsson, Carin (2005), *Kor och människor: Nötkreatursskötsel och besättningsstorlekar på torp och herrgårdar 1850-1914*, Hedemora, Gidlund

Jones, Susan D. (2003), *Valuing Animals: Veterinarians and their Patients in Modern America*, Baltimore, Johns Hopkins University Press

Korkman, Nils (1988), 'Svensk nötkreatursavel under ett sekel', *Kungl. Skogs- och Lantbruksakademiens tidskrift*, supplement 20, pp. 225-38

Lundgren, Elin (2001), 'Professionalisering inom djursjukvård 1890-2001: En undersökning i naturvetenskaplig medicin vs. kvacksalveri och alternativmedicin', unpublished bachelor thesis, Linköping University

Martiin, Carin (2011), 'Following Trails of Cattle through the Landscape: Dairy Cows were Key Players in Changes to Land', Antonson, Hans and Jansson, Ulf (eds.), *Agriculture and Forestry in Sweden since 1900: Geographical and Historical Studies*, Stockholm, Royal Swedish Academy of Agriculture and Forestry, pp. 198-216

——— (2017), 'From Farmer to Dairy Farmer: Swedish Dairy Farming from the Late 1920s to 1990', *Historia Agraria*, 73, pp. 7-34

Morell, Mats (2011), 'Agriculture in Industrial Society, 1870-1945', Myrdal, Janken and Morell, Mats (eds.), *The Agrarian History of Sweden: From 4000 BC to AD 2000*, Lund, Nordic Academic Press, pp. 165-213

Nordblom, Bengt (1982), 'Veterinärmedicinens utveckling', Månsson, Ingmar et al. (eds.), *Vårt lantbruksuniversitet: En bok till Lennart Hjelm*, Uppsala, Swedish University of Agricultural Sciences, pp. 37-42

Rendel, Jan (2003), *Från byatjur till genteknik: En agrar- och vetenskapshistorisk studie av utvecklingen av svensk husdjursgenetik och husdjursavel under 1900-talet*, Stockholm, Royal Swedish Academy of Agriculture and Forestry

Risan, Lars Christian (2003). *Hva er ei ku? Norsk Rødt Fe som teknovitenskap og naturkultur*, Oslo, University of Oslo

Roos, Arne (2004), 'Kontrollföreningarnas betydelse för jordbrukets utveckling', Rydén, Reine (ed.), *Jordbrukarnas kooperativa föreningar och intresseorganisationer i ett historiskt perspektiv*, Stockholm: Royal Swedish Academy of Agriculture and Forestry, pp. 221-36

Palm, Kerstin (1981), 'Veterinäryrket förr och nu: En redogörelse med tyngdpunkt på distriktsveterinärens och dennes föregångares arbetsförhållanden', unpublished bachelor thesis, Stockholm University

Sayer, Karen (2019), 'The Changing Landscape of "Labour": Work and Livestock in Post-Second World War British Agriculture', *History*, 104, pp. 911-40

Smith-Howard, Kendra (2014), *Pure and Modern Milk: An Environmental History since 1900*, Oxford, Oxford University Press

——— (2017), 'Healing Animals in an Antibiotic Age: Veterinary Drugs and the Professionalism Crisis, 1945-1970', *Technology and Culture*, 58, 3, pp. 722-48

Sveriges veterinärförbund (1961), 'Styrelseberättelse', *Medlemsblad för Sveriges veterinärförbund*, 13 pp. 267-70

Theunissen, Bert (2012), 'Breeding for Nobility or for Production? Cultures of Dairy Cattle Breeding in The Netherlands, 1945-1995', *Isis*, 103, 2, pp. 278-309

Wilmot, Sarah (2007), 'From "Public Service" to Artificial Insemination: Animal Breeding Science and Reproductive Research in Early Twentieth-Century Britain', *Studies in History and Philosophy of Biological and Biomedical Sciences*, 38, 2, pp. 411-41

Woods, Abigail (2007), 'The Farm as Clinic: Veterinary Expertise and the Transformation of Dairy Farming, 1930-1950', *Studies in History and Philosophy of Biological and Biomedical Sciences*, 38, 2, pp. 462-87

——— (2012), 'Is Prevention Better than Cure? The Rise and Fall of Veterinary Preventive Medicine, *c.* 1950-1980', *Social History of Medicine*, 26, 1, pp. 113-31

——— and Matthews, Stephen (2010), '"Little, if at all, Removed from the Illiterate Farrier or Cow-leech": The English Veterinary Surgeon, *c.* 1860-1885, and the Campaign for Veterinary Reform', *Medical History*, 54, 1, pp. 29-54

DIEGO CONDE GOMEZ AND LOURENZO FERNANDEZ PRIETO

10. The role of the Veterinary profession in the improvement of livestock in Contemporary Spain (1910-1936)

I. Introduction. Improvements in livestock production and veterinarian profession in Spain

The process towards the transformation and improvement of livestock farming in Spain ran parallel to changes within the veterinary profession during the early decades of the twentieth century. Both trends emerged in the context of sweeping transformations in the fields of agriculture, livestock production and science that were linked to the second wave of industrialisation.[1] Among other things, this context was to entail the formal, professional and administrative segregation of agriculture and livestock farming.

This process must be understood as part of a corporative and collective action that cannot be explained as a sum of individualities, as it often has been in more conventional veterinary histories. In the Spanish case, we have observed that there is also an interesting generational component because this transformation and corporative self-affirmation was driven by a generation of veterinarians associated with the first members of the Board of Inspectors of Livestock Hygiene and Veterinary Health (*Cuerpo de Inspectores de Higiene Pecuaria y Sanidad Veterinaria*)

1 (Polanyi 1944), Pujol and others (2001), Fernández Prieto (2007, 2017).

Diego Conde Gomez • Royal Academy of Veterinary Sciences of Spain. Univ. Santiago de Compostela. HISTAGRA: diego.conde.gomez@gmail.com - https://orcid.org/0000-0001-9449-1397

Lourenzo Fernandez Prieto • Univ. Santiago de Compostela. HISTAGRA: lourenzo.fernandez@usc.es - https://orcid.org/0000-0003-3206-2428

From Breeding & Feeding to Medicalization, Rural History in Europe, 17 (Turnhout: Brepols, 2023), pp. 237–266
BREPOLS PUBLISHERS 10.1484/M.RURHE-EB.5.135028

in 1910, which in a more general sense can be framed within the so-called silver age of Spanish science.[2]

Here we focus on the circumstances that conditioned the work of this generation of veterinarians, as well as the collective and corporative actions that led to the development of regulations and improvements in the livestock sector that were related to the implantation of measures in favour of public and animal health. Firstly, the labour carried out by veterinarians in Spain at the beginning of the twentieth century is analysed to determine how far it is appropriate to conceive of this collective as having an identity of its own in the sense of being a historical generation. At the same time, an attempt is made to identify the different traits, activities and motivations that drove veterinarians and their profession as a whole, whose main objective between 1900 and 1936 was to bring about a transformation in livestock farming.[3]

Spain's abundant historiography of veterinary practice is dominated by positivism and apologetics more concerned with dates and individual facts than with the global reach and vision of the profession within its historical context, in contrast with the broader focus of agrarian historiography when it comes to other occupations like agricultural engineering (Cabo Villaverde and Bernárdez Sobreira 1996; Cabo Villaverde 2001; Florencio Puntas 2004; Pan-Montojo González 2005; Fernández Prieto 2007) or the historiography of sanitary and scientific development in general.[4]

As is the case for the veterinary historiographies of other countries in Europe and America (Smith 2013), the history of Spanish veterinary science has been constructed from a corporativist perspective by the small number of retired veterinarians for whom "these works are a source of enjoyment laden with love for their profession" (Cordero del Campillo 1984: 12). This tendency has

2 From 1902 to 1939 (Mainer 1981; Sánchez Ron 1999; Sánchez Ron and García-Velasco 2010; Sánchez Ron and Lafuente 2007). Characterised by the founding of organisations such as the Free Institute of Education (*Institución Libre de Enseñanza*) and the Committee for the Expansion of Scientific Inquiry and Research (*Junta para Ampliación de Estudios e Investigaciones Científicas*) that championed education, science and technology. The silver age coincides with the awarding of the Nobel Prize for Physiology or Medicine (1906) to Santiago Ramón y Cajal, the only Spaniard to receive a Nobel Prize in a scientific field, along with his distant disciple Severo Ochoa (1959).

3 Here an improvement in livestock farming is not only understood to refer to aspects related to zootechnics and the field of production, but also those changes that suppose an improvement in animal welfare and health, as well as the adherence of the final product to certain health standards. It represents the sub-discipline of veterinary medicine that deals with bettering livestock production through the application of knowledge in the areas of breeding, genetics, statistics, fodder, hygiene, fertilisers, pedigree, herd health management, surgery, artificial insemination, milking machines etc.

4 Veterinary history's long track record goes back to Sanz Egaña's *Historia de la Veterinaria Española* (Sanz Egaña 1941), followed by *Semblanzas veterinarias* (Cordero del Campillo and others 1973). This field of study was consolidated in 1996 with the founding of the Spanish Association of Veterinary History (*Asociación Española de Historia Veterinaria*). For a critical summary, see Moreno Fernández Caparrós (2002). On the topic of scientific development, see Laín Entralgo (1981), López Piñero (González Blasco and others 1979) or Sánchez Ron (1999).

been partially corrected by contributions with a more analytical focus such as recent monographies by Vives Vallés and Mañé Seró (2016, 2018, 2020), which apart from questioning preconceived and long-reiterated ideas, also enter into transversal dialogue with specialist international literature, thus achieving a far more global vision through interaction with other historiographies.[5]

The works of Vital Ruibérriz (1982) and Benito Hernández (2003) study the coexistence of *albeitería*[6] and veterinary practice, as well as how professional training came to be regulated through the creation of the Royal School of Madrid (1792), which was heavily influenced by the French veterinarian sector; this first school modelled itself on its equivalents in Lyon (1761) and Alfort (1765), as Salvador Velasco (2015) affirms. The context and purpose of these schools' formation is common to other European countries like Germany (Schönherr 1991; Mitsuda 2017), France (Hubscher 1999; Berdah 2012, 2018; Conseil Général Vétérinaire and Glavany 2006), England (Woods 2014, 2007; Woods and others 2018; Brassley 2000; Corley and Godley 2011), the Netherlands (1999), Norway (Elvbakken 2017) or Sweden (Bruno 2020; Hellberg 1990), and also the United States (Jones 2003; Steele 1976, 2000).

Regarding Spanish veterinary medicine in its European context at the time when formal training and qualifications were put into effect, we can refer to Bonnaud and Fortané's research (Bonnaud and Fortané 2021). For the first third of the twentieth century, the works of Etxaniz Makazaga (2003) and Conde Gómez (2015) shed light on the life and work of Gordón Ordás and Rof Codina respectively, as does the final volume of *Semblanzas veterinarias* (Dehesa Santisteban and others 2011).

II. The creation of the Board of Inspectors of livestock hygiene and Veterinary Health (1910)

It could be judged that the first group of Inspectors of Livestock Hygiene and Veterinary Health was the foundation for the modern National Veterinary Corps (*Cuerpo Nacional Veterinario*). That first group of inspectors branched off into other self-defined entities like the Veterinary Union (*Unión Veterinaria*) (1917-1923) and the National Veterinary Association (*Asociación Nacional Veterinaria*) (1923-1936). The full acceptance of the profession into the State Administration's ranks was a generalised aspiration among veterinarians around 1900 because it would suppose a level of job security that their predecessors had

5 In this sense, Karen Brown and Daniel Gilfoyle (2010) offer new perspectives from the point of view of globalisation that are still pending in European historiography. For more information on Spain see Gutiérrez García (2006a, 2006b, 2010) or Pumarola i Battle (2002).

6 The art of healing animals, especially horses. This term was used in Spain until the consolidation of veterinary medicine as a profession based on science rather than practical experience. See Cordero del Campillo and others (1996).

Figure 10.1 Photographs of the first Inspectors of Livestock Hygiene and Veterinary Health (1910).
Source: Asociación del Cuerpo Nacional de Veterinaria

lacked, for example municipal veterinarians who were often dependent on the decisions of local politicians. Although the board of livestock inspectors could not guarantee veterinarians complete economic liberty, the stability that membership offered did allow inspectors to balance their official duties with other tasks. In this way, an important cadre of veterinarians in Spain achieved an unprecedented level of professional consolidation, and from this position they strived to dignify the sector by improving its social consideration.

This was not an easy process. The Ministry of Development (*Fomento*), presided by Augusto González Besada, issued a Decree (25th October 1907) that set out regulations for state-wide Agricultural and Livestock Services. With respect to the latter, the General Inspection for Livestock Hygiene (*Inspección General de Higiene Pecuaria*) was established along with the Board of Inspectors – one for each province – of Livestock Hygiene and Health, Ports and Borders (*Cuerpo de Inspectores de Higiene y Sanidad Pecuarias y de Puertos y Fronteras*). The titles of its parliamentary supporters reveal the project's aristocratic tenor; the duke of Bailén and the marquis de la Frontera, at that time president and secretary of the General Association of Livestock Producers of the Kingdom of Spain (*Asociación General de Ganaderos del Reino*), which represented the interests of large landowners.

The new institution comprised of an inspector-in-chief (selected from a number of applicants who had to be high-ranking veterinary professors with a minimum of ten years' experience), 49 provincial inspectors and fifteen inspectors for ports and borders. The provincial inspectors' remit covered the study of common diseases among livestock in a given province, including their description and the identification of measures for their prevention and cure; responding to queries made by farmers and associations; and writing an annual report on the aforementioned labours. They were required to live in the provincial capital with an office in the corresponding Livestock Station or, if that were not possible, Agronomic Section. Their salary oscillated between 2,500 and 6,000 pesetas depending on where they were placed,[7] and the post was incompatible with the exercise of any other kind of regulated or unregulated profession. Sub-delegates of veterinary health, municipal health inspectors and municipal veterinarians were all obliged to collaborate with them in tasks related to livestock health control.

The Decree also specified that the board's personnel would be subjected to an entrance examination, which was duly convened by Royal Order (7th May 1908) with a total of 72 posts available. One year later an assessment panel was approved with members drawn from the ranks of veterinary professors and inspectors. 364 candidates sat the exam in the Faculty of Medicine in Madrid during September 1909. The eventual Inspectors of Livestock Hygiene and Veterinary Health were named by Royal Order (23rd February 1910), with Dalmacio García Izcara assigned the post of inspector-in-chief.[8]

Previously published research has considered the work of several inspectors whose professional and political careers are worthy of note. Such is the case for the first and second of the men on the list of candidates who passed the examination process: Félix Gordón Ordás was a prestigious republican politician who would later become president of the government in exile (1951-1960);[9] Juan Rof Codina, who has been well studied as a disseminator and promotor of improvements in Galician livestock farming;[10] or indeed other inspectors like Santos Arán (Etxaniz Makazaga 2003: 11-113; Gómez Piquer and Pérez García

7 For comparison, a university professor earned 3,000 pesetas per annum.

8 Other posts included José Díaz del Villar, Government Adviser for Health, doctor in Medicine and professor at the Court Veterinary School; Bonifacio Estrada Viloira, Veterinary Health Services Inspector for the Ministry of Governance; Patricio Chamón Moya, military veterinarian; Juan de Dios González Pizarro, professor at the Córdoba Veterinary School; and Juan Castro Valero, professor at the Madrid Veterinary School. Appendix 1 enumerates the full list of inspectors and their posts.

9 Among the studies that centre on Gordón, we cite the abovementioned work by Etxaniz Makazaga (2003), contributions by Cordero del Campillo (1973, 2003, 2004) and Pérez y Pérez and others (1986). At the same time, historians have repeatedly used Gordón's writings to glean information about the republican governments: Seco Serrano (1973), Thomas (1976), Alted Vigil (1993), Marco (1998) and De Hoyos Puente (2018a, 2018b).

10 Though previous studies do exist (Fernández Pousa 1968; Fernández Prieto 1985; Rodríguez Troncoso 2007; Ruiz Martínez 1973), recent works by Conde Gómez (2013, 2014, 2015) have definitively profiled Rof's professional activity.

2000: 374; Serrano Tomé 1973), Cayetano López (Galindo García 1973a; Séculi Brillas 1971), Jose Orensanz, and Sanz Egaña (Galindo García 1973b).

III. Veterinary network and transfer of knowledge

Taken as a whole, this generation of inspectors made it its mission to strengthen the public and political role of veterinarians, following in the wake of Molina Serrano and García Izcara and their efforts to dignify the profession. Making the most of their new position in the Administration, they took a series of actions that, as we shall see, would clash with the measures that had previously been put in place.

If the classical Greek idea is to be believed, there are always two simultaneously active generations who exert their full powers to influence the same questions and issues, but at different life stages and, as such, with different approaches. In the present case, the veterinarian generation led by Dalmacio García Izcara coexisted with the one that came together in the first Board of Inspectors with Félix Gordón Ordás, the most distinguished candidate from the selection process, at its head (Etxaniz Makazaga 2003: 96-97). The divergences between each generation, the former conservative and the latter progressive, explain their distinct ways of understanding their profession and the consequent tensions between them, which coincides with the ideas of Strauss and Howe (1991) and Kuhn (1962; 2012) about periods of conflict within the structures of scientific revolutions. This bipolarity between the old and new veterinary schools permitted a synergetic balance that fed into political and organisational gains following the proclamation of the II Republic in 1931 (Moreno Fernández Caparrós 2004: 152-53).

The activities of the Board of Inspectors began in the second decade of the twentieth century, in the context of the impact made by the second wave of industrialisation, with clearly defined challenges ahead, the first being related to the articulation of Spanish State bureaucracy.

> The Inspection for Livestock Hygiene and Veterinary Health has a mighty role to play in the intricate mechanism of Spanish social life. It is a question of life and death, to be or not to be, as Shakespeare would say. If the obstacles to its creation and betterment should become entrenched, and should the Inspectors be forced to live as yet another class of civil servants, they will waste away through atrophy. (Gordón Ordás 1910)

The second challenge was related to the normative and legal vacuum over questions such as the administrative response to epizootics, public health, food control systems, or livestock statistics, all of which were to be addressed by the new inspection corps and its inspector-in-chief in the capital, Dalmacio García Izcara.

Due to the lack of coordination from the central Administration or any common instructions for livestock inspectors, a parallel network emerged in the midst of the new generation which would serve to share knowledge and the various initiatives that inspectors were pushing ahead with in different parts of the State. In this sense, Gordón Ordás took advantage of a range of professional publications to exhort his co-inspectors not to fall into the trap of bureaucratic complacency, but rather to take an active role in making the necessary improvements to livestock production in their jurisdictions.

> Damn those inspectors of livestock hygiene who, removing all obstacles from their path, limit themselves to the giving and taking of health statistics. Blessed are those who, though they cannot sow, sow culture, which is bread for those who are poor in spirit as they are in body. Opening the eyes of farmers is how robust livestock is bred and the fatherland invigorated. (Gordón Ordás 1910)

An important instrument in this professional articulation was Gordón Ordás' founding in April 1911 of the Journal of Hygiene and Veterinary Health (*Revista de Higiene y Sanidad Veterinaria*), which together with its successor, the Journal of Livestock Hygiene and Health (*Revista de Higiene y Sanidad Pecuarias*), was to become the principal medium for the definition of policy, practice and professional cohesion among veterinarians. Its first issue emitted a declaration of intentions in this regard.

FOUR WORDS
Our mission and our means

> Possessed of the greatest enthusiasm, we inaugurate the JOURNAL OF HYGIENE AND VETERINARY HEALTH. Our primordial aim is the defence and improvement of veterinary personnel registered with the Health service. There we have a whole organisation that, despite its huge deficiencies, lends invaluable services to the public wealth; and that organisation, which includes Inspectors of Livestock Hygiene, veterinary sub-delegates and qualified veterinarians… lacks an agglutinating agent that would give it the indispensable cohesion to reveal itself as a homogenous whole. ('Cuatro palabras' 1911)

Starting from its fourth issue, the journal brought in a regular "Good Works" section (*Labor positiva*) to showcase livestock inspectors' activities and promote the imitation of these selected experiences and practices all across the Spanish State.

IV Promotion of Animal Health

One of the problems that the inspectors faced was how to deal with the persistent epizootics that ailed livestock across Spain. From the late nineteenth century onwards, there was a move towards higher levels of specialisation in livestock farming and a greater value being put on animals as a source of production (Fernández Prieto 1996; Bernárdez Sobreira 1997; Soto Fernández 2006; Domínguez Martín and Puente Fernández 2009), specifically in the Cantabrian north and Atlantic northeast of the Iberian Peninsula. This helped to raise levels of interest in upgrading animal healthcare (Conde Gómez 2014, 2015).

IV. 1. Campaigns to monitor and eradicate disease

The fight against diseases like foot-and-mouth, anthrax and tuberculosis became a priority because they were a formidable obstacle to the development of Spanish livestock farming. Foot-and-mouth disease was one of the main epizootics that affected livestock in Spain. Although data is available for the presence of foot-and-mouth in the country since 1848 ('Sección Científica. Administración Veterinaria' 1848), an order issued by the English government on the 16th May 1881 for animals arriving from the peninsula to be put down was what brought the problem to public attention. Among other consequences, said measure interrupted the active exportation of livestock from Galicia to England that had developed since the mid-nineteenth century (Carmona Badía 1982; Martínez López 2000). One sign of the pitiful state of State epizootics control is the fact that, as a result of the English government's order, the Civil Governor of A Coruña published a note in the provincial press to request any kind of information that people might have about the disease.

Just thirty years later, in 1910, with the creation of the Board of Inspectors new generalised measures began to be taken against what had become an endemic problem. Rof Codina, the brand-new inspector for A Coruña, initiated pre-prophylactic efforts to inform the public through the publication of leaflets by the Provincial Development Council (*Consejo Provincial de Fomento*) and newspaper articles about foot-and-mouth disease with broad descriptions of its etiology, pathology and prophylaxis.[11] The inspector in Barcelona, Francisco Sugrañes Bardagí, began to vaccinate animals against the disease, employing the same techniques that were used for sheeppox. At the same time, policies were proposed for the disinfection of transport and lazarettos where animals from infected areas could be quarantined ('Labor Positiva. Los Inspectores de Higiene Pecuaria En Acción' 1911a).

In the case of bacterial anthrax, certain vaccination programmes had been put in place in the past, for example in Navarre in 1882 (Camarero Rioja 2009a, 2009b), but the constitution of the Board of Inspectors marked the commencement of generalised prophylactic campaigns. Here it is worth mentioning Rof Codina's work in Galicia, which contributed innovative features to the campaigns, such as the commercialisation of vaccines, as well as the inclusion, participation and complicity of farmers through agrarian associations, thus allowing for a greater intensification and dissemination of these programmes as a weapon to fight against, control and eradicate the disease. Rof's work was soon recognised in light of the fact that his informative and educational efforts had definitively

11 (Rof Codina 1911a, 1911b, 1911c, 1911d, 1912b, 1921a, 1921b).

Figure 10.2. Juan Rof Codina in an anthrax vaccination campaign in A Coruña (1911) ('Los Progresos Del Campo Gallego' 1911: 22)

increased the number of farmers who had opted for prophylactic solutions instead of far less efficient remedies (Salgueiros 1910).

Preventing tuberculosis was more complex because it involved good hygiene practices in sheds, barns and stables; the tuberculinisation of animals; and the issues of milk conservation, transport and sale (Conde Gómez and Fernández Prieto 2019). For more information on each of these aspects, see Conde Gómez (2018), Conde Gómez and others (2013), Gutiérrez García (2004); Hernández Adell (2012), or Ponte Hernando (2016).

IV. 2. Dissemination in livestock hygiene

Another facet of the new veterinary inspectors' mission was to try to disseminate their proposals in order that 'old' practices, rooted more in tradition than rigorous science, might be abandoned. With this in mind, inspectors often published articles in the local press and information bulletins edited by institutions like livestock development committees or crop and livestock farmers' associations.

The reiteration of bibliographical references to Gordón Ordás (Etxaniz Makazaga 2003) and Rof Codina (Conde Gómez 2015), as the main promotors of pedagogy in the field of livestock farming during the first half of the twentieth century undervalues the actions taken by a large part of the veterinarian community in the same timeframe. This considerable labour was carried out principally in the small and medium holdings of the peninsular north. The contributions of Francisco Abril Bocas in Oviedo, Fernández Turegano in Cuenca, Joaquín Castellanos in Teruel,[12] Martín Lázaro Calvo en la frontera de Tui or Xavier Prado *Lameiro* in Ourense (Prado Rodríguez *Lameiro* 1918), were qualitatively just as important as Rof's or Gordón's. Therefore, they help us to comprehend the territorial breadth of the inspectors' activities, as well as the influence of their profession on public opinion and the world of production.

Thanks to multiple publications aimed at farmers and an intense campaign of in-person educational talks and conferences in agrarian associations, inspectors managed not only to raise health standards in livestock production, but also had a positive effect on how society perceived the usefulness of their profession. And in their eyes, the second issue was no less important than the first.

The farmers themselves praised the work of livestock inspectors, as can be seen in the case of Agustín Fornells in Fuerteventura, on the Canary Islands.

> As farmers, we have admired don Agustín Fornells for his constancy, calm demeanour and tenacity, with which, scorning the fiercest of personal attacks, for months and months he has worked to bring prestige and dignity to the veterinary profession, which, in this province, is more necessary than anywhere else, for the majority of our animals are under the empirical care of individuals who are wholly ignorant of the most rudimentary elements of zoology, and who offer services that in their ignorance they believe to be beneficial, yet in reality not only diminish our livestock capital, but are true affronts and dangers to good sanitary practice. ('Labor Positiva. Los Inspectores de Higiene Pecuaria En Acción' 1912)

V Improvements in livestock zootechnics and productivity

Starting in 1910, livestock inspectors in all regions of the State initiated plans to raise productivity in cattle farming. There were two main reasons for this. On the one hand, this was one of a number of late-stage responses at the local level to a broader crisis in livestock production, which itself was part of the great turn-of-the-century agrarian crisis (Carmona Badía 1982; Carmona Badía and de la Puente Fernández 1988; Fernández Prieto 1992). On the other, both milk and meat consumption increased from 1910 onwards due to growing urban demand, particularly in northern Spain, and to new recommendations in the field of nutrition. This rise in demand also spurred farmers themselves to act as promotors of the inspectors' improvements as a strategy to maximise their output and income.

The new initiatives were first implemented in the peninsular north, where livestock farming was more advanced. An example of the importance of specialist journals can be found in how they were used to publicise methods for the

12 ('Labor Positiva. Los Inspectores de Higiene Pecuaria En Acción' 1911a; 1911b)

Figure 10.3. Livestock contest in Carballo. 1911. Attendants witness the veterinarians Ignacio Varela, Santos Romero and Rof Codina judging a Galician bullock. (Rof Codina 1912a: 220)

valuation and categorisation of animals that were being pioneered in Gipuzkoa (Rof Codina 1914) and Galicia (Conde Gómez 2014), which later served as a template for the progress of other breeds like the Tudanca cattle of Cantabria and Asturias ('Labor Positiva. Los Inspectores de Higiene Pecuaria En Acción'. 1913) through crossbreeding (Luque 1911; Luque and Pastor 1913; Mendizabal Aizpuru and others 2005).

> We have applied, in both Contests held in the years 1911 and 1912, the Lydtin method, lightly modified with the points method (that is used for Contests in Switzerland and England), and which for several years now has been utilised in Gipuzkoa and then Galicia, Navarre, Biscay and other provinces, with an infinity of advantages over judging by eye alone. (Enriquez 1913)

VI. The Republican Veterinary generation and the creation of the General Direction for Livestock farming (1931-1936)

The proclamation of the II Spanish Republic brought with it the materialisation of one of the key revindications of the 1910 veterinary generation: the establishment of a stratum within the Administration that would encompass all services pertaining to livestock production, which until that moment had been scattered between different Ministries (Public Instruction was charged with running Veterinary Schools; Governance handled veterinary inspection and public health; Development was responsible for livestock health, hygiene and improvement; meanwhile War organised military veterinary services). As early as 1904, during the I Livestock Producers' Congress in Madrid, a group of professors from the city's Veterinary School propounded the need for legislation to regulate livestock sanitary control, putting forward a project for the configuration of a national zootechnical sanitary service to oversee the development and conservation of the livestock industry (Asociación General de Ganaderos del Reino 1904). In the same vein, in two reports titled "Livestock Contests as a Means to Promote and Classify Livestock Production" (*Los Concursos de Ganados como medio de fomentar y clasificar la Ganadería*) and "The Galician Bovine Breed" (*La Raza Bovina Gallega*),[13] the second of which was awarded a prize by the General Association of Livestock Producers of the Kingdom of Spain, Rof Codina laid out his proposal for a Livestock Service that would bring together the Board of Inspectors; veterinary training; livestock development and publicity; Veterinary Schools; Livestock Stations; and military equine breeding centres – all under the same organisational umbrella.

During the 1922 National Veterinary Assembly, the Cordoban veterinarian Rafael Castejón insisted once again on the need for a General Directive for Livestock Farming (*Dirección General de Ganadería*) as part of the Ministry of Development, to which all veterinary services would be subordinated in the same manner as Rof Codina had proposed. The request for a General Directive would be repeated six years later at the I National Assembly of Veterinary Colleges, held between the 5th and 10th January 1928.[14]

In the month following the proclamation of the Republic, the General Directive for Livestock Farming and Industries finally materialised by Decree on the 30th May 1931. This represented a regrouping of services at the hands of Félix Gordón Ordás, the chief architect behind the move.[15] The General Directive monopolised all functions tied to the study and application of the production, exploitation, improvement, industrialisation, prophylaxis and treatment of animals that were previously dispersed among different Ministries, with the one exception of the Military Veterinarian Corps (*Cuerpo de Veterinaria Militar*), whose competencies were absorbed a few months later.[16]

The decision to implant a new ministerial department had a lot to do with some of these veterinarians' roles in the republican movement, and it came to fruition as veterinarians – as a group – took advantage of that moment of revolutionary rupture, the opening of a window of opportunity. In spite of the difficulties in its creation, it is significant that the new department and its

13 (Rof Codina 1915, 1917)

14 General Archive of the Administration. 1.01-331-61.

15 Etxaniz Makazaga (2003: 277-302), Moreno Fernández-Caparrós (2000), and García Partida (2009) have explained the context in which the most far-reaching reform experienced by the Spanish veterinary sector in the twentieth century took place.

16 For this transfer to take place a specific commission was convened ('Orden Designando La Comisión Que Se Indica Para Transferencia a Este Ministerio de Todos Los Servicios Referentes a Cría Caballar' 1931). In early August, the commission managed to coordinate the cession of animals, equipment, and facilities to the Ministry of Development.

concomitant restructuring process endured and remained in place under Franco – up until 1971 – with structures more or less unchanged since the II Republic. The central professional and political argument in favour of the General Directive for Livestock Farming was expressed in the cited Decree of 1931.

> The extraordinary importance of livestock farming, which constitutes one of the most solid pillars of the wealth of nations, has been recognised at the theoretical level by great minds and by all men of the countryside; but it has always occupied a subaltern position in the concerns of the Public Powers, as a secondary appendix to agriculture, with no recognition for the fact that both sources of wealth, though intimately related to each other, present distinct paths of development and require direction by experts of different cultural backgrounds, with shared knowledge in the field of biology of course, but of divergent branches that lead the expert in agronomy to be an agricultural technician, thanks to his study of plant biology, just as they lead the veterinarian to be a livestock technician, thanks to his study of animal biology.

Veterinarians desired to turn livestock farming into an industry, a new and modern economic motor. In this way, the veterinary profession was guaranteed a presence in the only sector where it had a chance of gaining strength, livestock farming, thus avoiding collisions with medics in the arena of Public Health, since a veterinarian could not compete with doctors in the struggle for directive posts (Etxaniz Makazaga 2003: 278). They were rather more direct in vying for influence at the expense of agricultural engineers because the Decree of 1931 effectively segregated their competencies – agriculture to the agronomists, livestock to the veterinarians – which fuelled corporativist tensions between the two professions.[17] Republican veterinarians justified the solution by arguing that the subordination of livestock to agricultural production prior to the II Republic had been the fault of the monarchy and aristocracy, and that large landowners' indifference to livestock farming had also condemned the veterinary profession to its subordinate position. In the end, the practical result of all this was that veterinarians attempted to close ranks with doctors and draw lines in the sand between themselves and agricultural engineers. A further Decree, on the 3rd June 1931, stipulated that the General Directive for Livestock Farming would be provisionally headed by the Sub-secretary of the Ministry of Development, who was none other than Gordón Ordás. On the same day, the first General Inspectors were also provisionally assigned: Santos Arán for Livestock Development, Research and Contrasting; and García Armendaritz for the General Inspection of Hygiene and Veterinary Health.[18]

17 For more information on corporativist disputes between veterinarians and agricultural engineers, see Conde Gómez (2015: 126).

18 ('Decreto disponiendo que las Inspecciones Generales de Higiene…' 1931)

Figure 10.4. Assembly of the National Association of Veterinarians (Asociación Nacional de Veterinarios). Madrid, 1932
Rof Archive. Biblioteca Intercentros de Lugo. University of Santiago de Compostela

As a result of the agricultural engineers' widespread questioning of and determined opposition to the General Directive for Livestock Farming, the Minister of Development, Álvaro Albornoz, was forced to submit the Decree to the consideration of the Constituent Assembly, elected in late June, before it could pass into law. Albornoz himself presented the draft bill to the Assembly. In his memoirs, Manuel Azaña recalls the agronomists' opposition as well as Albornoz and Ordás' political tussling to get the Assembly on side.

> ...the Sub-secretary of Development, Gordón, is a Veterinarian and there are some who say that in drafting that bill he has rendered a great service to his veterinarian colleagues, to the detriment of agricultural engineers (...); the Parliamentary Commission charged with analysing it did not give its approval, and it was necessary for the Minister, who announced his resignation if the project were to fall through, raised the possibility of a vote of no confidence; in the end the law was passed thanks to the dissenting vote of Gordón himself...

Once the Assembly had sanctioned the bill and it had passed into law,[19] the organisation of the General Directive for Livestock Farming was completed with a new Regulation (7th December 1931). This occurred merely three days after the law was passed and barely six months after the original Decree, which gives us a notion of how intense the process was. The High Council of Livestock Production (*Consejo Superior Pecuario*) was created as a central technical body, replacing the pre-existing Central Epizootics Committee (*Junta Central de Epizootias*) to take charge of the drafting of new regulatory projects and development plans. Furthermore, the High Council was granted the right to propose new legislation and administrative competencies within the General Directive for Livestock Farming.

VII. Conclusions

The process of transformation that was experienced at this time by the Spanish veterinary sector in an administrative, political, and corporative sense was brought about collectively by a generation of veterinarians that we have called the republican generation. Throughout the period of study, most practitioners were united by an interest in gaining prestige, respect and social dignification for the profession as a whole.

Veterinarians' active work to promote improvements in livestock production was not simply a question of professional responsibility, but was also a step towards redefining their image in society, as was the case in other countries (Hubscher 1999). They had recourse to a wide selection of strategies and made the most of all and any political opportunity to consolidate the role of the veterinarian in the realms of science and health on an equal footing to other professions like doctors, pharmacists, or agricultural engineers, who enjoyed greater social, intellectual and political recognition.

The influence that this generation wielded was the materialisation of the demands of past generations of veterinarians. Throughout the first third of the twentieth century, the profession as a collective equipped itself with tools and resources to achieve its objectives. In this sense, the constitution of internal information networks that were independent of the State Administration, and their deployment as organs of dissemination and corporative cohesion, permitted the development of diverse initiatives for improving livestock production and public health in all parts of the Spanish State. In any case, it should be pointed out that the defining characteristics of this veterinary generation were essentially identical to other generations that made up what is known as the silver age of Spanish science. And even accepting differences in the solutions that were chosen, their actions were not out of step with the initiatives being taken by the veterinary profession in other countries.

19 ('Ley disponiendo se agrupen…' 1931)

The most important of these is connected to the definition of livestock farming as a separate economic and political sphere from agriculture, as a way to reaffirm veterinarians' professional status. In other countries, this strategy was combined with a notable level of protagonism in the area of Public Health management (Schönherr 1991; Woods 2014).

In the Spanish case, the 1936 coup and the resulting Civil War, not to mention the long period of Francoist dictatorship, entailed the partial decapitalisation of the technical personnel, ideas and projects of the previous years, as happened in other fields. Outstanding members of the veterinary profession suffered persecutions and exile, while others continued to work and participate in administrative and scientific structures during the dictatorship. Despite all the administrative changes, the veterinary sector's professional definition did not undergo substantial modifications during the following decades.

A pending topic for future research is to carry out a comparative study of Spain and other European countries to confirm whether there were similarities in the existence and consolidation of the knowledge networks and corporativism herein described, and to determine their chronological differences. Therefore, it would be interesting to ponder the role played by the veterinary sector in neighbouring countries, to consider whether the abovementioned activities were coordinated by governments or whether they depended on individual or collective actions taken outside the remit of the Administration. Whatever the results, it is clear that the process of the affirmation of livestock farming as a differentiated productive sector in the context of the great transformations effected by the second wave of industrialisation is linked to the consolidation of the veterinary profession and its statal and professional configuration in Europe and America. The same can be said of the eradication of 'old peasant customs' that came to be seen as non-professional. What distinguished one country from another was the area of professionalisation and the competencies that were ceded to the veterinary profession, covering a wide range of options such as animal health, public health, epizootics control or genetic augmentation – worlds within worlds and as many possibilities as there were States.

Bibliography and sources

Legislation

Decreto disponiendo que las Inspecciones generales de Higiene y Sanidad pecuarias y de Sanidad veterinaria, radicantes, respectivamente, en los Ministerios de Economía y Gobernación queden incorporadas a la Dirección de Ganadería e Industrias pecuarias de este Ministerio, y nombrando, con carácter provisional, para los cargos que se indican, a los señores que se mencionan, Gaceta de Madrid, 05/06/1931, p. 1201.

Decreto disponiendo que todos los servicios relacionados con el estudio y aplicación de la producción, explotación, mejora, etcétera, etc., que actualmente están dispersos en los Ministerios de Instrucción pública, Gobernación, Economía y Guerra, con excepción de los que desarrolla en este último el Cuerpo de Veterinaria militar, se agrupen en una Dirección general de Ganadería e Industrias pecuarias dependientes del Ministerio de Fomento, Gaceta de Madrid, 31/05/1931, pp. 1046-47.

Ley disponiendo se agrupen en una Dirección general de Ganadería e Industrias pecuarias, dependiente de este Ministerio, todos los servicios relacionados con el estudio y aplicación de la producción, explotación, mejora, industrialización, profilaxis y tratamiento de los animales y sus productos, con excepción de los que desarrolla el Cuerpo de Veterinaria militar, Gaceta de Madrid, 04/12/1931, p. 1419.

Orden designando a los señores que se indican para que se hagan cargo del ganado, material y efectos correspondientes a los servicios de la Cría Caballar, y que serán entregados por la Comisión designada por el Ministerio de la Guerra, Gaceta de Madrid, 09/08/1931, p. 1119.

Orden designando la Comisión que se indica para transferencia a este Ministerio de todos los servicios referentes a Cría Caballar, Gaceta de Madrid, 25/06/1931, p. 1657.

Real Decreto Relativo a la Organización de los Servicios de Agricultura y Ganadería. 31/10/1907. Gaceta de Madrid, pp. 392-404.

Real Orden disponiendo se convoque a oposiciones para proveer 61 plazas de Inspectores de Higiene Pecuaria Provinciales, de Puertos y de Fronteras, y que se publiquen en la Gaceta los adjuntos Reglamentos y programa a que han de ajustarse dichas Oposiciones. 14/05/1908. Gaceta de Madrid, pp. 770-75

Real Orden nombrando a los señores que se expresan, Presidente y Vocales para juzgar las oposiciones a las plazas de Inspectores Provinciales de Higiene Pecuaria y de Sanidad Veterinaria. 15/06/1909. Gaceta de Madrid, p. 1469

Real Orden nombrando, en virtud de oposición, Inspectores de Higiene Pecuaria y Sanidad Veterinaria a los señores que se indican. 11/03/1910. Gaceta de Madrid, p. 516

Bibliography

Alted Vigil, Alicia. 1993. *El Archivo de La República Española En El Exilio, 1945-1977: (Inventario Del Fondo París)* (Madrid: Fundación Universitaria Española)

Asociación de General de Ganaderos del Reino. 1904. *I Congreso Ganadero* (Madrid)

Benito Hernández, Milagros. 2003. *Del Amanecer de Las Escuelas de Veterinaria En España: Aportaciones al Estudio de La Historia de La Veterinaria*, 1^{a} ed. (Moncada (Valencia): Universidad Cardenal Herrera-CEU ; Fundación Universitaria San Pablo-CEU)

Berdah, Delphine. 2012. 'Entre Scientifisation et Travail de Frontières: Les Transformations Des Savoirs Vétérinaires En France, XVIIIe-XIXe Siècles', *Revue d'histoire Moderne et Contemporaine (1954-)*, 59.4 (Societe d'Histoire Moderne et Contemporaine): 51-96

———. 2018. *Abattre Ou Vacciner: La France et Le Royaume-Uni En Lutte Contre La Tuberculose et La Fièvre Aphteuse (1900-1960)*, Collection 'En Temps & Lieux', 82 (Paris: Éditions EHESS)

Bernárdez Sobreira, Antonio. 1997. 'A Evolución Do Sector Pecuario Na Galicia Contemporánea: Especialización Productiva e Mercantilización Na Sociedade Rural (1865-1996)', *Semata: Ciencias Sociais e Humanidades*: 371-88

Bonnaud, Laure, and Nicolas Fortané. 2021. 'Being a Vet: The Veterinary Profession in Social Science Research', *Review of Agricultural, Food and Environmental Studies*, 102.2: 125-49 <10.1007/s41130-020-00103-1>

Brassley, Paul. 2000. 'Animal Health and Veterinary Medicine', in *The Agrarian History of England and Wales (1810-1914)* (Cambridge), pp. 587-93

Brown, Karen, and Daniel Gilfoyle (eds.). 2010. *Healing the Herds: Disease, Livestock Economies, and the Globalization of Veterinary Medicine*, 1st edn (Ohio University Press)

Bruno, Karl. 2020. 'Disciplining Cattle Reproduction: Veterinary Reproductive Science, Bull Infertility, and the Mid-Twentieth Century Transformation of Swedish Dairy Cattle Breeding', *Studies in History and Philosophy of Science Part A*, 84: 106-18 <10.1016/j.shpsa.2020.09.002>

Cabo Villaverde, Miguel 2001. 'Un Funcionario Particular, o Enxeñeiro Agrónomo En Galicia, 1900-1950', in *Estudios de Historia Das Ciencias e Das Técnicas: VII Congreso de La Sociedad Española de Historia de Las Ciencias y de Las Técnicas: Pontevedra, 14-18 de Setembro de 1999*, II vols (Pontevedra: Deputación Provincial de Pontevedra, Servicio de Publicacións), pp. 437-51

———, and Antonio Bernárdez Sobreira. 1996. 'Ciencia y Dictadura: La Investigación Agronómica En Galicia Durante El Primer Franquismo (1936-1950)', *Noticiario de Historia Agraria: Boletín Informativo Del Seminario de Historia Agraria*, 6.12: 119-40

Camarero Rioja, Fernando. 2009a. 'Navarra, pionera de la vacunación. I' <http://www.euskonews.com/0505zbk/gaia50503es.html>

———, Fernando. 2009b. 'Navarra, pionera de la vacunación. II' <http://www.euskonews.com/0506zbk/gaia50603es.html>

Carmona Badía, X. 1982. 'Sobre as Orixes Da Orientación Exportadora Na Producción Bovina Galega. As Exportacións a Inglaterra Na Segunda Mitade Do Século XIX', *Grial*, Anexo 1, Historia: 169-206

——— and Leonor de la Puente Fernández. 1988. 'Crisis agraria y vías de evolución ganadera en Galicia y Cantabria', in *La crisis agraria de finales del siglo XIX: (I Seminari Internacional d'historia de Girona)*, ed. by Ramón Garrabou i Segura (Crítica), pp. 181-211

Colegio Oficial de Veterinarios de Valencia. 1911. *Veterinarios Españoles* (Valencia: Imprenta de Soler, García y Compañia)

Conde Gómez, Diego 2013. 'Veterinaria e Mellora Pecuaria Na Galicia Contemporánea. O Papel de Juan Rof Codina'

——— 2014. *Canto Val Unha Vaca? Da Cuestión Agraria á Cuestión Pecuaria* (A Coruña: Deputación Provincial da Coruña)

——— 2015. *Juan Rof Codina: Renovación Na Veterinaria e Gandaría Galega Contemporánea* (Vigo: Xerais)

——— 2018. *Cuestiones Sobre La Mala Leche. Aportación de La Profesión Veterinaria a La Mejora Higénico Sanitaria de La Leche En España (1837-1931)* (Madrid: Real Academia de Ciencias Veterinarias de España)

Conde Gómez, Diego, José Manuel Cifuentes Martínez, and Lourenzo Fernández Prieto. 2013. 'Primeras Actuaciones Veterinarias Frente La Tuberculosis Bovina En Galicia (1900-1925). El Papel de Juan Rof Codina', in *Arte y Veterinaria. La Veterinaria En Las Ciudades. XIX Congreso Nacional y X Iberoamericano de Historia de La Veterinaria* (Madrid), pp. 321-26

Conde Gómez, Diego, and Lourenzo Fernández Prieto. 2019. 'Transformaciones Higiénico-Sanitarias Derivadas de La Especialización de La Producción Láctea Española, 1900-1970', in *Leche y Lecheras En El Siglo XX. De La Fusión Innovadora Organica a La Revolución Verde*, ed. by Lourenzo Fernández Prieto and Daniel Lanero Táboas (Zaragoza: Prensas Universitarias de Zaragoza), pp. 95-118

'Conferencia Sobre Higiene General'. 1924. *El Regional. Diario de Lugo* (Lugo), p. 2

Conseil Général Vétérinaire, and Glavany. 2006. *Histoire des services vétérinaires français* (Amicale du Conseil Général Vétérinaire)

Cordero del Campillo, Miguel. 1973. 'Félix Gordón Ordás (1885-1973)', in *Semblanzas Veterinarias: Volumen I* (León: Syva), pp. 287-333

———, Miguel. 1984. 'Prólogo', in *La Albeyteria Española En El Siglo XVIII* (Salamanca), pp. 11-13

———, Miguel. 2003. 'Félix Gordón Ordás y La Reivindicación Veterinaria', in *Conferencia Pronunciada En El Ciclo Conmemorativo Del I Centenario Del Colegio Oficial de Veterinarios de Madrid* (Madrid) <https://historiadelaveterinaria.es/wp-content/uploads/2003-Felix-Gordon-Ordas-y-la-reivindicacion-veterinaria.pdf>

———, Miguel. 2004. *Félix Gordón Ordás (1885-1973)* (León: Instituto Leonés de Cultura)

———, Miguel, Miguel Ángel Márquez, and Benito Madariaga de la Campa. 1996. *Albeytería, Mariscalía y Veterinaria: (Origenes y Perspectiva Literaria)* (León: Universidad de León, Secretariado de Publicaciones)

———, Miguel, Carlos Ruiz Martínez, and Benito Madariaga de la Campa. 1973. *Semblanzas Veterinarias: Volumen I* (León: Syva)
Corley, T. A. B., and Andrew Godley. 2011. 'The Veterinary Medicine Industry in Britain in the Twentieth Century', *The Economic History Review*, 64.3 (Wiley): 832-54
'Cuatro Palabras'. 1911. *Revista de Higiene y Sanidad Veterinaria*, 1.1: 1
De Hoyos Puente, Jorge. 2018a. 'Félix Gordón Ordás (1885-1973): la persistencia republicana en el exilio', in *Activistas, militantes y propagandistas: Biografías en los márgenes de la cultura republicana (1868-1978)* (Granada: Athenaica), pp. 375-94
———, Jorge. 2018b. 'Un embajador en Guerra: Gordón Ordás en México', *Revista Universitaria de Historia Militar*, 7.13 <http://ruhm.es/index.php/RUHM/article/view/393> [accessed 19 June 2018]
'Decreto Disponiendo Que Las Inspecciones Generales de Higiene y Sanidad Pecuarias y de Sanidad Veterinaria, Radicantes, Respectivamente, En Los Ministerios de Economía y Gobernación Queden Incorporadas a La Dirección de Ganadería e Industrias Pecuarias de Este Ministerio, y Nombrando, Con Carácter Provisional, Para Los Cargos Que Se Indican, a Los Señores Que Se Mencionan'. 1931. *Gaceta de Madrid*, p. 1201
Dehesa Santisteban, Francisco Luis, María Castaño Rosado, José Manuel Etxaniz Makazaga, and Joaquín Sánchez de Lollano Prieto. 2011. *Semblanzas Veterinarias: Volumen III* (Madrid: Consejo General de Colegios Veterinarios de España)
Domínguez Martín, Rafael, and Leonor de la Puente Fernández. 2009. 'Ganadería e Industrialización Láctea. El Complejo Ganadero Industrial En Cantabria En El Siglo XX': 143-79
'Ego Sum. Félix Gordón Ordás'. 1933. *Gutiérrez*, p. 1
Elvbakken, Kari Tove. 2017. 'Veterinarians and Public Health: Food Control in the Professionalization of Veterinarians', *Professions and Professionalism*, 7.2: e1806–e1806 <10.7577/pp.1806>
Enriquez, Carlos S. 1913. 'La Ganadería Montañesa. Los Concursos Pecuarios de Santander', *Revista de Higiene y Sanidad Veterinaria*, 3.9: 533-75
Etxaniz Makazaga, José Manuel. 2003. *Félix Gordón Ordás y Sus Circunstancias: Apuntes Para Su Biografía* (León: Fundación Vela Zanetti)
Fernández Pousa, Ramón. 1968. 'Juan Rof Codina y El Agro Gallego', in *Hombres de Galicia*, Temas Españoles, 486 (Madrid: Publicaciones Españolas), pp. 27-28
Fernández Prieto, Lourenzo. 1985. 'Unhas Notas Sobre a Vida e Obra de Juan Rof Codina (1874-1967)', in *Reformas Que Se Pueden Implantar En Galicia Para El Progreso de La Agricultura*, Edición facsímil (Sada (A Coruña): Ediciós do Castro), pp. 9-21
———, Lourenzo. 1992. *Labregos Con Ciencia: Estado, Sociedade e Innovación Tecnolóxica Na Agricultura Galega, 1850-1939* (Vigo: Edicións Xerais de Galicia)
———, Lourenzo. 1996. 'Notas Sobre a Gandería de Galicia Na Segunda Metade Do Século XIX', in *Humanitas. Estudios En Homenaxe ó Prof. Dr. Carlos Alonso Del Real* (Santiago de Compostela: Universidade de Santiago de Compostela), pp. 697-724
———, Lourenzo. 2007. *El Apagón Tecnológico Del Franquismo: Estado e Innovación En La Agricultura Española Del Siglo XX* (Valencia: Tirant lo Blanch)

———, Lourenzo. 2017. 'El Proceso Económico', in *Historia Contemporánea de España (Tomo I: 1808-1930)*, ed. by Jordi Canal (Barcelona: Taurus), I, pp. 687-739

Florencio Puntas, Antonio. 2004. *Ingenieros Agrónomos, Cambio Institucional e Innovación Tecnológica de La Agricultura Andaluza Contemporánea*, Sociología Agraria (Seville: Junta de Andalucía, Consejería de Agricultura y Pesca)

Galindo García, Francisco. 1973a. 'Cayetano López López (1860-1970)', in *Semblanzas Veterinarias* (León: Syva), I, pp. 335-63

———, Francisco. 1973b. 'Cesáreo Sanz Egaña (1885-1959)', in *Semblanzas Veterinarias* (León: Syva), I, pp. 265-86

García Partida, Paulino. 2009. *Creación de La Dirección General de Ganadería: Su Entorno Político* (Madrid: Ministerio de Medio Ambiente y Medio Rural y Marino, Centro de Publicaciones)

Gómez Piquer, José, and José Manuel Pérez García. 2000. *Crónica de 150 Años de Estudios Veterinarios En Aragón (1847-1997)* (Zaragoza: Institución Fernando el Católico, Excma. Diputación de Zaragoza)

González Blasco, Pedro, José Jiménez Blanco, and José María López Piñero. 1979. *Historia y Sociología de La Ciencia En España*, Alianza Universidad; 251 (Madrid: Alianza Editorial)

Gordón Ordás, Félix. 1910. 'Plumada', *Revista Veterinaria de España*, 4.11: 423-25

Gutiérrez García, José Manuel. 2004. 'La Tuberculosis Bovina Como Zoonosis En La España Contemporánea (1850-1950)' (unpublished Tesis Doctoral, Barcelona: Universitat Autònoma de Barcelona)

———, ———. 2006a. 'Meat as a Vector of Transmission of Bovine Tuberculosis to Humans in Spain: A Historical Perspective', *Veterinary Heritage: Bulletin of the American Veterinary History Society*, 29.1: 25-27

———, ———. 2006b. 'Milk as a Vector of Transmission of Bovine Tuberculosis to Humans in Spain: A Historical Perspective', *Veterinary Heritage: Bulletin of the American Veterinary History Society*, 29.2: 41-44

———, ———. 2010. 'Laboratory Medicine and the Identity Change of Veterinary Medicine in Spain at the Turn of the 20th Century', *Dynamis: Acta Hispanica Ad Medicinae Scientiarumque Historiam Illustrandam*: 239-60 <10.4321/S0211-95362010000100010>

Hellberg, Inga. 1990. 'The Swedish Veterinary Profession and the Swedish State', in *The Formation of Profession. Knowledge, State and Strategy*, ed. by Rolf Torstendahl and Michael Burrage (London: SAGE), pp. 174-85

Hernández Adell, Ismael. 2012. 'La Difusión de Un Nuevo Alimento: Producción y Consumo de Leche En España, 1865-1936'. (Universitat Autònoma de Barcelona)

Hubscher, Ronald. 1999. *Les Maîtres Des Bêtes: Les Vétérinaires Dans La Société Francaise, (XVIIIe-XXe Siècle)* (Paris: Odile Jacob)

Jones, Susan D. 2003. *Valuing Animals: Veterinarians and Their Patients in Modern America*, Animals, History, Culture (Baltimore: Johns Hopkins University Press)

Koolmees, Peter A. 1999. 'The Development of Veterinary Public Health in Western Europe, 1850-1940', *Sartoniana*, 12: 153-79

Kuhn, Thomas S. 1962. *The Structure of Scientific Revolutions* (Chicago: University of Chicago Press)

———, and Ian Hacking. 2012. *The Structure of Scientific Revolutions*, Fourth edition (Chicago; London: The University of Chicago Press)

'Labor Positiva. Los Inspectores de Higiene Pecuaria En Acción'. 1911a. *Revista de Higiene y Sanidad Veterinaria*, 1.6: 182-83

'Labor Positiva. Los Inspectores de Higiene Pecuaria En Acción'. 1911b. *Revista de Higiene y Sanidad Veterinaria*, 1.5: 144-48

'Labor Positiva. Los Inspectores de Higiene Pecuaria En Acción'. 1912. *Revista de Higiene y Sanidad Veterinaria*, 1.10: 310

'Labor Positiva. Los Inspectores de Higiene Pecuaria En Acción'. 1913. *Revista de Higiene y Sanidad Veterinaria*, 2.12

Laín Entralgo, Pedro. 1981. *Historia Universal de La Medicina*, VII vols (Barcelona: Salvat)

'Ley Disponiendo Se Agrupen En Una Dirección General de Ganadería e Industrias Pecuarias, Dependiente de Este Ministerio, Todos Los Servicios Relacionados Con El Estudio y Aplicación de La Producción, Explotación, Mejora, Industrialización, Profilaxis y Tratamiento de Los Animales y Sus Productos, Con Excepción de Los Que Desarrolla El Cuerpo de Veterinaria Militar'. 1931. *Gaceta de Madrid*, p. 1419

'Los Progresos Del Campo Gallego'. 1911. *Vida Gallega*: 22

Luque, Jesús. 1911. 'Por La Ganadería', *Revista de Higiene y Sanidad Veterinaria*, 1.4: 111-13

———, and Francisco Pastor. 1913. 'La Ganadería En La Provincia de Guipúzcoa', *Revista de Higiene y Sanidad Veterinaria*, 3.6: 321-

Mainer, José-Carlos. 1981. *La Edad de Plata (1902-1939): Ensayo de Interpretación de Un Proceso Cultural* (Madrid: Cátedra)

Marco, José María. 1998. *Manuel Azaña: Una Biografía* (Barcelona: Planeta)

Martínez López, Alberte. 2000. 'Perspectiva Histórica Da Gandería Galega: Da Complementariedade Agraria á Crise Da Intensificación Láctea (1850-1995)', in *Terra e Progreso: Historia Agraria Da Galicia Contemporánea.*, ed. by Lourenzo Fernández Prieto (Vigo: Edicións Xerais de Galicia), pp. 353-81

Martínez Morás, Fernando. 1917. 'Los Hombres de Valía. Un Bienhechor de Galicia', *La Voz de Galicia* (A Coruña), pp. 1-2

Mendizabal Aizpuru, J. A., J. R. Ibarbia Barreras, and José Manuel Etxaniz Makazaga. 2005. 'Aportaciones a La Historia de La Raza Vacuna Pirenaica. Paradigma de La Zootecnia Española', *Archivos de Zootecnia*: 39-50

Mitsuda, Tatsuya. 2017. 'Entangled Histories: German Veterinary Medicine, *c.* 1770-1900', *Medical History*, 61.1 (Cambridge University Press): 25-47 <10.1017/mdh.2016.99>

Moreno Fernández Caparrós, Luis Ángel 2002. 'Aportación a La Historia de La Inseminación Artificial Ganadera En España: Su Significado En El Desarrollo Pecuario y La Repercusión Económica En El Período 1931-1971' (Madrid: Universidad Complutense de Madrid)

——— 2004. 'Causas Que Llevaron a La Adopción e Implantación de La Inseminación Artificial Ganadera En España', in *Temas de Historia de La Veterinaria*, ed. by José Manuel Cid Díaz (Murcia, Spain: Universidad de Murcia), pp. 125-56

——— 2000. 'Los Antecedentes Históricos de La Dirección General de Ganadería', in *Temas de Historia de La Veterinaria II* (Murcia: Universidad de Murcia), pp. 83-124

'Orden Designando La Comisión Que Se Indica Para Transferencia a Este Ministerio de Todos Los Servicios Referentes a Cría Caballar'. 1931. *Gaceta de Madrid*, p. 1657

Pan-Montojo González, Juan Luis. 2005. *Apostolado, Profesión y Tecnología: Una Historia de Los Ingenieros Agrónomos En España* (Madrid: Asociación Nacional de Ingenieros Agrónomos)

Pérez y Pérez, Dionisio, Miguel Cordero del Campillo, Jose del Rio Rodriguez, and Paulino García Partida. 1986. *Homenaje a Félix Gordón Ordás En El Centenario de Su Nacimiento* (León: Universidad de León)

Polanyi, Karl. 1944. *The Great Transformation*, (New York; Toronto: Farrar & Rinehart, Inc.)

Ponte Hernando, Fernando. 2016. *Contribución a La Salud Pública y a La Lucha Antituberculosa de Una Figura Olvidada: El Dr. D. Marcelino Ramírez García (1864-1940) Veterinario Militar y Médico* (Santiago de Compostela)

Prado Rodríguez 'Lameiro', Xavier. 1918. 'Dificultades Que Entraña El Diagnóstico de Las Enfermedades Infectocontagiosas de Los Ganados y Necesidad de Que Los Ganaderos Conozcan Las Reglas Que Deben Poner En Práctica Para Facilitarle' (presented at the Semana Agrícola de Santiago, Santiago de Compostela)

Pujol Andreu, Josep, Manuel González de Molina, Lourenzo Fernández Prieto, Domingo Gallego Martínez, and Ramón Garrabou i Segura. 2001. *El Pozo de Todos Los Males: Sobre El Atraso En La Agricultura Española Contemporánea* (Barcelona: Crítica)

Pumarola i Battle, Martí. 2002. 'Meat Sanitary Control in Spain during the XIXth Century', *Historia Medicinae Veterinariae*, 27.1-4: 259-71

Rodríguez Troncoso, Manuel. 2007. *Juan Rof Codina. Manescal Exemplar* (Prat de LLobregat: Quaderns del Prat)

Rof Codina, Juan. 1911a. *Cartilla Contra La Glosopeda (Gripo)* (A Coruña: Consejo Provincial de Fomento)

———, Juan. 1911b. 'Glosopeda', *Prácticas Modernas e Industrias Rurales*, 9.191: 8-10

———, Juan. 1911c. 'La Glosopeda En Galicia', *La Voz de Galicia* (A Coruña), p. 1

———, Juan. 1911d. 'La Glosopeda Benigna En Bovinos', *La Voz de Galicia* (A Coruña), p. 1

———, Juan. 1912a. 'Estudio de 50 Vacas Gallegas Del Concurso de Ortigueira', *Revista Veterinaria de España*, 6.6: 208-22

———, Juan. 1912b. 'Hoja de Vulgarización Científica Para Evitar y Combatir La Glosopeda I', *Barbero Municipal*, p. 3

———, Juan. 1914. 'Concurso Nacional de Ganados. Especie Bovina. Estudio de Los Principales Grupos Indígenas de España', *Revista de Higiene y Sanidad Veterinaria*, 3.10-11: 627-723

———, Juan. 1915. *Los Concursos de Ganados Como Medio de Fomentar y Clasificar La Ganadería. Memoria Premiada Por La Asociación de Ganaderos Del Reino* (Madrid: Asociación General de Ganaderos)

———, Juan. 1917. *La Raza Bovina Gallega. Memoria Premiada Por La Asociación de Ganaderos Del Reino* (Madrid: Asociación General de Ganaderos)

———, Juan. 1919. 'El Comercio de Los Productos Agrícolas y Pecuarios En Galicia. Cómo Se Hace y Cómo Debe Organizarse', *Revista de Higiene y Sanidad Pecuarias*, 9.1-2: 1-35

———, Juan. 1921a. 'La Glosopeda Se Extingue', *El Ideal Gallego* (A Coruña), p. 6

———, Juan. 1921b. 'La Glosopeda Se Recrudece', *El Ideal Gallego* (A Coruña), p. 2

Ruiz Martínez, Carlos. 1973. 'Juan Rof Codina (1874-1967)', in *Semblanzas Veterinarias I* (León: Syva), pp. 195-218

Salgueiros, Juan. 1910. 'Una Buena Labor', *Prácticas Modernas e Industrias Rurales*, 8.184: 276-77

Salvador Velasco, Angel. 2015. *El inicio de la veterinaria en España* (Seville: Colegio Oficial de Veterinarios de Sevilla)

Sánchez Ron, José M., and José García-Velasco (eds.). 2010. *100 JAE: La Junta Para Ampliación de Estudios e Investigaciones Científicas En Su Centenario: Actas Del II Congreso Internacional Celebrado Los Días 4, 5 y 6 de Febrero de 2008* (presented at the Congreso Internacional '100 JAE: la Junta para Ampliación de Estudios e Investigaciones Científicas en su centenario', Madrid: Fundación Francisco Giner de los Río, Institución Libre de Enseñanza : Publicaciones de la Residencia de Estudiantes)

———, José M., and Antonio Lafuente (eds.). 2007. *El Laboratorio de España: La Junta Para La Ampliación de Estudios e Investigaciones Científicas (1907-1939)* (Madrid: Sociedad Estatal de Conmemoraciones Culturales : Residencia de Estudiantes)

———, José Manuel. 1999. *Cincel, Martillo y Piedra: Historia de La Ciencia En España (Siglos XIX y XX)* (Madrid: Taurus)

Sanz Egaña, Cesáreo. 1941. *Historia de La Veterinaria Española: Albeitería-Mariscalería-Veterinaria* (Madrid: Espasa Calpe)

Schönherr, W. 1991. 'History of Veterinary Public Health in Europe in the 19th Century', *Revue Scientifique Et Technique (International Office of Epizootics)*, 10.4: 985-94

'Sección Científica. Administración Veterinaria'. 1848. *Boletín Mercantil e Industrial de Galicia*, pp. 1-3

Seco Serrano, Carlos. 1973. *Historia de España*, 3ª, VI vols, ed. by Lluis Pericot García (Barcelona: Instituto Gallach de Librería y Ediciones)

Séculi Brillas, J. 1971. 'Evocación Del Academico Honorario Don Cayetano López y López', *Anales de Medicina y Cirugía*, 51.224: 135-39

Serrano Tomé, Vicente. 1973. 'Santos Arán San Agustín (1880-1970)', in *Semblanzas Veterinarias* (León: Syva), pp. 257-64

Smith, Donald F. 2013. 'Lessons of History in Veterinary Medicine', *Journal of Veterinary Medical Education*, 40.1: 2-11 <10.3138/jvme.1112.04>

Soto Fernández, David. 2006. *Historia Dunha Agricultura Sustentábel: Transformacións Produtivas Na Agricultura Galega Contemporánea* (Santiago de Compostela: Consellería do Medio Rural)

Steele, J. H. 1976. 'Veterinary Public Health in the United States, 1776 to 1976', *Journal of the American Veterinary Medical Association*, 169.1: 74-82

——— 2000. 'The History of Public Health and Veterinary Public Service', *Journal of the American Veterinary Medical Association*, 217.12: 1813-21

Strauss, William, and Neil Howe. 1991. *Generations: The History of America's Future, 1584 to 2069*, 1st Quill ed (New York: Quill)

Thomas, Hugh. 1976. *Historia de La Guerra Civil Española* (Barcelona: Grijalbo), ii

Vital Ruiberriz de Torres, Pedro L. 1982. 'Historia de La Ciencia Veterinaria Española: Del Antiguo Régimen al Liberalismo 1792-1847' (Madrid: Universidad Complutense de Madrid)

Vives Vallés, Miguel Ángel, and María Cinta Mañé Seró. 2016. *La veterinaria grecorromana* (Cáceres: Universidad de Extremadura)

———, ———, and ———. 2018. *El inicio de la medicina animal: del Neolítico a la cultura grecorromana* (Cáceres: Universidad de Extremadura)

———, ———, and ———. 2020. *Historia de la veterinaria medieval en Asia Central y Extremo Oriente*, Historia de la Veterinaria (Cáceres: Imanguxara), iii

Woods, Abigail. 2007. 'The Farm as Clinic: Veterinary Expertise and the Transformation of Dairy Farming, 1930-1950', *Studies in History and Philosophy of Science Part C: Studies in History and Philosophy of Biological and Biomedical Sciences*, Between the farm and the clinic: agriculture and reproductive technology in the twentieth century, 38.2: 462-87 <10.1016/j.shpsc.2007.03.009>

———. 2014. 'Science, Disease and Dairy Production in Britain, *c.* 1927 to 1980', *Agricultural History Review*, 62.2: 294-314

———, Michael Bresalier, Angela Cassidy, and Rachel Mason Dentinger. 2018. *Animals and the Shaping of Modern Medicine: One Health and Its Histories.* (Place of publication not identified: Springer International)

Appendix

Board of provincial inspectors of livestock hygiene and veterinary health, ports and borders

Table 10.1. List of the first group of Inspectors of Livestock Hygiene and Veterinary Health, Ports and Borders (Colegio Oficial de Veterinarios de Valencia 1911)

POST	NAME	LOCATION	PROVINCE
First-class Inspectors of Livestock Hygiene and Veterinary Health	Mr. Félix Gordón Ordás	Madrid	Madrid
	Mr. Juan Rof Codina	A Coruña	A Coruña

POST	NAME	LOCATION	PROVINCE
	Mr. Santos Arán San Agustín	Seville	Seville
	Mr. Cayetano López y López	Barcelona	Barcelona
	Mr. José Orensanz y Moliné	Valencia	Valencia
	Mr. Celestino-Luis Jiménez Vicente	Granada	Granada
	Mr. Juan Bautista Montserrat Foncuberta	Cádiz	Cádiz
	Mr. Cesáreo Sanz Egaña	Málaga	Málaga
Second-class Inspectors of Livestock Hygiene and Veterinary Health	Mr. Publio Fructuoso Cordeque Gómez	Zaragoza	Zaragoza
	Mr. Salvador Esteban Martí Güell	Murcia	Murcia
	Mr. José Ródano Gómez	Toledo	Toledo
	Mr. Balbino Sanz García	Valladolid	Valladolid
	Mr. Emilio Aramburu Ibáñez	Alicante	Alicante
	Mr. Francisco Abril Brocas	Oviedo	Oviedo
	Mr. Juan Bort Cerdán	Burgos	Burgos
	Mr. José Mª Beltrán Monferrer	Córdoba	Córdoba
Third-class Inspectors of Livestock Hygiene and Veterinary Health	Mr. José García Buela	Pontevedra	Pontevedra
	Mr. Félix Núñez Méndez	León	León
	Mr. Martín Ciga Lecuna	Bilbao	Biscay

POST	NAME	LOCATION	PROVINCE
	Mr. Tomás Rota Minondo	Valcarlos border post	Navarre
	Mr. Jesús Luque Artó	Donostia	Gipuzkoa
	Mr. Pascual Luna López	Pamplona	Navarre
	Mr. Diego María Martín Ortiz	Ciudad Real	Ciudad Real
	Mr. Joaquín Castellanos García	Albacete	Albacete
	Mr. Manuel Prieto Briones	Salamanca	Salamanca
	Mr. Emiliano Sierra Sierra	Jaén	Jaén
	Mr. Antonio Bosch Miralles	Palma	Baleares
	Mr. Enrique Arcniega Cerrada	Soria	Soria
	Mr. Domingo Aisa Sánchez de Ávila	Huesca	Huesca
	Mr. Luís Núñez Herrero	Ávila	Ávila
	Mr. Félix Fernández Turégano	Cuenca	Cuenca
	Mr. Pelayo Jiménez de la Torre	Cáceres	Cáceres
	Mr. Victoriano López Guerrero	Badajoz	Badajoz
	Mr. Juan Verdaguer Estrach	Girona	Girona
	Mr. Arturo Anadón Piris	Lleida	Lleida
	Mr. Mateo Arciniega y Añastro	Vitoria-Gasteiz	Álava
	Mr. Esteban Gabín Abadía	Canfranc border post	Huesca

POST	NAME	LOCATION	PROVINCE
	Mr. Guillermo Manuel Moreno Amador	Huelva	Huelva
	Mr. Carlos Díez Blas	Teruel	Teruel
	Mr. Carlos Santiago Enríquez	Santander	Santander
	Mr. Francisco Pastor Calvo	Irún border post	Gipuzkoa
	Mr. Froilán Fernández Silva	Zamora	Zamora
	Mr. Javier Prado Rodríguez (Prado "Lameiro")	Ourense	Ourense
	Mr. Rufino Portero López	Segovia	Segovia
	Mr. Juan Miralles Más	Castellón	Castellón
	Mr. Felipe Gómez Chamorro	Logroño	Logroño
	Mr. Lázaro Lechuga Millán	Almería	Almería
	Mr. Ángel Martín Puebla	Guadalajara	Guadalajara
	Mr. Pedro Rosell Vilá	Puigcerdá border post	Girona
	Mr. Fidel Ruiz de los Paños Corbacho	Palencia	Palencia
	Mr. Agustín Fornells Plana	Las Palmas	Canary Islands
	Mr. Antonio Panés Rodríguez	Tarragona	Tarragona
	Mr. Niceto-José García Armendáritz	Lugo	Lugo
	Mr. Antonio Moraleda Burillo	Cartagena border post	Murcia

POST	NAME	LOCATION	PROVINCE
	Mr. Martín Lázaro Calvo	Tui border post	Pontevedra
	Mr. Andrés Benito García	Port Bou border post	Girona
	Mr. Santiago Herrero González	Valencia de Alcántara border post	Cáceres
	Mr. Román Ergueta Sanz	Verín border post	Ourense
	Mr. José Rubio García	Villanueva del Fresno border post	Badajoz
	Mr. Francisco Castillo Estremera	La Fresgeneda border post	Salamanca
	Mr. Severo Curiá Martínez	Benasque border post	Huesca
	Mr. Protasio García Salmerón y Rodríguez	Alcañices border post	Zamora
	Mr. Teodoro José Moreno Amador	Farga de Moles border post	Lleida
	Mr. Manuel Moreno Amador	Paymogo border post	Huelva

KRISTINA NORDEUS

11. When science is not enough

Antibiotic Growth Promoters and the precautionary principle within the European Union

I Introduction

Antibacterials made their way into both human and veterinary medicine in the early 20th Century (see Aminov 2010 for review). The first antibacterials were synthetic substances, like Salvarsan (1909) and Prontosil (1935). They were soon supplemented with naturally occurring antibiotics – the first being penicillin which was discovered by chance in 1928 and used clinically from the 1940s. Signs of antibiotic resistance started appearing soon after the substances came into use in human and veterinary medicine.

The positive effect of antibiotics at subtherapeutic concentrations on growth was soon discovered, and AGPs were approved for use in feed for production animals in Europe in the early 1950s (Castanon 2007; Cogliani and others 2011). This was considered a 'technical use' as opposed to medical use, i.e., administering higher concentrations of antibiotics to treat or prevent disease (Greko, 2010, p. 232). Even so, their use was not uncontroversial from a resistance perspective. Following a severe outbreak of food-borne salmonellosis of bovine origin in Great Britain, a committee charged with investigating the use of antibiotics in animal husbandry and veterinary medicine was appointed (Bellerby 1970, p. 50). In 1969, the so-called Swann Committee issued a report concluding that 'the administration of antibiotics to farm livestock, particularly at sub-therapeutic levels, poses certain hazards to human and animal health'. The Committee made several recommendations – most notably that only substances with little or no application as therapy in human or veterinary medicine should be used as growth promoters (Joint Committee on the use of Antibiotics in Animal Husbandry and veterinary Medicine 1969). In the wake of this report, some antibiotic substances were removed from use as AGPs in Europe, such as tetracycline and penicillin (Castanon 2007; Cogliani and others 2011), but the practice persisted.

Kristina Nordeus • Department of Urban and Rural Development, Agrarian and Rural History, Swedish University of Agricultural Sciences

From Breeding & Feeding to Medicalization, Rural History in Europe, 17 (Turnhout: Brepols, 2023), pp. 267–286
BREPOLS PUBLISHERS 10.1484/M.RURHE-EB.5.135029

Based on official and working documents, as well as scientific literature, this paper looks at the transformation of the EU legislation on AGPs, starting from the accession of Sweden in 1995, and what role Sweden played in this process. The focus is primarily on the main contributions by Swedish actors, and events in the institutions of the European Union. As such, it does not provide an exhaustive account of all events that were part of the process leading to the ban. While the issue of antibiotics in the food chain in EU member states have been described thoroughly elsewhere (e.g., Kahn 2016, Kirchelle 2020), this paper provides new insight into Swedish efforts and developments in the EU institutions, as well as on the process of EU decision making.

The second part gives an overview of the Community legislation governing AGPs, and the process leading to the Swedish ban, as well as the outcome of the Swedish membership negotiations with the EU. The following parts deal with Swedish efforts to introduce a similar ban at the EU-level, and developments in the EU institutions, as well as the opposition of such a ban. The final discussion aims at analysing why a ban was finally introduced, and how the process differed from the Swedish ban a decade earlier.

The introduction and codification of the precautionary principle into Community food safety legislation following food crises in the Union, most notably the 'mad cow disease', was important for this process. The precautionary principle emerged as an explicit concept within environmental science in the 1970s (Gee and others 2001). The purpose was to establish a policy tool to be able to deal with potentially serious threats to health or the environment before the risk could be properly assessed and quantified.

Another important contextual process was the US-EU hormone dispute. Mainly due to consumer concerns and adverse impact on the single market, the EU introduced bans on the use of hormones for growth promotion in animal production in the 1980s. The US retaliated by imposing duties on EU-imports. The issue resulted in a WTO dispute between the US and the EU, where in 1997 the WTO Dispute Settlement Panel found the hormone ban in breach of the SPS Agreement since it was not based on risk assessment evaluating impacts on human health. The ruling was upheld by the WTO Appellate Body in 1999 (Kastner and Pawsey 2002; Kleinman and Kinchy 2003).

II. AGPs in Sweden and the EU

The view that AGPs were different from therapeutic antibiotics was reflected in the preamble of the legal act governing feed additives in the EU, Directive 70/524/EEC, where it was stated that 'antibiotics should also be regarded as additives since, when used in small quantities, they have a physiological nutritional

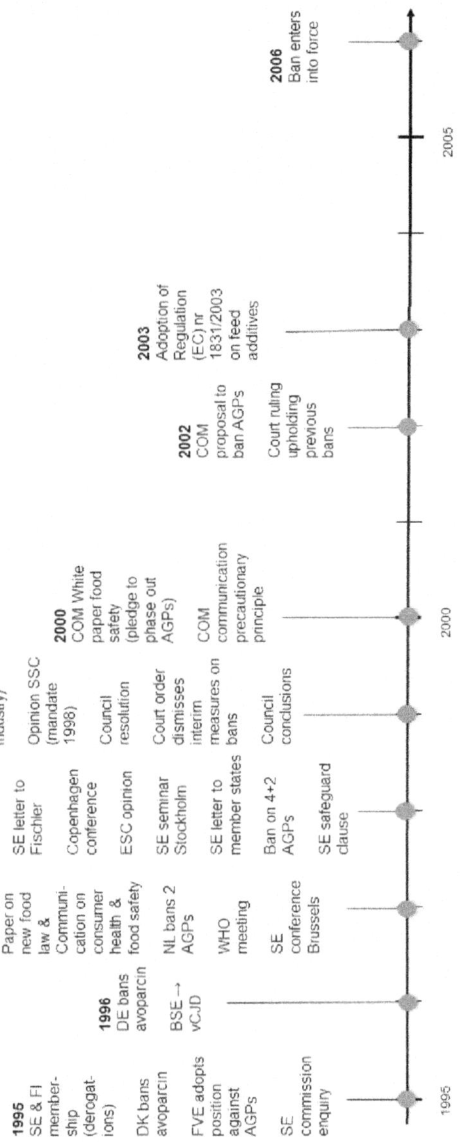

Figure 11.1. Timeline

effect, although when used in large quantities they have a medicinal effect'.[1] According to Article 6 of the Directive, substances should only be authorised for use if 'at the level permitted in feeding-stuffs, treatment or prevention of animal disease is excluded' and if 'for serious reasons concerning human or animal health its use must not be restricted to medical or veterinary purposes'. As Butaye and others (2003) point out, the Swann report formed the basis for these provisions.

Article 7 of the Directive contained a 'safeguard clause', allowing member states to unilaterally suspend the use of an additive for a period of four months if they believed it might endanger animal or human health. The Council of the European Union should then decide, on proposal from the Commission, whether the additive should be removed from the list or its use be restricted.[2]

In Sweden the rules governing AGPs were tightened in the late 1970s. There was some hesitation about the use of AGPs among involved Government agencies, but with these stricter rules for authorisation and the removal of some substances, the practice could continue. AGPs were routinely incorporated in almost all pig and broiler feed and their use was increasing drastically, even after the introduction of stricter rules. In 1981, one of Sweden's leading daily newspapers published a series of articles on the topic of Swedish agriculture. At that time Sweden's agricultural market was heavily regulated, with fixed prices for several products, problems with surplus production and high food prices. The cooperative farmer organisations, under the umbrella of the Federation for Swedish Farmers (LRF), controlled most of the agri-market. Among other things, the articles highlighted the extensive use of antibiotics in Swedish animal production. This sparked an intensive debate regarding farm animal welfare and the use of antibiotics. From fear of losing consumer confidence, LRF adopted a policy advocating restrictive use of antibiotics in 1981. In 1984, the LRF general assembly voted in favour of requesting a compulsory ban on AGPs and there were also parliamentary bills on the same subject. In 1985, a new feed law was passed and the use of antibiotics in animals, for any other reason than to cure, prevent or alleviate symptoms of disease, was banned. The ban entered in to force in 1986 (Nordéus 2019; Nordéus, forthcoming). Sweden became the first country to regulate withdrawal of AGPs in food animal production (Cogliani and others, 2011).

Not all actors were pleased with the Swedish ban. It coincided with the introduction of a new strict animal welfare act in 1988 and these two pieces of legislation forced the sector to undergo an extensive transformation (see

1 Council Directive of 23 November 1970 concerning additives in feeding-stuffs (70/524/EEC).

2 Article 11 in amended versions of the Directive. The European Commission consists one commissioner from each member state and has sole right of initiative, i.e. to put forwards proposals. The Council of the European Union consists of Government ministers from each member state, according to the policy area to be discussed. The European Parliament consists of directly elected Members (MEPs). Before 1992, the Council held the legislative powers, but in 1992 co-decision between Council and Parliament was introduced and further strengthened in 1999 and now most European laws are adopted jointly by the two institutions.

also Waluszewski, chapter 12 in this volume). Nearly a decade after the ban was introduced and Sweden was preparing for EU-membership, some degree of consensus regarding the ban had been reached. This was manifested at a seminar on antibiotic use and resistance, hosted by cooperative industry associations. The report from the seminar stated that although the ban on AGPs had initially been a measure aimed at consumers, the benefits of a restrictive use of antibiotics in form of reduced development of resistance were becoming increasingly clear (*Lantmännen Foderutveckling AB* and *Scans djuromsorgsnämnd* 1993). To prepare for membership negotiations, the Swedish Board of Agriculture compiled arguments to support the Swedish negotiation positions. Regarding AGPs it was stated that a non-restrictive use of antibiotics would pose risks to human and animal health, and the environment. Furthermore, measures introduced by farmers to maintain production levels after the ban had resulted in better animal environments, feed and management, but had also come at a cost to farmers and consumers. Despite the cost it was still 'public opinion in Sweden [...] that antibiotics [...] should only be used for medicinal purposes after prescription by a veterinary surgeon'.[3]

In the 1994 Treaty of Accession between EU member states and Austria, Finland, and Sweden, the three countries were granted several exemptions from Directive 70/524/EEC (Official Journal of the European Communities C 241 1994, pp. 330-31). Sweden could maintain its ban on AGPs until 31 December 1998. Within this time limit a decision would be taken on requests for adaptation presented by Sweden in accordance with the safeguard clause: either the EU would change its rules or Sweden would have to lift its ban. Finland was similarly allowed to maintain its ban on certain AGPs.[4]

III. Working together to defend the Swedish model

Sweden joined the EU in 1995 and this marked the start of a period of intense lobbying on the topic of AGPs, involving various Swedish actors. Even if the Swedish ban was introduced as a measure to increase consumer confidence, it was legally justified by concern for human and animal health – although the risk of resistance development could not be quantified (*Regeringens proposition* 1984/85: 149). Given the limited scientific basis for the Swedish ban, it was an arduous task to provide scientific evidence as to why the ban should be maintained in Sweden and introduced in EU legislation. Consequently, the first four years as a member of the EU were characterized by extensive efforts to compile scientific evidence of, and spread knowledge about, the connection between use of AGPs and development of resistance to antibiotics used in human and veterinary therapy.

---

3 Document in author's possession.

4 Finland could maintain its ban on avoparcin for dairy cows, tylosin phosphate and spiramycin, as well as antibiotics with corresponding effects, until the end of 1997.

In 1995, Swedish Minister of Agriculture Winberg commissioned a two-year enquiry charged with compiling scientific findings that could serve as grounds for the Swedish application. The enquiry resulted in a 350-page report in English.[5] Among the main conclusions of the report, it stated that antibacterial feed additives had favourable economic effects, but from a long-term perspective the effects were questionable, especially regarding animal welfare and animal health. At levels permitted in the Directive, AGPs could be used for the treatment or prevention of disease which was in violation of the Directive. Also, some substances were toxic to humans or animals. Furthermore, the risk of increased resistance associated with the use of AGPs was far from negligible, with potentially serious consequences for animal and human health. Antibacterials not presently used therapeutically in humans or animals could become valuable templates for future drugs. The overall conclusion was that the benefits did not outweigh the risks, and that antibiotics should be restricted to medical and veterinary purposes (Government Official Reports 1997:132, p. 1).

In November 1997 the report was presented at a conference in Brussels arranged by the Ministry (Swedish Ministry of Agriculture 1998). Invitations were issued to EU ministries, farmers' unions, consumer organisations and animal welfare groups, as well as to representatives of the feed additives industry and other interested parties and there were over 200 attendees. The brochure 'Can we use less antibiotics?' was produced by the Ministry of Agriculture (1997) and distributed to the participants. It contained information about antibiotic use and development of antimicrobial resistance, as well as the Swedish ban and its effects in terms of animal health problems and increased costs. It was however stated that 'over time, the Swedish approach pays for itself'. It provided information about the 'Swedish model' of animal husbandry: the need for a horizontal approach to animal welfare, housing conditions, management, preventive health work and antibiotic use. As a testimony to the collective approach, the brochure was prepared by the Ministry, in collaboration with the Board of Agriculture and 'representatives of numerous organizations' and featured quotes from scientists, governmental agencies, consumer organisations and industry representatives. The outcome of the conference was documented in a booklet in which the new Minister of Agriculture Åhnberg expressed hope that 'common understanding [had been reached] on the importance of restrictive use of antibiotics' (Swedish Ministry of Agriculture 1998).

The only possibility for Sweden to maintain its ban on AGPs, after the four-year exemption period, was to convince the EU to introduce the same rules. Consequently, Sweden submitted an application to the European Commission in February 1998, requesting an adaptation of the EU legislation in accordance with

---

5 Sweden had been granted other exemptions regarding feed additives. These feed additives, mainly coccidiostats (veterinary medicines approved as feed additives for prevention of coccidiosis mostly in poultry), also fell within the remit of the report but they are outside the scope of this paper.

the Swedish ban.[6] The report of the commission enquiry constituted the scientific justification for such a ban.

Later that year, the Ministry also arranged a two-day seminar and study visit in Stockholm for representatives from EU ministries and the European Commission. The aim was to spread information about AGP-free animal production, and the speakers included representatives from industry, academia, and state agencies (Ministry of Agriculture 1999).

The issue of AGPs was also gaining momentum outside of Sweden and Finland. In many of these European initiatives, however, there was often a Swedish presence. Representatives from the Swedish Veterinary Association worked hard to raise awareness about the issue within the Federation of Veterinarians of Europe (FVE). In November 1995 the FVE General Assembly voted on a Danish proposal, in favour of adopting a statement that antibiotics should only be used therapeutically, and not as AGPs. The vote was preceded by an animated debate and the Assembly was divided. In the preparatory discussions in the Union of European Veterinary Practitioners, French and British delegates were among those opposed to the proposal, and at the General Assembly, the strongest resistance came from the Federation of European Veterinarians in Industry and Research. The lack of scientific evidence was the main argument against the position (Tronstad 1995).

A WHO (World Health Organisation) meeting on the topic of the medical impact of antimicrobial use in food animals, held in October 1997, resulted in recommendations which stated that 'increased concerns regarding risk to public health resulting from the use of antimicrobial growth promoters indicate that it is essential to have a systematic approach towards replacing growth-promoting antimicrobials with safer non-antimicrobial alternatives'. Sweden was represented by veterinary professors Franklin and Wierup who spoke about the status of antibiotic resistance in Swedish animal production (WHO, 1997).

In 1998, the Economic and Social Committee of the European Communities (ESC) approved an opinion on antibiotic resistance as a threat to public health. The chairman of the Swedish Consumers' Association, Turid Ström, initiated the process and acted as rapporteur for the opinion. It was prepared by an interdisciplinary group and covered topics in both human and veterinary medicine. According to the opinion, administration of antibiotics to animals could select for resistant bacteria which could spread to humans via the food chain. The use of antibiotics in animals should be limited to veterinary medical purposes, and AGPs should be replaced with safer, non-antimicrobial alternatives (Economic and Social Committee of the European Communities 1998).

---

6 Mentioned in a Governmental Decision dated 22 December 1998, document in author's possession.

IV. Developments in the EU Legislation

In 1997 the use of the AGP avoparcin was temporarily banned through the Commission Directive 97/6/EC.[7] In 1995 and 1996 respectively, Denmark and Germany had prohibited the use of avoparcin within their territories, arguing that avoparcin caused resistance to other glycopeptides used in human medicine. In its opinion in May 1996, the Scientific Committee on Animal Nutrition had concluded that it was not possible to establish cause and effect, and consequently it was not necessary to preserve the use of glycopeptides exclusively for human medicine. The Commission on the other hand concluded that 'in this climate of uncertainty it is preferable to show extreme caution' and prohibited the use of avoparcin as an interim protective measure, to be re-examined before 1999 (Commission 1997a).

Further changes were in the making. In the late 1990s, it had become evident that the food safety system within the EU was flawed (Ugland and Veggeland 2006; Vos 2000). In 1996 scientific evidence indicated that that the prions causing bovine spongiform encephalopathy (BSE or mad cow disease) could be transmitted to humans via the food chain and cause the deadly neurological variant Creutzfeldt-Jakob Disease (vCJD). Even though the use of specified bovine offal was banned in all human food as a precautionary measure in 1989, it was found in 1995 that the ban was not being strictly enforced (Ainsworth and Carrington 2000). In response to these failures, the Commission published a Green Paper on European Food Law in April 1997, stating that the BSE crisis had highlighted the need for a new food policy to ensure that only foodstuffs that were safe, wholesome and fit for consumption, could be placed on the market. Major efforts would be deployed to ensure that the most recent and complete scientific advice would be considered, and a precautionary approach would be the rule in case of insufficient scientific evidence (Commission 1997b). The Commission also issued a communication on consumer health and food safety where it announced actions to reinforce the way the Commission obtained and made use of scientific advice (Commission 1997c). The responsibility for the food and veterinary committees, including the reformed scientific committees that had previously been dispersed within the Commission, would be placed under the authority of the Commissioner for Consumer Policy and Health Protection and General Directorate (DG) XXIV. Furthermore, the Multidisciplinary Committee would become the Scientific Steering Committee (SSC) (Ugland and Veggeland 2006; Commission 1997c).

In April 1998 the European Parliament adopted a resolution on the Commission Green paper, where it asked the Commission 'to review Community legislation on the use of antibiotics in animal feed in order to avoid any use other than for therapeutic purposes' (Official Journal of the European Communities C

---

7 Commission Directive 97/6/EC of 30 January 1997 amending Council Directive 70/524/EEC concerning additives in feedingstuffs.

104 1998). It was followed by another resolution on the topic of antibiotics in animal feed (Official Journal of the European Communities C 167 1998). In the debate preceding its adoption, Swedish Members of European Parliament Olsson (Farmers' Party) and Schörling (Socialdemocratic Party) advocated a ban on AGPs based on the precautionary principle. The Commissioner for Agriculture, Rural Development and Fisheries, Franz Fischler, stated that a ban would have to be founded on solid scientific evidence and that such evidence was not yet available.[8]

In May 1998 a multidisciplinary working group was set up under the new Scientific Steering Committee and was mandated to scientifically evaluate the prevalence and development of antimicrobial resistance, examine its implications for human and animal health, and make recommendations based on scientific evidence. The 'use/misuse of feed additives' was listed among issues for consideration.[9] Swedish veterinarian Professor Martin Wierup was appointed as one of two vice-chairmen of the working group.[10]

As the end of Sweden's exemption period was approaching, Minister of Agriculture Åhnberg sent a letter to Commissioner Fischler in September 1998. Åhnberg accounted for recent developments and ongoing processes in the EU and pleaded for Fischler to find a suitable solution to the AGP issue; her suggestion was to let Sweden maintain its ban in the short term, and to introduce a similar ban at the EU level in the longer term. According to Åhnberg, it would be most unfortunate if Sweden would have to abandon its AGP policy. Apart from the public health considerations and the scientific evidence Sweden had presented, Swedes perceived the keeping of animals in environments where a continuous supply of antibiotics was necessary to be unethical, and stakeholder support for the Swedish ban was unanimous.[11]

In September 1998 a conference on 'the Microbial Threat', initiated by the EU Chief Medical Officers, and covering both human and animal use of antibiotics, was held in Copenhagen. On the topic of AGPs, it was stated that:

> most of those at the conference considered the use of antimicrobials for growth promotion was not justified and that it was essential to have a systematic approach towards replacing growth promoting antimicrobials with safer non-antimicrobial alternatives including better farming practice. Others thought that it was essential to conduct a full risk assessment before taking

8 *EP-Veckan: 15-05-98, Antibiotika i djurfoder* (https://www.europarl.europa.eu/press/sdp/pointses/sv/1998/p980515.htm#24, accessed 12 April 2023).

9 Minutes of the meeting of 14-15 May 1998 (https://food.ec.europa.eu/system/files/2020-12/sci-com_ssc_out21_en.pdf, accessed 12 April 2023).

10 Minutes of the meeting of 25-26 June 1998 (https://food.ec.europa.eu/system/files/2020-12/sci-com_ssc_out23_en.pdf, accessed 12 April 2023).

11 Letter dated 1998.09.02, *Brev från Åhnberg till Kommissionen, Franz Fischler, angående användningen av antibiotika som tillväxtbefrämjare i djurfoder, önskemål om möte*, dnr. Jo1998/01981-1.

> any further decisions. (Danish Ministries of Health and Food, Agriculture and Fisheries 1998)

In December 1998 six of the remaining AGPs were banned: antibiotics bacitracin zinc, spiramycin, virginiamycin and tylosin phosphate, through Council Regulation (EC) No 2821/98, and synthetic growth promoters carbadox and olaquindox, through Commission Regulation (EC) No 2788/98. Here, again, the Commission went against the opinion of the Scientific Committee on Animal Nutrition.

For the four antibiotics, references were made to: the Swedish application for adaptation of the EU legislation which was still pending; the Finnish ban on tylosin phosphate and spiramycin; and the Danish ban on virginiamycin. The ban was justified by the risk of cross-resistance (spiramycin, tylosin phosphate and virginamycin), as well as the substance being used in human medicine (spiramycin and bacitracin zinc). As for the earlier avoparcin ban, these bans were said to be for the interim. Despite the Swedish efforts, the Commission concluded that there was no evidence supporting the request from Sweden to ban the four antibiotic substances that remained in use – monensin sodium, salinomycin sodium, flavophospholipol and avilamycin – since they were not used in human or veterinary medicine, and/or did not cause cross-resistance. For the two synthetic growth promoters carbadox and olaquindox, references were also made to the Swedish application as well as to the prohibition of carbadox in the Netherlands (1997). These substances were banned on the grounds of being genotoxic.[12]

Just before the two pieces of legislation were adopted, the re-appointed Minister for Agriculture Winberg wrote letters to her ministerial colleagues on the issue of AGPs and the Swedish application. It was clear from the Commission's draft proposals that it wanted Sweden to adapt to the EU legislation: to allow the use of the four remaining AGPs on its territory after December 1998. Winberg asked her colleagues to support Sweden's request to be able to maintain its national ban, at least until the scientific committees had finished their ongoing assessment. During the year that followed civil servants from the Ministry, and often also from the Swedish Veterinary Institute, flew across Europe to meet with colleagues in other member states. Eleven such meetings were held. Several arguments in favour of maintaining the Swedish ban were presented: that it would be beneficial to maintain Sweden AGP-free as a reference country to measure resistance, that it would be impossible to move backwards after all the expensive investments that had been made, and that the public opinion of the EU would become even more negative if Sweden was forced to remove its ban. The Swedish representatives also referred to the precautionary principle and reiterated technical arguments from the report of the Swedish commission enquiry. From the written responses from the member states, and reports from the meetings, it can be concluded that

---

12 Genotoxic substances damage the genetic information within a cell causing mutations, which may lead to cancer.

ten member states were against a total ban, but six out of these could support Sweden's request to maintain the ban on its own territory. Four member states were in favour of a total ban.[13]

The Commission did not however change its stance on the four remaining AGPS, and the requirement for Sweden to adapt to existing legislation remained. Consequently, the Swedish Government decided to apply the safeguard clause in Article 11 of Directive 70/524/EEC in December 1998 and maintain its national ban.[14]

V. The tipping point

In May 1999, the Scientific Steering Committee issued its opinion on antimicrobial resistance, requested by the Commission due to great concern over the implications for human and animal health. One of the recommended measures in the opinion concerned AGPs. It stated that:

> the use of agents from classes which are or may be used in human or veterinary medicine [...] should be phased out as soon as possible and ultimately abolished. Efforts should also be made to replace those antimicrobials promoting growth with no known risk of influencing intestinal bacterial infections by non-antimicrobial alternatives. [...] Meanwhile, it should be reiterated to manufacturers and farmers that the continuous feeding of AMGPs to food animals for the purpose of disease prevention is a contravention to EU regulations and represents misuse; more effective enforcement measures should be adopted. (European Commission 1999)

In 1999 member states became increasingly vocal on the topic of AGPS in the Council (Official Journal of the European Communities C 195 1999), finally inviting the Commission to 'propose as soon as possible necessary changes on the basis of Directive 70/524/EEC on additives in feeding stuffs in order to gradually replace antimicrobial growth promoters by non-antimicrobial alternatives and to promote changes in animal production practices that will improve animal health and welfare'.[15]

In January 2000 the Commission made a pledge in its White Paper on food safety to phase out AGPs as part of its broad strategy against antibiotic resistance (European Commission 2000a). In February 2000 the Commission adopted a communication on the precautionary principle (European Commission 2000b). The precautionary principle would cover 'cases where scientific evidence is

---

13 Letters and reports from meetings from the Governments archives, dnr. Jo1999/00142 (1-13).

14 Press release from the Ministry of Agriculture, dated 1998.12.22, *Jordbruksdepartementet Pressmeddelande – Regeringen utlyser skyddsklausulen*. Document in author's possession.

15 Minutes from the 2236th Council meeting – AGRICULTURE – Brussels, 14 December 1999 (https://ec.europa.eu/commission/presscorner/detail/en/PRES_99_412, accessed 12 April 2023).

insufficient, inconclusive or uncertain and preliminary scientific evaluation indicates that there are reasonable grounds for concern that the potentially dangerous effects on the environment, human, animal or plant health may be inconsistent with the high level of protection chosen by the EU' (European Commission 2000c). The precautionary principle was codified in Article 7 of the new food law in 2002.[16]

A proposal for a regulation on additives for use in animal nutrition, including a ban on AGPs, was issued by the Commission in March 2002 (European Commission 2002). According to Commission information from negotiations in the Council on the 22 April 2002 (C/02/98), 16-20 December 2002 (C/02/399), 17-18 March 2003 (C/03/76) and 22 July 2003 (C/03/215), most member states seem to have welcomed, or at least accepted the ban, and only Austria voted against the proposal.[17]

Article 11 of the new Regulation that was adopted in 2003 stated that 'antibiotics, other than coccidiostats and histomonostats, may be marketed and used as feed additives only until 31 December 2005; as from 1 January 2006, those substances shall be deleted from the Register'.[18]

VI. The opposition – scientists and industry

In the case of the Swedish ban, LRF taking a stand meant that most of the agricultural sector had to follow suit, given the cooperative organisation of Swedish agriculture (Nordéus 2019, Nordéus, forthcoming). At the European level however opposition from both industry and the scientific community was fierce, given that vast economic interests were at stake.

In 1997 Professor Viaene from University of Ghent published the report 'The Swedish animal production – could it be applied across the European Union?' and the answer was 'no'. According to the report, Swedish animal production was not relevant from an EU-perspective. It was characterised by small scale units and limited output, and the Swedish market had been protected during the years following the ban. The ban on AGPs and the new Animal Welfare Act had led to lowered production efficiency and increased costs, increasing the economic burden on consumers and farmers. Viaene claimed that the use of antibiotics in feed remained high in Sweden and it was not evident that the ban had affected

16 Regulation (EC) No 178/2002 of the European Parliament and of the Council of 28 January 2002 laying down the general principles and requirements of food law, establishing the European Food Safety Authority and laying down procedures in matters of food safety.

17 Available at https://ec.europa.eu/commission/presscorner/detail/en/PRES_02_98; https://ec.europa.eu/commission/presscorner/detail/en/PRES_02_399; https://ec.europa.eu/commission/presscorner/detail/en/PRES_03_76 and https://ec.europa.eu/commission/presscorner/detail/en/PRES_03_215, all accessed 12 April 2023.

18 Regulation (EC) No 1831/2003 of the European Parliament and of the council of 22 September 2003 on additives for use in animal nutrition.

antibiotic resistance. From an EU-perspective, it was essential for farmers to have access to all cost-reducing technologies, both to compete on the global market, and to provide consumers with high quality food at affordable prices (Viaene, 1997).

In 1999 a conference was organised by the World Organisation for Animal Health (OIE) and the Food and Agriculture Organization of the United Nations (FAO) with support from, among others, FEDESA and FEFANA, the European animal health and feed additives industry associations. The aim was to find a model for risk assessments of antibiotic substances used in animal production, and to discuss the proper use of antibiotics in animals and monitoring of resistance. The conference gathered 280 participants from 29 countries, but no invited speaker was critical of the use of AGPs. Even so, according to the report of the Swedish Ministry of Agriculture from the meeting the outcome of the conference was better than expected.[19] A report from the Heidelberg Appeal Nederland Foundation (HAN),[20] requested by FEFANA, was launched at the conference. It resembled the Swedish report from 1997 but drew very different conclusions (Bezoen and others 1999). The human health risk related to the use of AGPs could not be properly assessed for lack of data and there was no evidence of AGPs compromising therapeutic use of related antibiotics in humans, nor that use of AGPs led to an increase of infectious diseases in humans. Furthermore, a comprehensive multidisciplinary research effort was needed to properly assess all aspects of AGPs, including beneficial effects on animal health and welfare. In the following years the report was cited by other researchers opposed to the ban, such as Acar and others 2000.

The issue of AGPs polarized the scientific community and this can be illustrated by a debate in the Journal of Antimicrobial Chemotherapy. An international group of scientists, initially convened as an advisory board by the Animal Health Institute (a U.S. association of manufacturers of animal health-care products), published a review article on the topic of AGPs (Phillips and others 2004). The main conclusion of the review was that 'even if resistant pathogens do reach man, the clinical consequences of resistance may be small' and that the application of the precautionary principle was a non-scientific approach. Following its publication, Peter Collignon (2004) from the Microbiology Department of the Canberra Hospital wrote in a letter to the editor that he 'was dismayed to see that Phillips et al. were defending what nearly all microbiologists know are grave abuses of antibiotics – their continued use in feed as growth promoters'.

The pharmaceutical industry was keen to keep their products on the market. When partial bans on AGPs were introduced in 1998, two AGP producers, Pfizer Animal Health SA/NV (Case T-13/99) and Alpharma Inc. (Case T-70/99),

---

19 Report from Ministry of Agriculture and LRF memorandum '*Vitbok*', in author's possession.

20 An organisation (1993-2008) mainly active in criticising Dutch efforts in the environmental area for being unnecessary. In 2008 HAN became a part of *De Groene Rekenkamer* (https://groene-rekenkamer.nl/han/, accessed 25 April 2023).

applied to the Court of First Instance for interim measures and annulment of the legislation, claiming that the bans were not scientifically grounded. Pfizer was supported by the Spanish pork and beef breeders' associations, as well as by FEDESA and FEFANA, and two Belgian farmers. The Council of the European Union was supported by the European Commission as well as Denmark, Finland, Sweden and, at a later stage, the United Kingdom. The same member states, except for Denmark, participated in the Alpharma case as well. In June 1999 the President of the Court of First Instance ordered that their applications for interim measures were dismissed, and in the judgement from September 2002, the Court dismissed the applications for annulment of the legislation.

VII. Conclusion

When Sweden became a member of the EU, Swedish animal husbandry had recently gone through a great transformation, driven by the ban on AGPs in 1986, and the new Animal Welfare Act of 1988. This transformation came at a cost, especially for farmers, but in the middle of the 1990s the sector had rallied around the holistic approach of the 'Swedish model' of animal husbandry. Another important change was the transition from an almost closed national market to the common market of agricultural products, and the competition it entailed. Specializing in providing the citizens of the Union with safe and healthy food could be a means to compete on that market (Konsumentberedningen 1993, p.2). Consequently, it was important for Sweden to be able to maintain its ban on AGPs. The accession negotiations were successful in this aspect and according to a Commission representative, part of the success was already achieved by this exemption since it would force the Commission and the Council to deal with the topic. It was becoming increasingly clear that the use of AGPs was of concern to the Nordic member states, especially Finland and Sweden.[21]

The common market is one of the pillars of the EU and barriers to trade should be avoided. Maintaining the national ban after the exemption period was not possible and consequently it was necessary to produce scientific evidence to support the introduction of an EU ban. The Swedish efforts in this regard were quite remarkable. One important part of this was the involvement and interest of the Swedish Ministers of Agriculture, Olsson (Farmers' Party) during the accession negotiations, and Åhnberg and Winberg (Socialdemocratic Party) during the country's first years as an EU-member. Perhaps even more important however was the commitment of certain civil servants and researchers working at the Ministry, governmental agencies, farmers' organisations and in academia. Another important aspect was the structural approach to this work. A commission enquiry was put in place early in the process to compile the much-needed scientific

21 Civil servant from Commission (DG SANTE), interviewed on 13 September 2017.

evidence, and various Swedish actors helped spread the information within their EU contexts. The importance of staying on the same message within EU policy work has been highlighted previously (Ludlow 2005).

The Swedish and Finnish exemptions placed the issue of antibiotic resistance and AGPs on the agenda and initiated the process in the EU. Furthermore, former Commissioner Fischler believes that Sweden served as an example to show that animal protection without AGPs could be successful.[22] Once in motion other parallel processes were probably equally as important. Shortly after Sweden and Finland became members, other member states started applying the safeguard clause to ban certain AGPs; the issue was expanding well beyond the two most recent, small member states. Similarly, the WHO meeting in 1997, with its recommendations on AGPs, seems to have been a key event as it provided advocates of prudent use of antibiotics with important arguments.

The Commission was initially reluctant to ban the use of AGPs. According to Fischler, the initial standpoint of the Commission was that any bans on AGPs had to be based on scientific evidence, and the scientific mainstream position at the time was that the risk of antibiotic resistance associated with AGPs could be limited by differentiating between therapeutic antibiotics and AGPs. With time, consumer and animal welfare organisations became increasingly powerful and more scientific evidence was presented. The Commission had to address this and regulating the issue at community level would serve the functioning of the common market, as well as increase the trust in EU food safety and quality.[23]

The hormone dispute within the WTO set the scene for the discussions on AGPs and may have influenced the Commission's view on the issue by highlighting the importance of scientifically sound policymaking. At the same time, according to a Commission representative, the hormone dispute may have drawn some attention away from the ongoing discussions on AGPs, allowing it to fly somewhat under the radar both within EU and internationally. Furthermore, the outfall of the BSE crisis was still being handled and the EU wanted to be a world-leader in food safety. Dealing with the issue of AGPs went well with these ambitions.[24]

The importance of such parallel events in the Union cannot be stressed enough. The BSE scandal and other food scares of the 1990s exposed flaws in European food safety, which ultimately led to a reorganisation of the entire system. The focus shifted from the functioning of the market to consumer safety (Ugland and Veggeland 2006; Vos 2000), and the precautionary principle was introduced in food legislation. Furthermore, the Scientific Steering Committee was created and charged with coordinating opinions where the input of several scientific committees was needed. This fundamental change paved the way for the phasing out of AGPs, even when the risks associated with their use could not be

22 Personal message, 23 January 2018.
23 Personal message, 23 January 2018.
24 Civil servant from Commission (DG SANTE), interviewed on 13 September 2017.

properly quantified. A comparison between the initial bans in 1997 and 1998, and the final ban in 2003, provides a good illustration of this shift. The earlier bans were also of a precautionary nature, but they were motivated by concrete risks such as cross-resistance. The substances for which no such risks were apparent were left in use. The passing of time also had an effect – with ongoing research and monitoring, scientific findings increased with time.

Vast economic interests were at stake and these influenced actors taking part in the process. The pharmaceutical industry wanted to keep their products on the market. For farmers' organisations and others, it was important that European agriculture remained competitive on the world market. Even though the EU bans were, at least to a certain extent, scientifically grounded, we see the same debate as in connection to the Swedish ban (Nordéus, forthcoming). Those opposed to banning AGPs claimed that the bans were 'unscientific' and cautioned for a devaluation of science. This polarization of the scientific community neatly illustrates a dilemma in applying the precautionary principle – how do you know if you are being careful or if you are just hampering progress?

Bibliography and references

Acar, J., Casewell, M., Freeman, J., and Goosens, H. (2000), 'Avoparcin and virginiamycin as animal growth promoters: a plea for science in decision-making', *Clinical Microbiology and Infection* 6, 9, pp. 477-82

Ainsworth, C., and Carrington, D. (2000), 'BSE disaster: the history', *New Scientist*, 25 October 2000

Alpharma Inc. v. Council of the European Union. Case T-70/99, Court of First Instance. 30 June 1999 and 11 September 2002

Aminov, R. (2010), 'A brief history of the antibiotic era: Lessons learned and challenges for the future', *Frontiers in Microbiology*, 1:134. DOI: 10.3389/fmicb.2010.00134

Bellerby, J. (ed.) (1970), *Factory farming: a symposium*, London: The British Association for the Advancement of Science

Bezoen A., van Haren, W., and Hanekamp, J. (1999), 'Emergence of a Debate: AGPs and public health', Heidelberg Appeal Nederland Foundation. Available at: https://www.researchgate.net/publication/261286820_Emergence_of_a_Debate_AGPs_and_Public_Health, accessed 12 April 2023

Butaye, P., Devriese, L., and Haesebrouck, F. (2008), 'Antimicrobial growth promoters used in animal feed: Effects of less well known antibiotics on gram-positive bacteria', *Clinical Microbiology Reviews* 16, 2, pp. 175-88

Castanon, J. (2007), 'History of the use of antibiotics as growth promoters in European poultry feed', *Poultry Science*, 86, pp. 2466-2471

Cogliani, C., Goossens, H., and Greko, C. (2011), 'Restricting antimicrobial use in food animals: Lesson from Europe', *Microbe*, 6, 6, pp. 274-79

Collignon, P. (2004), 'Antibiotic growth promoters', *Journal of Antimicrobial Chemotherapy*, 54:1, 272. https://doi.org/10.1093/jac/dkh266

Danish Ministries of Health and Food, Agriculture and Fisheries. 1998. 'The Copenhagen recommendations. Report from the invitational EU conference on the microbial threat. Copenhagen, Denmark, 9-10 September 1998'

Economic and Social Committee of the European Communities. 1998. 'Resistance to antibiotics as a threat to public health' CES 567/98 fin E/o

European Commission. 1997a. 'Press release 30 January: Ban on the antibiotic "Avoparcin" in animal feed' (IP/97/71)

European Commission. 1997b. 'The general principles of food law in the European Union. Commission Green Paper', COM(97) 176 final

European Commission. 1997c. 'Communication from the Commission – Consumer health and food safety', COM(97) 183 final

European Commission. 1999. 'Opinion of the Scientific Steering Committee on anti-microbial resistance 28 May 1999'

European Commission. 2000a. 'White paper on food safety', COM (1999) 719 final

European Commission. 2000b. 'Communication from the Commission on the precautionary principle', COM(2000) 1 final

European Commission. 2000c. 'Commission adopts communication on precautionary principle', IP/00/96

European Commission. 2002. 'Proposal for a regulation of the European Parliament and of the Council on additives for use in animal nutrition', COM(2002) 153 final

Gee, D., MacGarvin, M., Stirling, A., Keys, J., Wynne, B., and Vaz, S. (2001), *Late lessons from early warnings: the precautionary principle 1896-2000*, Harremoës, P. (ed.), Luxembourg, Office for Official Publications of the European Communities

Government Official Reports 1997:132, Ministry of Agriculture. 'Report from the Commission on antimicrobial feed additives'. Stockholm: Nordstedts Tryckeri AB

Greko, C. (2010), 'Antibiotikaresistens'. In Appelgren, L-E., Jämte, I., and Östensson, K. (Eds.), *Veterinär – yrke i förvandling*, Sveriges Veterinärförbund, pp. 230-38

Joint committee on the use of antibiotics in animal husbandry and veterinary medicine. 1969. 'Report presented to Parliament by the Secretary of State for Social Services, the Secretary of State for Scotland, the Minister for Agriculture, Fisheries and Food and the Secretary of State for Wales by Command of her Majesty'. London: Her Majesty's Stationery Office

Kahn, L. (2016), *One Health and the politics of antimicrobial resistance*, Baltimore, Johns Hopkins University Press

Kastner, J., and Pawsey, R. (2002), 'Harmonising sanitary measures and resolving trade disputes through the WTO-SPS framework. Part I: a case study of the US-EU hormone-treated beef dispute', *Food Control* 13, pp. 49-55

Kirchelle, C. (2020), Phyrric progress, *The history of antibiotics in Anglo-American food production*, New Brunswick, Rutgers University Press

Kleinman, D., and Kinchy, A. (2003), 'Boundaries in science policy making: Bovine growth hormone in the European Union', *Sociological Quarterly*, 44:4, pp. 577-94

Konsumentberedningen. 1993. 'Faktablad nr 16, Djurskydd och djurhälsa i Sverige och i EG'

Lantmännen Foderutveckling AB and Scans djuromsorgsnämnd. 1993. 'Antibiotikaanvändning och Resistensutveckling, Seminarium, Uppsala, 29 september 1992'

Ludlow, N. (2005). 'The making of the CAP: towards a historical analysis of the EU's first major common policy' [online], London: LSE Research Online. Available at: http://eprints.lse.ac.uk/2415, accessed 12 April 2023

Nordéus, K. (2019), 'The Swedish ban on antibiotic growth promoters: The role of the institutions'. In *Leche y lecheras en el siglo xx – De la fusión innovadora orgánica a la Revolución Verde* L.F. Prieto and D.L. Táboas, (eds), pp. 251-86. Zaragoza: Prensas de la Universidad de Zaragoza

———. Forthcoming. 'The Swedish ban on antibiotic growth promotors – the role of involved actors and consequences of the ban'

Official Journal of the European Communities C 241, 29 August 1994, p. 9-404

Official Journal of the European Communities C 104, 6 April 1998, p. 60. A4-0009/98 'Resolution on the Commission Green paper on the general principles of food law in the European Union (COM(97)0176 – C4-0213/97) (A4-0009/98)'

Official Journal of the European Communities C 167, 1 June 1998, p. 306. B4-0447, 0448, 0449, 0450, 0451, 0452, 0453 and 0454/98 'Resolution on antibiotics in animal feed'
Official Journal of the European Communities C 195, 17 July 1998. 'Council Resolution of 8 June 1999 on antibiotic resistance "A strategy against the microbial threat"', (1999/C 195/01)
Pfizer Animal Health SA/NV v. Council of the European Union. Case T-13/99 R, Court of First Instance 30 June 1999 and 11 September 2002
Phillips, I., Casewell, M., Cox, T., De Groot, B., Friis, C., Jones, R., Nightingale, C., Preston, R., and Waddell, J. (2004), 'Does the use of antibiotics in food animals pose a risk to human health? A critical review of published data', *Journal of Antimicrobial Chemotherapy*, 53, 1, pp. 28-52
Regeringens proposition 1984/85: 149 om lag om foder
Swedish Ministry of Agriculture, Food and Fisheries. 1997. 'Can we use less antibiotics?'
Swedish Ministry of Agriculture, Food and Fisheries. 1998. 'Today we defeat bacteria. What about tomorrow? Documentation from a Conference in Brussels 13 November 1997'
Swedish Ministry of Agriculture, Food and Fisheries. 1999. 'The Swedish model of animal production, information based on presentations given at a seminar held in Stockholm 3-4 September 1998'
Tronstad, A. (1995), 'Europas veterinärer vill förbjuda antibiotika i tillväxtbefrämjande syfte', *Svensk Veterinärtidning*, 16, pp. 738-39
Ugland, T., and Veggeland, F. (2006), 'Experiments in food safety policy integration in the European Union', *JCMS*, 44, 3, pp. 607-24
Viaene, J. (1997), 'The Swedish animal production system: Could it be applied across the European Union?', State University of Ghent
Vos, E. (2000), 'EU food safety regulation in the aftermath of the BSE crisis', *Journal of Consumer Policy*, 23, 3, pp. 227-55
WHO 1997. 'The medical impact of antimicrobial use in food animals. Report of a WHO meeting. Berlin, Germany, 13-17 October 1997', WHO/EMC/ZOO/97.4

ANNICK JACQ

12. Too much antibiotics in aquaculture too?*

The difficult emergence of a public problem in France retraced through actors' discourses

I. Introduction

The discovery and industrial development of penicillin and streptomycin in the 1940s, followed by the discovery of numerous new antibiotics, opened an era in which antibiotics appeared to have vanquished deadly bacterial infectious diseases like pneumonia and tuberculosis. However, resistance to antibiotics (known as antimicrobial resistance, or AMR) rapidly emerged and spread to the point where it is now considered to be a major health threat worldwide (Bud, 2007; Levy, 1992).

Despite early warnings by medical scientists, it took many years for the problem of AMR to be recognized in France as a 'public problem' that needed to be tackled at the global level through a combination of national policies and international efforts (Arquembourg, 2016a; 2016b). In November 2001, the EU adopted a recommendation on the prudent use of antimicrobial agents in human medicine, and in 2002 the first awareness campaign was launched by the French Ministry of Health and the national health insurance system that was called *Les antibiotiques, c'est pas automatique* ('antibiotics are not automatic').[1] Several plans aimed at preserving antibiotic effectiveness followed, progressively integrating the

* This work was carried out as part of the 'Amagri project' funded by the ANR grant ANR-18-CE03-0001. I am especially grateful to Delphine Berdah for useful discussions, encouragements, comments, suggestions and setting up several of the interviews, to Aurélien Féron for discussions and comments, and to Carin Martiin and two anonymous reviewers for careful reading of the manuscript and suggestions for improvements.

1 The Council of the European Union, *Council recommendation of 15 November 2001 on the prudent use of antimicrobial agents in human medicine*. https://eur-lex.europa.eu/eli/reco/2002/77/oj (last consulted on 26/03/2023).

Annick Jacq • Groupe d'Histoire et de Diffusion des Sciences d'Orsay, Université Paris-Saclay, UR-Études sur les Sciences et les Techniques

From Breeding & Feeding to Medicalization, Rural History in Europe, 17 (Turnhout: Brepols, 2023), pp. 287–312
BREPOLS PUBLISHERS 10.1484/M.RURHE-EB.5.135030

dimension of both human and veterinary usage, to finally refer to the notion of *One Health'* promoted by the World Health Organization (WHO) in 2015.[2]

The emergence and spread of AMR have been linked to the inappropriate use of antimicrobial drugs, not only to treat diseases in humans, but also in livestock. Several authors have recently retraced the history of antibiotic use and its regulation in animal husbandry in the US, UK (also in comparison with Sweden) and France, highlighting the economic and socio-cultural dimensions of the problem and the role played by veterinarians and the pharmaceutical industry (Kirchelle, 2018; 2020; Berdah, 2021; McKenna; Begemann et al. 2018; Fortané, 2016; Finlay and Marcus, 2016; Corley and Godley, 2011).

The development of aquaculture in the 1970s and 1980s also led to the massive use of antibiotics, but for decades this practice was little questioned and drew much less attention from historians (Lulijwa et al., 2020). This article is an initial attempt to understand the specific dynamics of the debate over antibiotic usage in aquaculture compared to agriculture in France. In the first part, I provide elements of context regarding the danger of AMR linked to the usage of antibiotics in animal husbandry, the history of the transformation of French aquaculture since the 1970s, and the role played by antibiotic use in this transformation and its specificity compared to livestock farming. In the second part, using the French newspaper *Le Monde* and the monthly journal of the consumer association *Que Choisir* as a door to the 'public space', I show that antibiotic usage in French aquaculture has been largely overlooked as a possible cause of AMR, thus failing to emerge as a 'public problem' (Sheppard-Sellam, 2019; Arquembourg, 2016b; Dewey, 2010). In the third part, I investigate whether this reflects an absence of controversies among aquaculture actors, or results from a more complex situation. I do so by analyzing the positioning over time of various actors, such as scientists and veterinarian experts, governmental institutions and regulatory agencies, and the profession itself, drawing on the content of various media in which these actors express their views, including professional journals, scientific academic literature, institutional reports, and for the most recent period, websites. In addition, several semi-structured interviews were conducted, especially with veterinarians specializing in aquaculture. The results show that the lack of attention paid to aquaculture as a potential source of AMR in the public space does not reflect an absence of controversies among aquaculture actors. These controversies are largely framed by a broader debate regarding intensive agriculture.

---

2 Ministère des Solidarités et de la Santé. https://solidarites-sante.gouv.fr/prevention-en-sante/les-antibiotiques-des-medicaments-essentiels-a-preserver/des-antibiotiques-a-l-antibioresistance/article/l-antibioresistance-concerne-les-hommes-mais-aussi-les-animaux-et-l (last consulted on 26/03/2023).

II. The rise of antibiotic usage and AMR in agriculture and aquaculture in France 1950-2020

II-1. Antibiotics in agriculture and the rise of AMR

As shown by several scholars, the use of antibiotics has been an integral part of the 'modernization' of animal husbandry since 1945 in the US. While the intensification of livestock production favored disease development, the discovery that the addition of antibiotics to feed could promote animal growth made their use an essential component of the zootechnical infrastructure of modernized, intensified post-war agriculture (Bud, 2017: Chapter 7th; McKenna, 2017; Fortané, 2016; Kirchhelle, 2018, 2020). This practice was soon imported to France under pressure from the French government and the pharmaceutical industry. Delphine Berdah has shown that the creation of this new market allowed the pharmaceutical industry to valorize the byproducts of antibiotic production, which could be incorporated into feed, thus decreasing the cost of antibiotics needed for medical usage (Berdah, 2021).

It was recognized very early that antibiotic use could lead to the selection of antibiotic resistance in animal and human pathogens, but this had little effect on public policies. In 1969, the Swann report, produced in the United Kingdom, found "that the administration of antibiotics to farm livestock poses certain hazards to human and animal health since it has led to the emergence of strains of bacteria which are resistant to antibiotics", and recommended that antibiotics administered in human medicine should not be used as growth promoters.[3] Although the Swann report is considered to be a turning point with regard to putting the risk posed by animal AMR on the political agenda, it would take more than 45 years before the European Union banned, in 2006, the usage of antibiotics as growth promoters, one major recommendation of the report.[4]

In France, the monitoring of veterinary medicine usage was established in 1999 after a call by the World Organization for Animal Health (OIE) to fight AMR. From then on, the French Food Safety Agency (AFFSA)[5] produced regular reports on antibiotic sales for veterinary usage, including an analysis by species, with data provided for fish farming.[6] Another marker of the construction of AMR in animals as a public problem was the restructuration of Resabo, a research

3 Michael Meredith Swann, Report of the joint committee on the use of antibiotics in animal husbandry and veterinary medicine, (London, UK, Her Majesty's Stationary Office, 1969).

4 See chapter 11 by Kristina Noredeùs in this volume for the EU ban of antibiotic usage as growth promoters.

5 Agence Française de Sécurité Alimentaire des Aliments, later to become ANSES, Agence Nationale de Sécurité Sanitaire de l'Alimentation, de l'Environnement et du Travail (Agency for Food, Environmental and Occupational Health and Safety).

6 AFFSA *Suivi des ventes de médicaments vétérinaires contenant des antibiotiques en France en 2004* (2004). https://pop.anses.fr/fr/system/files/ANMV-Ra-antibiotiques2004.pdf (last consulted on 26/03/2023).

network established in 1982 to monitor AMR in bacteria isolated from bovines. In 2001, Resabo became Resapath, a true surveillance system of AMR in animals. Initially restricted to bovines, poultry and pigs, Resapath later extended its scope to collect data on sheep, goats, rabbits, horses, cats and dogs… and finally fish in 2007 (Fortané, 2015).

II-2. brief history of French aquaculture

FAO defines aquaculture as the 'farming of aquatic organisms including fish, mollusks, crustaceans and aquatic plants'.[7] Here, we will restrict ourselves to aquatic animal farming. It is a millennia-old activity, with Chinese rearing carps as early as 2000-1000 BC and tilapias being reared in tanks over 4000 years ago in Egypt.[8] Oysters have been farmed since Roman antiquity, especially in Mediterranean lagoons. In Europe, carps were completely domesticated in the Middle Ages and carp farming in ponds and tanks was an important resource for monasteries and nobles' estates. Trout, essentially rainbow trout, were domesticated in the 19th century by two fishermen who developed a method for their artificial reproduction. *Bouchot* culture, an intertidal growth technique still used today that involves binding mussels to nets on wooden poles, is said to have been invented in the 13th century.[9]

From a small scale, traditional activity, French aquaculture was significantly transformed between the 1970s and 2000. The framework used to develop aquaculture resembled the one that presided over the transformation of traditional agriculture 20 years earlier, namely improving productivity through technological innovations relying on science (Bonneuil and Thomas, 2009). Public policies aiming to develop aquaculture were motivated by the need to feed a growing population, provide additional sources of income to local populations and respond to the problem of the dwindling supply of fish due to overfishing. French aquaculture is distinguished by the existence of two legal regimes, one for the inland domain, covering freshwater management, including fresh water (or continental) fish farming, and another for the maritime domain and marine water usage. Fresh water production mostly involves trout, while marine production includes oysters, mussels and a number of marine fish species such as turbot, sea bass, sea bream and sturgeon. The National Institute for Agriculture, Food and Environment (INRAE, previously INRA), which was created in 1946 to contribute to the post-war modernization of agriculture, is in charge of backing up continental aquaculture. Four hydrobiology stations were integrated into INRA

7 https://www.fao.org/fishery/en/collection/aquaculture (last consulted on 26/03/2023).

8 FAO *History of Aquaculture* (1988) https://www.fao.org/3/ag158e/AG158E01.htm (last consulted on 26/03/2023).

9 https://www.poisson-aquaculture.fr/lhistoire-de-la-pisciculture/; https://www.alimentarium.org/en/knowledge/history-aquaculture; http://eprints.cmfri.org.in/10665/1/6.%20Asokan.pdf (last consulted on 3/06/2023).

in 1964, including the Paris *Station d'Hydrobiologie Continentale*, a laboratory established for the diagnosis of bacterial and parasitic diseases of fish (Denis, 2014; Hedrick et al., 2013). For the marine domain, the *Centre National pour l'Exploitation des Océans* (National Center for the Exploitation of the Oceans, CNEXO), an applied research institution, was created in 1967. One of its missions was to support marine aquaculture development, including the identification of appropriate sites, choices of species, and the conception, realization and running of aquaculture facilities.[10] In 1984, CNEXO fused with ISTPM (Scientific and Technical Institute for Maritime Fisheries) resulting in the creation of IFREMER (French Research Institute for Sea Exploitation). In term of veterinary usage and sanitary regulations, both fresh water and marine water activities fall under the responsibility of ANSES, itself accountable to the French Ministry of Agriculture and Food (*Ministère de l'Agriculture et de l'Alimentation*).[11]

French aquaculture is characterized by three elements:

* The top product is bivalve mollusks, with 80 000 t/y of oysters and 62 000 t/y of mussels to compared to 60 000 t/y and 40 000 t/y respectively in 1970, which puts it in second place in Europe (first oyster producer) (Tanguy, 2008).[12]
* Continental aquaculture is dominated by the production of trout, 96% of which are rainbow trout. Trout production was 39 500 t in 2019 according to the professional organization CIPA, making France the leading European producer of freshwater fish.[13]
* The development in the 1980s of marine fish farming. The reproductive cycle of several sea fish was mastered, leading to the development of hatcheries. New marine species were domesticated, such as sea bass, turbot and sea bream. In the 1990s, the sturgeon reproduction cycle also was mastered, allowing the development of the very lucrative production of caviar, making France one of the world's leaders. In addition, France has an important activity of eggs and fry production, both for shellfish (oysters) and fish, the huge majority of which is exported across Europe. France also tried to take advantage of its overseas territories to develop tropical shrimp production in Tahiti and in New-Caledonia with the help of the CNEXO but has faced difficulties linked to the emergence of pathologies (Académie des sciences, 2003).

---

10 https://archives.ifremer.fr/Histoire-des-organismes/CNEXO (last consulted on 26/03/2023).

11 In 2020, for the first time since 1991, a *Ministère de la mer*, (Ministry of the Sea) was created to which the direction of fisheries and aquaculture is now attached, after previously being part of the Ministry of Agriculture.

12 Comité National de la Conchyliculture; http://www.cnc-france.com/ (last consulted on 26/03/2023).

13 CIPA: Comité Interprofessionnel des produits de l'aquaculture (Interprofessional committee for aquaculture products) https://www.poisson-aquaculture.fr/chiffres-cles/ (last consulted on 26/03/2023).

The zootechnical progress made by INRA and IFREMER in terms of feed quality and control of reproduction thus played an essential role in the development of French aquaculture from the 1970s to the end of the 1990s.

Continental freshwater aquaculture is characterized by a large number (~300) of businesses varying greatly in size, with some very small. Seventy percent of production is located in three regions: the North, South West (Aquitaine) and Brittany. Marine fish production is much more concentrated, with 36 farms and five companies specializing as hatcheries and nurseries.[14]

Shellfish production involves around 4 000 producers, distributed all along the French coast, with many local specificities. In the 1990s, the control of oyster reproduction as well as other shellfish species, mostly clams, led to the rapid development of hatchery/nursery activities.

These transformations of French aquaculture, with increased productivity and the apparition of new farmed species, had important consequences on antibiotic usage.

II-3. Specificity of antibiotic usage in aquaculture compared to animal husbandry

As in agriculture, the intensification of aquaculture farming activities led to an increased density of animals compared to the situation in the wild, favoring contagion and possibly decreasing animal immune defenses (Doumenge, 1984). The cause of some infectious diseases (but not all) are pathogenic bacteria that can be fought using antibiotic treatments. In an aquatic environment, with a high density of animals, diseases can spread very rapidly and the use of antibiotics is mostly metaphylactic, meaning they are administered in the feed to an entire batch of fish as soon as a few fish present clinical signs. Antibiotic-laced feed is dispersed in the water and accumulates in the sediment, particularly as sick fish do not eat. The water itself is contaminated, raising the question of wastewater treatment (Ducateau, 2015: 60-62). Overall, one of the major problems of antibiotic usage in fish farming is that the dissemination of antibiotics in the environment is much faster in water compared to a terrestrial environment.

In the case of oysters, which are filter feeders eating phytoplankton present in sea water, the usage of antibiotics to treat bacterial infections in adult oysters farmed in open water is not practical, but the control of reproduction in fish and shellfish hatcheries has generated new antibiotic usage. Reproduction and outgrowth of larvae and juveniles are very sensitive steps, potentially encouraging the preventive and even systematic use of antibiotics. In the case of oysters, the traditional method of spat collection from the marine environment has been partially replaced by the use of spats produced in hatcheries. In the 1980s, IFREMER developed the production of triploid oysters, which are sterile, grow quickly and

14 CIPA/FranceAgrimer *Les filières Pêche et Aquaculture en France* (2017).

are 'fat' all year round, a quality preferred by customers. All triploid spats are produced in hatcheries. Oyster farmers also may prefer to obtain spats of diploid oysters from hatcheries rather than through collection in the wild.

Another distinguishing feature is a prescription practice known as 'the cascade'. As only five antibiotics currently have obtained marketing authorization (MA) for fish pathologies, veterinarians are allowed under certain conditions to choose another antibiotic, for instance one known to be efficient for the same pathogen in another animal, but which does not have an MA for fish. Antibiotics with an authorization for fish in France are oxytetracycline, florfenicol (from the same family as chloramphenicol), oxolinic acid, flumequine (a fluoroquinolone), and the sulfadiazine/trimethoprime combination. Other antibiotics that can be used through the cascade process are amoxicilline, difloxacine, enrofloxacine, and erythromycin (Michel, 2018: 439). In 2014, ANSES advised that the metaphylactic use of fluoroquinolones in fish farming should be regulated and all preventive (prophylactic) usage of antibiotics should be completely banned.[15]

Since the 1975 law on the use of pharmaceuticals in veterinary medicine, antibiotics, like other veterinary drugs, may be administered only upon prescription by a veterinarian, although under certain conditions, prescription without clinical examination has been possible since 2007.[16] This possibility is conditioned on regular visits of the farm by the veterinarian in charge of the sanitary follow-up. According to the Ministry of Agriculture and Food, less than 20 veterinarians specialize in fish farming in France, and they are mostly present in the Aquitaine region,[17] although fish farms are numerous and dispersed across the country. Of the four national veterinary schools in France, only two provide training in aquaculture. This situation, making it difficult to combine proximity and expertise in prescription practices, is considered as potentially detrimental to good sanitary management (Février, 2013: 9; 18).

For the years 2014-2015, the amount of antibiotics used in fish farming represented about 0.49% of the total used in veterinary medicine, which may seem negligible. However, if one considers the amount of antibiotics sold per live weight of fish, it is in the same order of magnitude as what is used for other farmed animals (62.53 mg/kg of live fish to compare with 42.61 for poultry and 65.09 for pigs). Another indicator is the Animal Level of Exposure to Antimicrobials, or ALEA, which divides the treated live weight by the total farmed population live weight. The ALEA for fish in this same period was 0.214, much less than that of

15 ANSES (2014) *Risques d'émergence d'antibiorésistances liées aux modes d'utilisation des antibiotiques dans le domaine de la santé animale.* p. 207 https://hal-anses.archives-ouvertes.fr/anses-01955006 (last consulted on 26/03/2023).

16 *Décret n° 2007-596 du 24 avril 2007 relatif aux conditions et modalités de prescription et de délivrance au détail des médicaments vétérinaires et modifiant le code de la santé publique (dispositions réglementaires)* (Decree 2007-596 of April 24th 2007 relative to the conditions and terms of prescription and retail distribution of veterinary medicine, modifying the public health code -regulatory dispositions-).

17 https://fichiers-publics.agriculture.gouv.fr/dgal/ListesOfficielles/SPA7_VS_AQU.pdf (last consulted on 26/03/2023).

poultry (0.783) or pigs (0.837), but within the same range as bovines (0.297).[18] It is also important to note that antibiotics are not used as growth promoters in fish farming.

III. Antibiotic usage in aquaculture as a potential source of AMR: not a public concern (1951-2020)?

One important aspect of what can be defined as a 'public problem' is its visibility in the public sphere as a problem that requires political action (Arquembourg, 2016b). Jocelyne Arquembourg emphasizes: "Thus, a public problem emerges from problematic situations that become objects of public attention as soon as they are an object of debate" (Arquembourg, 2016a: 8).

Although the presence of a problem in a given media (or even several media) is surely not sufficient to warrant the existence of an issue in the public space, a study of the press can provide an indication of the emergence of such an issue, and whether the problem has or has not met its public (Arquembourg, 2016b; Marres, 2005; Kirshhelle, 2020).[19] In addition, it can provide information about the way the problem is framed. Here, I used two different types of written media: the newspaper *Le Monde*, published daily since 1944, and considered as a reference, and the monthly magazine *Que Choisir* published by the main French consumer association of the same name. One goal of the analysis was to determine whether and when, in these media, antibiotic usage in aquaculture was considered as a possible problem in terms of AMR compared to the usage in animal husbandry.

III-1. AMR in the newspaper Le Monde

The content of a corpus of 245 articles retrieved with the keywords *résistance aux antibiotiques* (antibiotic resistance) was analyzed and classified in terms of the potential sources of AMR identified, namely: misuse in human medicine, misuse in animal husbandry, use of genetically modified organisms (GMOs) in agriculture, misuse in aquaculture or none of these (other). An article could correspond to several categories if several potential sources of AMR were mentioned in the article. The number of articles on the subject published each year and the results of this analysis are presented on Figure 12.1.

The figure clearly shows three periods. The first period corresponds to the 1950s and early 1960s. The apparition of resistant bacteria is mentioned as early as 1951. In the following years, the accent is put on the discovery of new antibiotics every time a resistance problem emerges. The message hence was that there was

---

18 ANSES, *Suivi des ventes de médicaments vétérinaires contenant des antibiotiques en france en 2015. Rapport annuel* (2016) https://www.anses.fr/fr/system/files/ANMV-Ra-Antibiotiques2015.pdf (last consulted on 26/03/2023).

19 Regarding the limits of such sources, see Claas Kirshhelle's comments *in* (Kirshhelle, 2020, p. 24).

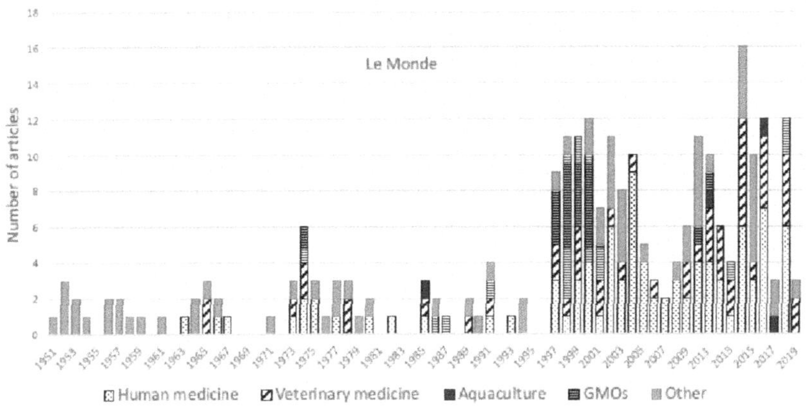

Figure 12.1. Problematization by the French daily newspaper *Le Monde* of antimicrobial resistance over the years (until 2019).

nothing to worry about.[20] Christophe Gradmann has shown, in the case of the pharmaceutical company Bayer, how the appearance of AMR was used by the company as an opportunity to create a market for new drugs (Gradmann, 2016). The second period starts in 1963, and is characterized by a steady number of articles criticizing the overuse of antibiotics in human medicine, both in ambulatory and hospital settings, but also pointing out the responsibility of veterinary usage in the rise of AMR. The first mention of a potential link of AMR with animal husbandry is found in a 1965 article by Professor Hugues Gounelle, a member of the French Academy of Medicine, criticizing the use of antibiotics as growth factors and food preservatives.[21]

Starting in the 1980s, when genetically modified crops arrived on the market, the debate revolved around GMOs as a potential source of AMR due to the resistance gene that they generally carry. Although initially a concern, the European Food Safety Agency (EFSA) concluded that the possibility of an actual transfer of an antibiotic resistance gene from a GM plant to a pathogenic bacterium was very unlikely, although there was a minority dissenting opinion.[22]

In parallel, with the development of the organic food movement, not using antibiotics in animal husbandry became one of the hallmarks of healthy food.

20 This period has been qualified as 'heroic' (*l'épopée des antibiotiques* – the epics of antibiotics) by Jocelyne Arquembourg (2016b).

21 A detailed analysis of articles published by *Le Monde* on antibiotic resistance is presented in (Arquembourg, 2016 b).

22 See for instance https://www.efsa.europa.eu/en/news/efsa-evaluates-antibiotic-resistance-marker-genes-gm-plants (published in 2009, last consulted on 26/03/2023).

A third period starts in 1997, with an increased number of articles on the subject, discussing possible causes (and hence remedies) to the problem of AMR. This is also, as we will see below, the period when public authorities started to address the AMR threat through specific policies.

Out of these 245 articles, only four question the usage of antibiotics in aquaculture in terms of AMR. The first, which appeared in 1985 and was dedicated to the development of salmon production in Norway, briefly mentions the increased use of antibiotics linked to the intensification of production.[23] The next one published in 2011, mentioned, among other examples, the appearance of a multiresistant *Salmonella* strain that could have emerged in Egypt due to the massive usage of antibiotics in aquaculture and poultry farms.[24]

In 2016, one article focused on salmon farms in Chile and the last article in our set, published in 2017, examined the debate around triploid oysters on the occasion of the end of the IFREMER patent. One factor behind criticism of oyster reproduction in hatcheries is the requirement for antibiotic treatment.[25]

III-2. Consumers as a public? The vision of AMR through the Que Choisir *magazine*[26]

The *Union Fédérale des Consommateurs-Que Choisir* is one of the major independent consumer associations in France. Besides testing and comparing marketed products and services, it produces wider analyses and propositions for public actions, leads campaigns and organizes collective actions. Created in 1961, it publishes a monthly magazine that bears the same name. I identified and analyzed a corpus of 37 issues containing at least one article dealing with the problem of antibiotic misuse. Fourteen of these, the first one in 1972, denounced medical misuse, mostly over-prescription or non-respect of the legislation by pharmacists, with references to the risk of AMR. One article in 1995 on the subject of biotechnology mentions a potential risk linked to the presence of a resistance gene in GM food.

Fifteen articles criticized the heavy use of antibiotics in animal husbandry, the first one in 1973. Two types of risks were highlighted: antibiotics can select microbial antibiotic resistance that could move from animal pathogens to human pathogens, and residues in consumed food can be toxic for the consumer (for instance, chloramphenicol can cause cytopenia, a potentially fatal diminution of

23 'L'essor des fermes marines (The expansion of marine farms)' *Le Monde*, 26 November 1985.

24 'Le boom des bactéries résistantes aux antibiotiques' *Le Monde*, 30 August 2011. In Egypt, aquaculture is often 'integrated'. Farms are raising both poultry and fish, chicken manure being used as pond fertilizer.

25 'Le saumon chilien veut s'alléger en antibiotiques', *Le Monde*, 5 August 2005; 'Huître ou ne pas huître ?', *Le Monde*, 12 December 2017.

26 This analysis is restricted to the articles published in the *Que Choisir* magazine, that were retrieved by scrolling the Table of Contents of all issues published between 1961 and 2020.

blood cells). The use of chloramphenicol to treat food-producing animals was finally banned in the EU in 1994,[27] and the problem of residues was considered as having been solved by the imposition of a waiting period between treatment and product marketing and the definition of maximum residue limits (MRL). Progressively, an emphasis was put on the risk of AMR. One important article was published in 1988 (*L'abus d'antibiotiques dans la viande* 'Overuse of antibiotics in meat production'), followed by articles on the same theme every two or three years between 1997 (with an article announcing an imminent ban of avoparcin by the EU) and 2015.[28] In 2015, one can read:

> the growing resistance of bacteria to antibiotics calls for ambitious action on the part of public authorities to fight against their excessive use in livestock farming. Last year, while *Que Choisir* was publishing worrying results on poultry meat (QC N° 523), French authorities were yielding to the lobbying of veterinarians and farmers most focused on productivity…

This analysis shows that slightly before 2000, antibiotic resistance linked to overuse of antibiotics in animal husbandry became an important concern of the association, moving away from the problem of contamination by residues, and that in 2015, it was well on the agenda of *Que Choisir*.

In contrast, our corpus comprised only five articles mentioning antibiotics in the context of aquaculture. The presence of antibiotic residues in seafood was a recurring theme, with the positive results of testing for antibiotic traces in shrimp imported from Thailand in 1982, and a positive finding in 2020 of forbidden nitrofurane in shrimp imported from Vietnam.[29] This use of antibiotics was not framed in terms of AMR. A 1991 article mentions the use of antibiotics in aquaculture to cure diseases with no note of criticism. In 1994, another article criticizes overmedication, linked to the likeliness of diseases due to high animal density, without explicitly mentioning AMR as a possible consequence. Only the most recent article, about shrimp and published in 2020, spells out:

---

27 (UE) N°37/2010 https://eur-lex.europa.eu/eli/reg/2010/37(1)/oj (last consulted on 26/03/2023).

28 Avoparcin was used as a growth promoter for pigs and poultry. Its use was forbidden in 1997 in the EU because avoparcin resistance determinants also confer resistance to vancomycin, a clinically important antibiotic in humans. Some of the titles were 'Les microbes font de la résistance (Microbes resist)'; 'Antibio et santé: une efficacité en péril (Antibiotics and health, an endangered efficiency)'; 'Les animaux malades de l'élevage (Animals sick from husbandry practices)'.

29 Shrimp imported from South-East Asia, especially Thailand and China have been repeatedly found contaminated by antibiotics, with a specific concern about nitrofurane and chloramphenicol, forbidden in the EU after 1994. Antibiotics can be added to the water of the shrimp ponds, but can also be used as preservative. A series of several positive findings in different European countries, especially after 2000, have led to some articles in the general press: Frédéric Pons, 'Bruxelles bloque la crevette chinoise', *Libération*, 1 January 2002. https://www.liberation.fr/futurs/2002/02/01/bruxelles-bloque-la-crevette-chinoise_392372/. Charles Haquet and Géraldine Meignan 'Alerte sur les produits alimentaires chinois', *L'Expansion*. 30 August 2007. https://www.lexpress.fr/economie/alerte-sur-les-produits-alimentaires-chinois_466303.html (last consulted on 26/03/2023).

> As in human medicine, the inappropriate use of antibacterial compounds accelerates the development of resistance genes, eventually making antibiotics useless

Nonetheless, UFC-Que Choisir actions against the inappropriate usage of antibiotics in animal husbandry also targeted aquaculture. In 2003, it submitted a request to AFFSA to review the risks linked to veterinary usage, a request that led to a report published in 2006 that included some data on aquaculture practices and acknowledged that fish farming was especially at risk of disseminating antibiotics in the environment. The final recommendations by AFFSA were, however, general, and did not distinguish fish from other farmed species.[30] In 2016, the association participated in a campaign led by Consumers International to support propositions of the European Parliament to the European Commission to better control antibiotic usage in livestock farming.

IV. Beyond the media: actors' discourses

This analysis of the content of *Le Monde* and *Que Choisir* confirms the emergence in the public space in the late 1990s-early 2000s of concerns regarding animal husbandry as a potential source of AMR. In contrast, although the development of aquaculture since the 1970s also had led to a similar overuse of antibiotics, it does not appear to be similarly problematized. Does that mean that no controversy existed regarding aquaculture? In this third part, I analyze the discourses of various actors regarding a possible role of aquaculture in AMR.

IV-1. Academic scientists and AMR: between science and politics

In the AMR literature, one comes across the notions of antibiotic misuse, overprescription, growth factors, prophylactic and metaphylactic usage, and so on. These potential causes of the rise of AMR first had to be constructed scientifically. The very notion of the existence of an AMR epidemic had to be objectivated scientifically before it had a chance of becoming a problem that deserved political attention. In a 2014 paper published in *Nature Reviews Microbiology*, 30 scientists called for 'Tackling antibiotic resistance', making the following recommendations:

> Non-therapeutic use of antibiotics must be discontinued. A total restriction on both human-approved compounds and their structural derivatives as routine feedstocks in animal feed, agriculture, or fisheries must be enforced. At the present time, more than 50% of the antibiotics produced are employed

30 AFFSA, '*Usages vétérinaires des antibiotiques, résistance bactérienne et conséquences pour la santé humaine*' (2006). https://www.anses.fr/fr/system/files/SANT-Ra-ABR.pdf (last consulted on 26/03/2023).

> as animal feeds to promote growth. This practice has been ongoing since the 1950s, despite efforts to prevent such non-prescription use by regulatory authorities, such as the UN and WHO. Is there convincing evidence that these supplements are of use when good animal husbandry is common practice? Appropriate treatment for sick animals must, of course, be maintained. Therapeutic use of antibiotics in domestic pets should be by veterinarian prescription only. (Bush et al., 2011: 9)

This article was highly cited, underscoring its impact. The authors also noted the use of antibiotics in fisheries, which was unusual, as most critics of agricultural antibiotic usage at the time tended to focus on poultry and hog farming.

Most of the authors were well-known microbiologists[31] with interests in mechanisms of antibiotic resistance and resistance gene transfer between bacteria. Coming from molecular microbiology basic science, they used this expertise to develop molecular epidemiology, tracing resistance determinants from strains to strains. They mostly worked with laboratory strains or clinical isolates, with some exceptions. For instance, Tim Walsh analyzed strains isolated both from animals and patients. They typically published their work in the American Society of Microbiology (ASM) journal, Antimicrobial Agents and Chemotherapy (AAC), although Wolfgang Witte of the Koch Institute in Germany, who works on pathogenic multiresistant strains of *Staphylococcus aureus*, also published in *Veterinary Microbiology*.

One particularly prominent figure among the authors is Stuart B. Levy (1938-2019). Following the publication of the Swann report, the US Food and Drug Agency (FDA) created a task force that came up with similar conclusions and recommendations as the report, leading to a counter-offensive by the pharmaceutical and agriculture industries. Through the Animal Health Institute, they recruited Stuart Levy, a microbiologist working on bacterial mobile genetic elements carrying antibiotic resistance genes in the gut bacterium *Escherichia coli*. They wanted him to carry out a study that they hoped would disprove the possibility of a migration of resistance from antibiotic treated animals to humans (McKenna, 2017: 108-22). The results of the study were a disappointment to its sponsors. In 1976, Stuart Levy published a paper showing that antibiotics present in animal feed could lead to the appearance of drug-resistant microbes, not only in the normal bacterial flora of treated animals, but also in non-treated animals living on the farm and in non-treated farm workers (Levy-FitzGerald-Macone, 1976: 40-44). The US Agriculture Department and pharmaceutical industries strongly rejected these findings, and it took decades (in 2017) before antibiotics were banned as growth promoters in the US and that preventive antibiotic use was

31 Amongst them, we can cite Karen Bush, Patrice Courvalin, Julian Davies, Jeffrey H. Miller, Timothy R. Walsh and Helen I. Zgurskaya. Their scientific interests, trajectories, publications and type of research were determined through a Pubmed search.

put under the control of veterinarians.[32] Stuart Levy became a strong advocate of what he called 'prudent use of antibiotics'. His academic research was also often relevant in term of public policies, for instance when he showed that selecting for triclosan resistance (a widely used biocide found in many cosmetics) also selected for cross resistance to antibiotics. In 1981, he created the international non-profit group Alliance for the Prudent Use of Antibiotics and continued to both work on antibiotic resistance mechanisms and crusade for better public policies.[33] He regularly published articles for a wider audience warning about the dangers of the way antibiotics were being used, both in human medicine and in agriculture.

Scientific work and campaigning by academic scientists such as Stuart Levy and Patrice Courvalin at the Pasteur Institute, in Paris, France contributed to shaping a perception of antibiotics as 'societal drugs' and AMR as an 'ecological problem', notions that will be integrated later in the 'One Health' approach (Levy, 1997: 1-9).

IV-2. The fisheries experts: antibiotic usage as a zootechnical practice

In the 1970s, the French government launched policies to develop aquaculture. For freshwater species, INRA (now INRAE) was in charge of supporting fish farmers. Pierre de Kinkelin, a veterinarian, was recruited in 1964 and trained in microbiology at the Pasteur Institute before taking the charge in 1968 of the newly founded fish-pathology laboratory at Jouy-en-Josas, near Paris, to work mostly on viral hemorrhagic septicemia. He then recruited Christian Michel, also a veterinarian who had trained in microbiology at the Pasteur Institute, to work on bacterial diseases. In 1975, de Kinkelin was appointed Secretary General of the OIE Fish Diseases Commission (now the Aquatic Animal Health Standards Commission) and served as the Commission's President from 1988 to 1991. In 1965, the fish farmers association, *Fédération Nationale de la Pisciculture,*[34] started publishing the journal *La pisciculture française* intended for a professional readership. On the academic side, an international peer-reviewed journal, *Aquaculture,* was created in 1972 to publish results from applied research carried out in academic laboratories and dedicated research institutes. In contrast, the role of *La pisciculture française* was to provide advice to fish farmers based on research published in academic journals and the experience of colleagues. De Kinkelin was a frequent contributor to this French professional magazine, in contrast with his small number of publications in academic journals (13 listed in the Web of Science), which reflected the applied orientation of the laboratory, whose main

32 Maryn McKenna, 'After years of debate, the FDA finally curtails antibiotic use in livestock', *Newsweek*, 13 January 2017 https://www.newsweek.com/after-years-debate-fda-curtails-antibiotic-use-livestock-542428 (last consulted on 26/03/2023).

33 Obituary Stuart Levy, *The Lancet*. https://www.thelancet.com/journals/lancet/article/PIIS0140-6736(19)32273-1/fulltext (last consulted on 26/03/2023).

34 National Federation of Fish Farming.

role was to help aquaculture farmers to better control farmed fish diseases thanks to scientific progress. He never published in *Aquaculture,* but in virology and fish disease journals.

Early reports on antibiotic usage in fish farming, such as those published in the 1970s in *Aquaculture,* dealt mostly with zootechnical considerations, such as which antibiotic to use for which disease, at which concentrations, preventively, or therapeutically? What are the effects of water pH, salinity, etc., on antibiotic efficiency? Antibiotic usage then contributed to stabilizing a zootechnical arsenal that also included water treatment, feed quality, temperature, etc., allowing increased productivity by reducing mortalities despite a higher density of fish. The key word 'Antibiotic-resistance' did not appear until 1991. In *La pisciculture française,* several articles discussed the technical difficulties of using antibiotics, and their high cost, which led to advice to restrict their use to hatcheries or crayfish production (because of their high added value). The problem of the apparition of AMR was only considered in term of therapeutic failure. In term of possible harmful consequences for human health, the focus was again on residues in consumed food.

In 1987, the risk of restrictions on the use of antibiotics due to European legislation is mentioned:

> Generally speaking, the ability to treat fish is seriously threatened by regulations being set up in the EEC (that already exist in other countries). Antibiotics are particularly targeted[35]

However, de Kinkelin two years later acknowledged potential problems:

> Vibrio belonging to different species are bacteria, which, in association, most often attack shrimp. This results in the massive distribution of antibiotics, possibly preventively, but in warm water and in presence of bacteria potentially pathogenic to humans, in water as well as in the shrimp, the risk of inducing bacterial resistance in human pathology becomes very real.[36]

Another veterinarian, who contributed to the initial development of shrimp farming in Ecuador in the late 1970s, says:

> Question: But you don't systematically add antibiotics?
> Answer: Yes, in this setup. Could we do without? Once again, by decreasing… (*implied: decreasing farmed shrimp density*). That said, I can tell you that it has no impact on the life of the post-larvae or the adult shrimp. None.[37]

However, according to him, such usage concerned only the early period of shrimp farming development (in the 1970-1980s). Later on, Ecuadorian farmers found ways to avoid using antibiotics by adding probiotics.

---

35 Pierre de Kinkelin (1987) in *La Pisciculture française d'eau vive et d'étang saumâtre et marine,* 92.
36 Pierre de Kinkelin (1989) in *La Pisciculture française d'eau vive et d'étang saumâtre et marine,* 98, p.10.
37 A consulting veterinarian specializing in shrimp farming. Interviewed 15 June 2016.

For him, it was a zootechnical question:

> (*Regarding antibiotic usage*:) This has become routine, we don't think about it, for instance, that is the way it happened in the Philippines, with the same circuit. [...] But before this, we still need to have the antibiotic resistance monitoring in place, things like that.

Altogether, this interview highlights two points. Antibiotic usage was problematized in term of zootechnical considerations (high cost, potential sanitary problems of residues in animal flesh), and antibiotic resistance in term of therapeutic failure, not in term of possible consequences for human global health. In addition, regulations were judged too restrictive: *Si vous respectez le droit, les normes, ce n'est quasi pas possible* ("if you follow the law, the rules, it's nearly impossible").[38]

IV-3. The science of AMR in aquaculture: controversy in the academic world

In the 1990s, several articles dealing with the potential risks linked to aquaculture were published in academic journals. The characterization of genetic elements at the DNA sequence level could be used to trace how AMR migrates from a fish or a shrimp pathogen, where it had been selected by an antibiotic treatment, to a human pathogen. Initial reports of such examples in the late 1980s and 1990s multiplied in the 2000s (DePaola et al., 1988: 1861-1863; Kim-Aoki, 1993: 103-09; Kim-Nonaka-Suzuki, 2004: 147-56). As in the debate over agriculture usage, authors were fundamental microbiologists, who researched the molecular mechanisms underlying resistance as well as resistance gene circulation, using molecular tools. The results were published in microbiology rather than aquaculture journals.

On the other side, a fish pathologist, David James Alderman, who worked at the Center for Environment, Fisheries and Aquaculture Science (CEFAS, UK) came to the rescue of aquaculture in 1998:

> Research has shown that antibiotic resistance can be transferred experimentally from a fish pathogen to a potential human pathogen. However, the important question is not the existence of a risk but the magnitude of that risk. Relevant data are in short supply but it would appear that the risk to public health from antimicrobial use in aquaculture is probably very low. (Alderman- Hastings, 1998)

Felipe Cabello, a microbiologist at New York Medical College, reviewed in a 2006 paper several examples of potential transfer of AMR determinants from aquaculture to human pathogens, such as *Salmonella enterica* serotype Typhimurium DT104 (Cabello, 2006). Previous work claiming that the multiresistance of a *Vibrio cholerae* strain responsible for the 1991 Latin American epidemic came

38 Veterinarian. Interviewed 15 June 2016.

from an Ecuadorian shrimp farm also was cited. The ensuing controversy involved Peter Smith, a microbiologist at the National University of Ireland, Galway, and a specialist in infectious diseases control in aquaculture who was serving as editor of the disease section of the journal *Aquaculture*, where he published a rebuttal of Cabello's paper (Smith, 2007).

More recently, another scientific article proposed that the multiresistance determinants observed in a *Salmonella* strain that initially emerged in Egypt could have come from aquaculture, given the heavy use of antibiotics in the 1990s in integrated farms rearing both chickens and fish (Le Hello et al., 2011).

Hence, since the 1990s, an increasing number of scientific articles were published which questioned the risks posed by the use of antibiotics in aquaculture for the development of AMR in humans. Examples of antibiotic resistance determinants present in aquaculture farms were multiplying in the literature, leading scientists to search for alternatives to antibiotics to fight diseases, such as phage therapy and essential oils. However, not all specialists were convinced. De Kinkelin's colleague and co-author, Christian Michel, drew this conclusion in his recent book *Gestion de la santé des poissons* (Michel, 2018: 439).

> Aquaculture does not escape from the criticisms periodically expressed by eco-activist groups, consumer and medical, for whom the unique solution one can think of is to ban the usage in animals of all anti-infectious drugs of interest for human treatment.

It is now possible to identify two groups of experts/scientists taking sides in this debate. On the one side, microbiologists, working on antibiotic resistance mechanisms from a fundamental view point, well recognized by the academic community, insist on the risk posed by intensive aquaculture as a source of AMR, and campaign for stricter regulations. On the other side, applied microbiologists and aquaculture experts, often veterinarians, working in applied laboratories to support aquaculture development, or directly for farms or feed producers, closer to the profession, minimize the risk and adopt a zootechnical framework, where the usage of antibiotics has to be considered as much with regard to the economic viability of fish and shellfish farming as to the highly hypothetical risk for human health.

IV-4. From science to regulatory actions: the difficult emergence of a public problem

In the mid-1990s, Denmark discovered that the increase in animals of enterococci resistant to avoparcin, an antibiotic used as a growth promoter in poultry and pigs, correlated with an increase in hospital settings of enterococci resistant to vancomycin, an antibiotic used only in humans but from the same family as avoparcin. This led Denmark, Germany and Norway to ban its usage as a growth factor. In 1997, following the *precaution principle*, the European Commission temporarily banned the addition of avoparcin to animal feed (Fortané, 2016;

Berger, 2016). According to Tristan Berger, the avoparcin crisis was a key event in putting animal antibiotic resistance on the European political agenda, making it a public problem (Berger, 2106, Arquembourg, 2016a).

In 2002, the national French health insurance system organized a campaign based on the slogan *les antibiotiques, c'est pas automatique* (antibiotics, it's not automatic) targeting doctors and patients, which led to a 24% reduction of antibiotics in 2005 but did not address the question of usage in animals (Carlet-Le Coz, 2015: 46). As already mentioned, starting in 1999, AFFSA produced regular reports on antibiotic sales for veterinary use.[39] The 2006 AFFSA report, in addition to the survey results, discussed the role of antibiotic use in animal farming in AMR development and the potential consequences for public health.[40] Known mechanisms of antibiotic resistance and transfers were discussed, based on the scientific literature. For aquaculture, reports of the presence of resistance to florfenicol and tetracycline in fish pathogens, two antibiotics which are authorized for fish, were cited.

After 2007, the surveillance of AMR in fish pathogens also was included under the Resapath surveillance network. However, data collection (strains isolation and antibiogram to determine resistance) in fish appeared problematic. In the 2015 report, one can read:

> Currently compiled data do not allow the presentation of AMR results due to the low number of isolates at the bacterial species level and the uncertainty surrounding the representativeness of the data and the antibiogram methodology for certain bacteria such as *Aeromonas salmonicida*.[41]

Since the Resapath surveillance network relies heavily on practicing veterinarians, the small number of veterinarians specializing in aquaculture may partly explain these difficulties (Fortané, 2015). In 2014, an ANSES report on the risk of AMR due to veterinary usage included several sections dedicated to the problematic in aquaculture.[42] This report was part of the first governmental program to reduce the risk of AMR in veterinary medicine, Ecoantibio 2012-2017.[43] In the proposed actions, none specifically concerned aquaculture, or fish farming, the only mention being to signal a coming action by the OIE. Later on, in the document presenting the second plan (2017-2021), aquaculture remains absent,

39 AFFSA *Suivi des ventes de médicaments vétérinaires contenant des antibiotiques en France en 2004* (2004). https://pop.anses.fr/fr/system/files/ANMV-Ra-antibiotiques2004.pdf (last consulted on 3/06/2023).

40 AFFSA, *Usages vétérinaires des antibiotiques, résistance bactérienne et conséquences pour la santé humaine* (2006). https://www.anses.fr/fr/system/files/SANT-Ra-ABR.pdf (last consulted on 26/03/2023).

41 *Aeromonas salmonicida* is a major pathogen encountered in *Salmonicidae* aquaculture.

42 ANSES, 2014.

43 Ecoantibio 2012-2017: https://agriculture.gouv.fr/plan-ecoantibio-2012-2017-lutte-contre-lantibioresistance (last consulted on 26/03/2023).

although three research projects addressing problems specific to fish farming will be financed by the program.[44]

It thus seems that in France, the problem of AMR linked to aquaculture started to emerge in the regulatory arena in the mid-2000s, but without becoming a specific target of public policies. Present in reports from regulatory agencies in charge of the problem of AMR (for instance, Février, 2013), it was absent in reports addressing conditions for the development of aquaculture. On the contrary, such reports, generally commissioned by the Ministry of Agriculture, complain regularly about too many regulations, restricting the possibility to fight diseases in aquaculture and compromising the sector's economic viability. In this framework, addressing risks resulting from seafood farming activities could be done only from the unspecific inclusion of seafood in the generic 'Animal' category.

IV-5. Fish and shellfish farmers response to the perception of growing concerns in the public

In France, fish and shellfish farmers are organized into national committees that reflect more or less the administrative division between the marine and the continental water domains and the different sectors of aquaculture (shellfish farming, fresh water and marine water production). Oyster, mussel and other shellfish farmers are members of the *Comité National de la Conchyliculture* (CNC) and are further organized at the regional level, with the existence in concerned regions of a *Comité régional de la conchyliculture* (CRC, regional committees). In the case of fish farming, whether fresh water or marine fish, the organizations representing the different categories of actors (farmers, feed producers and fish transformers) are regrouped into the *Comité Interprofessionnel des Produits de l'Aquaculture* (CIPA). The structuration as well as the committee missions and functioning are highly institutionalized and defined by law.[45] State representatives are present in these committees, which are placed under the control of the Ministry of Agriculture. Adhesion of all members of the profession is mandatory and the committees must be consulted on all matters concerning the sector for which they are competent. CIPA, created in 1997, was officially recognized by public authorities at the national and European level in 1998. Its missions include interactions with customers, promoting aquaculture and its products, representing

---

44 Ecoantibio 2: https://agriculture.gouv.fr/le-plan-ecoantibio-2-2017-2021 (last consulted on 26/03/2023).

45 Respectively '*Le Comité national des pêches maritimes et des élevages marins (CNPMEM)* (National committee for marine fisheries and marine farming) (https://www.comite-peches.fr/)'; 'Le comité national de la conchyliculture (CNC) (http://www.cnc-france.com/)'; '*Comité Interprofessionnel des Produits de l'Aquaculture (CIPA)* (Interprofessional committee for aquaculture products)' (https://www.poisson-aquaculture.fr/le-cipa/). See also CGEDD and CGAAER *Comités en charge des pêches, de la conchyliculture et de la pisciculture : bilan et perspectives* (2015).

the professions' interest to public authorities, and managing environmental and sanitary constraints.

Around 2000, CIPA engaged in a reflection on the theme of 'sustainable aquaculture' in response to the need to improve their image with customers, but more specifically spurred by big supermarket chains following the Bovine Spongiform Encephalopathy sanitary crisis.

Concretely, this translated into the setting-up of a certification scheme for sustainable aquaculture, the quality charter *Aquaculture de nos régions*, in collaboration with WWF, and with the support of public authorities, especially the Ministry of Agriculture. In this charter, the accent is on freshness, quality and sanitary safety and more broadly on the reduction of the environmental impact of the aquaculture industry, reflecting the new preoccupations of customers.

> Reducing the environmental footprint of aquaculture has become a high priority as part of the drive for greater sustainability in modern society [...] Equally of note is the heightened environmental awareness among aquaculture professionals, who are now concerned about the image their business enjoys with consumers and the general public. (Le Gouvello and Simard, 2017: 8)

A new auto-diagnostic tool, IDAqua®, was developed in 2006 (and finalized in 2008) in collaboration with several research organizations (INRAE, IFREMER) to help farmers determine the sustainability of their farms. Usage of antibiotics is one specific item among many others in the sanitary section, and is incorporated in the calculation of the final score. In addition, CIPA was a partner in the Ecoantibio 2012-2017 program.

The use of antibiotics in hatcheries is a controversial subject, openly discussed on videos and documentaries available on the internet. The CEO of one of the major French oyster hatcheries, producing triploid oysters, forcefully claims in a documentary that antibiotics must be used at critical stages of the rearing cycle:

> When we are growing oyster larva, we use an antibiotic for one day at three precise stages. That's it. Anyone who says, 'I don't use antibiotics in rearing', is a quack. Either he is a liar, or he is a very bad farmer who produces very bad quality products.[46]

V. Conclusion

According to Jocelyne Arquembourg, the problem of the risks posed by AMR to human health emerged in the public arena around 2000 in France, meeting two criteria of what constitutes a public problem, namely visibility in the public sphere

46 https://www.youtube.com/watch?v=yCe4akQDKXE= *Triploïde : enquête sur les huîtres génétiquement trafiquées* (2016) (last consulted on 23/03/2023).

and the call for political action (Arquembourg, 2016b). This public problem was further fragmented into several problems, each the object of a specific public policy. The 2002 campaign, *Les antibiotiques, c'est pas automatique* targeted the doctor/patient relationship as the main cause of antibiotic misuse. At the same time, several reports and surveillance plans contributed to constructing the usage in animal husbandry as a public problem, but it was only with the first Ecoantibio plan, in 2011, that this usage became the object of public actions.

Antibiotics are not used only in human medicine and animal husbandry, but also in aquaculture, a usage that accompanied the transformation of this activity in the 1970s, and one following the model set up in the 1950s for agriculture. Despite attempts by some scientists to question the use of antibiotics in aquaculture, especially after the 1990s, the debate over its role in AMR remained mostly confined to the scientific and expert arena in France. Opposition between two types of expertise can be seen. On the one side, there are microbiologists working on mechanisms of bacterial resistance and transfer. On the other, there are veterinarians working in public research laboratories but with a mission to support aquaculture development. The latter are close to actors in the profession, and have adopted their zootechnical framework and concerns for the economic viability of the activity. In terms of public actions, reports from regulatory agencies mobilize the scientific data provided by the first category of experts, whereas reports in favor of aquaculture development adopt the framework of the veterinarians. Hence, two logics are at odds: the economic logic of development of aquaculture and the sanitary logic aimed at protecting human health. These two logics fail to interact together to generate a coherent public policy.

This does not mean that French aquaculture is outside all regulatory processes, or that it is not subject to public policies. Indeed, both farmers and veterinarians keep complaining about the complexity of the rules and advocate for more flexibility. However, in terms of AMR, if the aquaculture question has gradually come onto the political agenda, it was by extending the question of antibiotic usage in animal husbandry. Fish and shellfish are legally part of 'Animals', and their farming is subject to the same veterinary laws, in particular regarding prescriptions and sanitary regulations, including antibiotic prescriptions. This extension of animal husbandry regulations to the domain of aquaculture has led veterinarian experts to join the debate, first to claim 'their right to cure', but also paving the way for a progressive acknowledgement of the problem. Alerts and regulations coming from international organizations such as OIE and the EU also likely have played a role in helping the question of AMR and aquaculture to emerge, although the importance of this role must be studied further.

On the side of seafood farmers, narratives defending efficient zootechnical practices, where antibiotic usage is presented as both unavoidable and harmless, coexist with recent efforts by the profession to address perceived consumer concerns. However, what the consumers' narrative attacks is the unsustainability of intensive farming rather than the AMR problem *per se*. In this framing of the debate, for seafood as well as for animals, antibiotic usage becomes the hallmark of a certain type of intensive, industrialized farming, together with pesticides, herbicides and chemical fertilizer utilization. Reciprocally, reduced usage is associated with organic food production and quality labeling systems, set up by the profession, or certain sectors of the profession, if possible with the collaboration of environmental organizations such as WWF and consumer associations. In this framework, the problem of AMR linked to antibiotic usage in aquaculture has yet to meet its broader public.

Bibliography

Académie des sciences (2003) '*L'Aquaculture marine. Exploitation et surexploitation des ressources marines vivantes. Rapport sur la science et la technologie*', 17, pp. 269-79

Alderman, D.J. and Hastings T.S. (1998) 'Antibiotic use in aquaculture: development of antibiotic resistance- potential for consumer health risks'. *International Journal of Food Science and Technology*, 33, pp. 139-55

Arquembourg, J. (2016a) 'Les Paradoxes de l'Antibiorésistance', *Questions de Communication*, 29, pp. 7-13

——— (2016b) 'L'antibiorésistance en France, du risque à la menace pour la santé publique. Analyse d'un processus inachevé de constitution d'un problème public', *Questions de Communication*, 29, pp. 29-47. https://journals.openedition.org/questionsdecommunication/10380

Begemann, S. (2018) 'How political cultures produce different antibiotic policies in agriculture: a historical comparative case study between the United Kingdom and Sweden'. *Sociologica Ruralis*, 58, 4, pp 765-85

Berdah, D. (2021) 'Pour une autre histoire de la « modernisation » des pratiques d'élevage : des antibiotiques dans les rations pour stabiliser les systèmes sociotechniques pharmaceutiques industriels français et britannique après 1945' in *Histoire des Modernisations Agricoles au XXᵉ siècle*, eds Lyautey, M., Humbert, L. and Bonneuil C. Rennes: Presses Universitaires de Rennes: pp. 101-20

Berger, T. (2016) 'Doit-on s'inquiéter de l'usage d'additifs antibiotiques en élevage ? Evolution des questionnements liés à l'antibiorésistance animale' *VertigO – la revue électronique en sciences de l'environnement*, 16, 3. https://www.erudit.org/en/journals/vertigo/2016-v16-n3-vertigo03075/1039996ar/

Bonneuil, C. and Thomas, F. (2009) *Gènes, pouvoirs et profits. Recherche publique et régimes de production des savoirs de Mendel aux OGM*. Versailles, Lausanne: Quae; Fondation Pour Le Progrès De L'homme

Bud, R. (2007) *Penicillin: Triumph and Tragedy*. Oxford: Oxford University Press

Bush, K. (2011) 'Tackling antibiotic resistance'. *Nature Rev Microbiol*, 9, pp. 894-96

Cabello, F. (2006) 'Heavy use of prophylactic antibiotics in aquaculture: a growing problem for human and animal health and for the environment'. *Environmental Microbiology*, 8, pp. 1137-1144

Carlet, J and Le Coz, P. (2015) *Together, let's save antibiotics. Report from the special working group for keeping antibiotics effective* Paris, Ministère des Affaires Sociales, de la Santé et des Droits des femmes

Corley, T. and Godley, A. (2011) 'The veterinary medicine industry in Britain in the twentieth century', *Economic History Review*, 64 (3), pp. 832-54

Denis, G. (2014) 'Une histoire institutionnelle de l'Institut national de la recherche agronomique (Inra) – Le premier Inra (1946-1980)'. *Histoire de la recherche contemporaine* [On line, 2014], III, 2. https://journals.openedition.org/hrc/802; DOI : https://doi.org/10.4000/hrc.802

DePaola, A. (1988) 'Phenotypic and genotypic characterization of tetracycline- and oxytetracycline-resistant *Aeromonas hydrophila* from cultured channel catfish (Ictalurus punctatus) and their environments' *Appliedl Environmental Microbiology*, 54, pp. 1861-1863
Dewey, J. (1927) '*Le public et ses problèmes*' transl. from English USA by Joëlle Zask, Paris: Gallimard, 2010
Doumenge, F. (1984) 'L'aquaculture française: bilan et perspectives'. *in* thematic issue: *La France et la gestion du milieu marin et côtier, Norois*, 121, pp. 77-96. https://www.persee.fr/doc/noroi_0029-182x_1984_num_121_1_7364
Ducateau, P.F (2015) 'Pisciculture : Bilan sur les modes d'élevage et perspectives en matière de durabilité environnementale. Exemple du saumon atlantique en Norvège' (unpublished veterinarian doctoral thesis, Ecole Nationale Vétérinaire d'Alfort. theses.vet-alfort.fr/telecharger.php?id=2010
Février, J. (2013) 'Un cadre pour conduire une politique de santé animale dans la filière aquacole. Conseil général de l'alimentation, de l'agriculture, et des espaces ruraux'. *CGAAER*, 12107. p. 9, 18. https://agriculture.gouv.fr/sites/minagri/files/documents//CGAAER_12107_2013_Aquaculture_cle8f91d3.pdf
Finlay, M. and Marcus, A.I. (2016) 'Consumer terrorists: Battles over agricultural antibiotics in the United States and Western Europe'. *Agricultural history*, 90, 2, pp. 146-72
Fortané, N. (2015) 'La surveillance comme dispositif-frontière. La triple ontologie des bactéries résistantes d'origine animale' *Revue d'Anthropologie des Connaissances*, 9, pp. 265-90
——— (2016) 'Le problème public de l'antibiorésistance en élevage: essai de généalogie et caractérisation' *Questions de Communication*, 29, pp. 49-66
Gradmann, C. (2016) 'Re-Inventing infectious disease: antibiotic resistance and drug development at the Bayer company 1945-80', *Medical History*, 60,2, pp. 155-80
Hedrick, R. (2013) 'Obituary – Pierre de Kinkelin Pelletan'. *Journal of Fish Diseases*, 36, pp. 975-76
Kim, E.H. and Aoki, T. (1993) 'Drug resistance and broad geographical distribution of identical R plasmids of *Pasteurella piscicida* isolated from cultured yellowtail in Japan'. *Microbiology Immunology*, 37, 103-09
Kim, S.R., Nonaka, L. and Suzuki, S. (2004) 'Occurrence of tetracycline resistance genes *tet*(M) and *tet*(S) in bacteria from marine aquaculture sites'. *FEMS Microbiology Letters*, 237 147-56
Kirchhelle, C. (2018) 'Pharming animals: a global history of antibiotics in food production (1935-2017)', *Palgrave Communications*, 4, pp. 96-109
——— (2020) *Pyrrhic Progress: the History of Antibiotics in Anglo-American Food Production*. (Critical Issues in Health and Medecine, English Edition eBook) Piscataway: Rutgers University Press
Le Gouvello, R. and Simard, F. (eds.)(2017). *Durabilité des aliments pour le poisson en aquaculture : Réflexions et recommandations sur les aspects technologiques, économiques, sociaux et environnementaux*. Gland, Suisse : UICN, et Paris, France : Comité français de l'UICN. 296 p. 8

Le Hello, S. , (2011) 'International spread of an epidemic population of *Salmonella enterica* serotype Kentucky ST198 resistant to ciprofloxacin'. *Journal of Infectious Diseases*, 204, pp. 675-84

Levy, S.B., FitzGerald G.B., and Ann B. Macone (1976) 'Changes in intestinal flora of farm personnel after introduction of a tetracycline-supplemented feed on a farm'. *New England Journal of Medicine*, 260, 40-44

——— (1992) *The Antibiotic Paradox: How Miracle Drugs are Destroying the Miracle*. New York: Plenum 2002

——— (1997) 'Antibiotic resistance: An ecological imbalance' in *Antibiotic resistance: origins, evolution, selection and spread*. Ed. by Derek J. Chadwick and Jamie Goode. Book series: Ciba foundation symposia, 207, 1-9

Lulijwa, R., Rupia E.J. and Alfaro A.C. (2020) 'Antibiotic use in aquaculture, policies and regulation, health and environmental risks: a review of the top 15 major producers'. *Reviews in Aquaculture*, 12, pp. 640-63

Marres, N. (2005) 'Issues spark a public into being: a key but often forgotten point of the Lippmann-Dewey debate' *Making Things Public*, Latour, B. and Weibel, P. eds. Cambridge MA: MIT Press, pp. 208-17

McKenna, M. (2017) *Big chicken. The incredible story of how antibiotics created modern agriculture and change the way the world eats* Washington, D.C.: National Geographic

Michel, C. (2018) *Gestion de la Santé des Poissons*. Coord. Versailles: Quae

Sheppard-Sellam, E. (2019) 'Problème public' in *Dictionnaire des politiques publiques: 5e édition entièrement revue et corrigée*. ed. by Laurie Boussaguet. Paris: Presses de Sciences Po, pp. 504-10 https://doi.org/10.3917/scpo.bouss.2019.01.0504

Smith, P. (2007) 'Antimicrobial use in shrimp farming in Ecuador and emerging multi-resistance during the cholera epidemic of 1991: A re-examination of the data'. *Aquaculture*, 271, pp. 1-7

Tanguy, H. (2008) *Rapport final de la mission sur le développement de l'aquaculture*. Ministère de l'agriculture et de la pêche/ Ministère de l'écologie, de l'énergie, du développement durable et de l'aménagement du territoire

ALEXANDRA WALUSZEWSKI

13. More research? Or more *use* of research and practical experiences?

On a systemic innovation compensating for regular use of antibiotics in Swedish pig meat production

I. Introduction and research design

What motivated Swedish farmers engaged in pig meat production to mobilise against routine use of antibiotics in the early 1980s, decades before antibiotic resistance stabilised as a health challenge on the global policy agenda (Thoms 2012; Kirchhelle, 2018; Chandler, 2020)? And why did 'insiders', farmers and the Swedish Farmers' Association, take on a role commonly represented by 'outsiders' such as interest groups, media debaters and activists (Van de Poel, 2000; Elzen et al., 2011)?

What was the content of the systemic change initiated in the pig meat setting, an area that conventionally, along with poultry, represents the highest consumption of antibiotics (Kirchhelle, 2018)? A systemic shift that eventually made Sweden the most restricted consumer of antibiotics for animals within the EU, and in Europe surpassed only by Norway and Iceland (not in EU but in the European Economic Area) (ESVAC 2020)? How was the cost of change distributed, when the main costs appeared in the producing setting and the main benefits in health care systems, in environment and society at large?

Finally, how does the systemic change relate to the contemporary research and innovation policy, and its great trust in independent market actors' ability to transform scientific knowledge advances to innovations? A policy recipe based on the idea that knowledge advances corresponding to economic and societal needs will, if they are made available to economic actors, be absorbed and transformed to innovations without contextual interference (Hasselberg, Y., Rider, S, Waluszewski, A, 2013: 5-11; Mirowski, 2011: 41-86; Waluszewski, 2011).

The aim of this chapter is to shed light over these questions concerning motives, content, and distribution of costs and benefits related to a systemic innovation able to compensate for routine group treatment of animals with

Alexandra Waluszewski • Department of Economic History, Uppsala University: alexandra.waluszewski@sts.uu.se

From Breeding & Feeding to Medicalization, Rural History in Europe, 17 (Turnhout: Brepols, 2023), pp. 313–336
BREPOLS PUBLISHERS 10.1484/M.RURHE-EB.5.135031

antibiotics, as well as how these patterns correspond to contemporary innovation policy. The chapter relates to literature on systemic characteristics of innovation (Van de Ven et al., 1999; Geels, 2004; Håkansson and Waluszewski, 2007 (eds.); Håkansson et al., 2009) and more precisely, systemic changes called forth by the antibiotic resistance challenge (Elzen et al., 2011; Van Boeckel et al., 2015; Baraldi et al., 2016; Kirchelle, 2018; Waluszewski, Baraldi, Ciabuschi, 2018; Chandler 2020). Furthermore, the study relates to literature on path dependencies due to investments in place (Utterback and Abernathy, 1975; Håkansson et al., 2009); that is interdependencies of material and social resources across organisational and legal borders (Håkansson and Waluszewski, 2002; Håkansson et al., 2009; Waluszewski, Baraldi and Ciabuschi, 2018). In studies of AMR, such embeddedness has been captured in terms of 'antibiotic infrastructure', referring to the physical and cultural interdependencies characterising production and use (Chandler, 2020; Kirchhelle, 2020: 38-45).

Finally, the chapter adds to literature on contemporary research and innovation policy (Slaughter and Rhoades, 2004; Grandin, Wormbs, Widmalm, eds. 2004; Rider, Hasselberg, Waluszewski, 2013) and more precisely, to policies on the antibiotic resistance challenge, commonly referred to as antimicrobial resistance, AMR[1] (Kahn, 2016; Begemann et al., 2018; Kirchelle, 2018; Chandler, 2020). Although contemporary policy recognises systemic characteristics of societal challenges, the basic analytical approach is still based the assumption of independent actors. Through information, individuals are supposed to be able to change use of this behaviour (Chandler, 2020; Waluszewski 2011). Through transfer of knowledge corresponding to identified social and economic needs, economic actors are supposed to transform it to innovative products and services, without interfering with social and material investments in place (Elzinga, 2004; Håkansson and Waluszewski, 2007; Rider and Waluszewski, 2015).

A process-oriented network approach was used as a theoretical point of departure (Håkansson et al., 2009:65-91), based on the fundamental notion that economic exchange has a qualitative content; that is, the exchanged resources are heterogeneous and dynamic, and the value is created in directly and indirectly related resource interfaces, creating cross-organisational interdependencies. These underpinnings have a close kinship with observations of interdependency made in the heterodox economics research field (see e.g.: Lawson, 2006; Marglin, 2008; Mirowski, 2011; Fourcade, et al., 2015). Furthermore, the notion of different return-of-investment patterns in using, producing settings and developing settings (Håkansson and Waluszewski, 2007) was used, with a special focus on how changes in the producing setting are related to facility systems, and how changes in the user setting are related to established user systems (Håkansson and Waluszewski, 2007:153-55). The study also recognised the distinction between

1 Antibiotics are used to prevent and treat bacterial infections, while the wider concept antimicrobials include drugs to treat other microbes, such as parasites, viruses and fungi. Antimicrobial resistance, AMR, and antibiotic resistance are often used synonymously, which is also the case in this chapter.

the changes occurring at an image level and on a material level (Håkansson and Waluszewski, 2002: 78-85; Czarniawska and Joerges, 1996).

Methodologically, the investigation of the emergence, content and distribution of costs and benefits related to the systemic change compensating for regular input of antibiotics focused on directly and indirectly connected resources and the actors behind them. The study, which is related to two international research projects,[2] relied on a combination of methods. Primary data was collected through interviews with insights into the producing, using setting and developing setting; that is pig meat farmers, representatives of the Swedish Farmers' Association (LRF), slaughter/food processing firms, retail industry and researchers engaged in veterinary medicine and agrarian history. Twenty interviews with 18 respondents were conducted; a few were interviewed twice.[3] (See appendix 1). Participating observations were conducted during a one-day workshop titled 'Antibiotics & Resistance, Pig Meat Production & Consumption', that took place in Uppsala in early May 2019. The aim of the workshop, designed by the author, was to gather expert views on different aspects of this topic.[4] Six key notes engaged in veterinary medicine, agrarian history, business studies, anthropology/health studies, sociology and a consumer organisation contributed, along with some 35 researchers/experts in the audience. (See appendix 1). Finally, a range of studies published in related research fields were used. STS studies, history of medicine, agrarian history, and industrial network studies provided insight into the role of antibiotics in animal-based food production over time, and on policies related to antibiotic use, production and development. Reports by ESVAC, WHO and FAO and similar bodies were used to gain data on animal antibiotic consumption and resistance patterns.

II. Modern animal production and the emergence of an antibiotic infrastructure

During the first decades of the 21st century, antibiotic resistance and the need for changing user patterns, as well as measures to support development and production of new drugs and diagnostics stabilised on the global policy agenda

2 The study is related to two international research projects. The first is a Nordic research collaboration, labelled 'Follow the Pig' where related investigations on the role of antibiotics are carried out in the Danish, Norwegian and Finnish pig meat production system. The second is a Swedish-Italian collaboration, where the emergence and content of these different pig meat production systems and the role of antibiotics are investigated. The author has also benefitted from being part of the research network related to UAC, Uppsala Antibiotic Centre, Uppsala University.

3 All interviews followed a semi-structured guideline, focusing on the most important changes on the image level and materialised level, from each respondent's point of view. The interviews with pig meat farmers included site visits.

4 The workshop, on May 8, 2019, was arranged by the author in cooperation with Uppsala University's Antibiotic Centre, UAC and the 'Follow the Pig' project, and was documented by a research assistant.

(Chandler, 2020; Waluszewski, Baraldi, Ciabuschi, 2018).[5] AMR was discussed at a high-level meeting at the United Nations Global Assembly in 2016. Through the 'tripartite alliance' of WHO, FAO and OIE, the complexity of the resistance was approached with the 'One Health' concept; that is, as a public health threat to the 'human-animal-ecosystem interface'.[6]

> WHO strongly recommends an overall reduction in the use of all classes of medically important antibiotics in food-producing animals, including complete restriction of these antibiotics for growth promotion and disease prevention without diagnosis.[7]

However, the step from policy declarations to systemic change is an issue that 'AMR champions' have recurrently expressed worries about (Chandler 2020). *If* global policy commitments do not result in changed use, antibiotics for animals; which currently represents about two thirds of the world's antibiotic consumption, is estimated to increase more than 50 percent until 2030, then to reach 200,000 tons per year (van Boeckel et al., 2017). The consumption is, to a large extent, based on production economy: as precautionary measures to combat diseases related to intense production of animals, to stimulate animal growth and shorten the production cycle (Van Boeckel et al., 2015, 2017; Thoms, 2012; Kahn, 2016). From the Anglo-American origin, antibiotic-intense production systems were exported to high, middle and even low-income countries, with the common denominator that antibiotics are embedded into every day feeding routines (Kirchhelle, 2018; Lhermie, Gröhn, Raboisson, 2017; Finlay, 2004).

Animal-based food production's exact contribution to AMR is difficult to pinpoint, but transmission of antibiotic-resistant bacteria of animal origin to humans can take place in several ways: through people working at or with farms, through food products and the environment (van Boeckel et al., 2015; Kirchelle, 2018). However, scientific knowledge about the resistance phenomenon is far from new.

When antibiotics became embedded as a regular additive in animal-based food production, resistance was already a recognised phenomenon. In fact, scientific knowledge about the intrinsic characteristics of bacterial mutation giving rise to resistance had been articulated in parallel with the launch of antibiotics since the late 1940s, both in research publications and the general media (Kirchhelle, 2018; Wise, 2007).

---

5 The pharmaceutical industry's radical disinvestments in R&D and production of new antibiotics had contributed to the awareness. From the around 30 pharmaceutical companies that had antibiotics on the R&D agenda in the late 1980s and early 1990s, the first decade of the 21st century counted a handful still engaged (Kinch et al., 2014; Waluszewski, Baraldi and Ciabuschi, 2018). The need to restrict the use of eventually successful new drugs implied the ordinary innovation logic was broken; maximizing sales could no longer be used to cover R&D costs (Outterson et al., 2015).

6 https://www.who.int/news-room/q-a-detail/one-health.

7 https://www.who.int/en/news-room/detail/07-11-2017-stop-using-antibiotics-in-healthy-animals-to-prevent-the-spread-of-antibiotic-resistance.

Considering the severe consequences of bacterial infections *before* antibiotics were launched in the end of the 1940s and early 1950s, the great attention directed to the therapeutic advantages is understandable, and also when they were launched these drugs were labelled 'magic bullets'.[8] Infectious diseases that previously were devastating for animal health and for farmers' economy suddenly became manageable.

When it rather rapidly became evident that, along with their therapeutic role, antibiotics could also be used as a production economy mean to decrease costs and increase output, a new growing application area emerged (Thoms, 2012; Kahn, 2016; Kirchhelle, 2018). In the wake of Stockstad and Jukes' groundbreaking findings on growth effects on chicken and pigs (Stockstad et al., 1949), practical trials and publications on antibiotics as feed additives flourished. As stated by Thoms (2012:189):

> The literature boomed accordingly: during the 1960s, 100,000 research papers were published on the growth effects of antibiotics in animals.

The findings fit hand in glove with the industrialisation of animal-based food production (Thoms, 2012; Kahn, 2016; Begemann et al., 2018). The embedding of antibiotics in daily feeding routines implied that the boundaries between therapeutic use and the production mean blurred (Kirchhelle 2018). Routine group treatment with antibiotics in feed or water became a mean to speed up growth; especially of pigs and poultry, to increase animal density and using space more efficiently, and reducing manual treatment of each animal (Nathan and Cars, 2014; Thoms, 2012; Kahn, 2016; Kirchhelle, 2018). Put briefly, antibiotics became a key ingredient in industrialised animal-based food production, or as Thoms (2012:182) puts it:

> Antibiotics played an important – though hitherto not fully acknowledged – role in this development, as they allowed for an intensification of farming and animal husbandry that had been unthinkable before.

A number of economic and political forces supported and benefitted from this development path. In the *policy sphere*, in both Western and Communist economies, facilitating animal-based food producers' access to antibiotics became a mean in the rationalisation of agriculture and increasing the supply of affordable food (Kirchhelle, 2018; Begemann et al., 2018). For the *retail industry*, antibiotic-based production systems constituted an important source of low priced animal-based food products, used as important tools to attract customer streams (Finlay, 2004; Kirchhelle, 2018). For the *pharmaceutical industry*, antibiotics for animals became a significant application area at the same time as residues could be used as supply to *feed producers* (Kirchhelle, 2018; Finlay 2004). Antibiotics emerged both as a veterinary product, supplied by a handful of US and European companies,

8 A metaphor launched by Paul Ehrlich around 1900, based on the idea that a drug could kill a specific microbe without harming the body at large.

and as a feed additive (Wise, 2007; Li, 2017; Kirchhelle, 2018). Over time, the animal pharma industry played a double role in relation to supplying antibiotics for animals. On one hand, animal pharma engaged in the development and production of vaccines, allowing decreased consumption of antibiotics for animals (Postma et al., 2015). On the other hand, animal pharma took a stand for the contemporary consumption levels, through lobbying groups and through individual statements (Kirchhelle, 2018).

In *pig meat production* the productivity was affected through increased growth and disease prevention. Additionally, the natural weaning process gradually occurring at 2-3 months of age, could be made as short as three weeks, if the infection sensitive piglets were given antibiotics. This, in turn, allowed increased use of the sow's reproduction ability (Li, 2017). As Li, (2017:1668) summarises:

> Antibiotics added to the diet can improve a sow's reproductive performance, conception rate, litter size and farrowing rate.

A purchaser at a Swedish slaughterhouse made a similar reflection:

> In order to increase productivity, the main thing to work with is shortening suckling periods and time between litters. Short suckling gives infections sensitive pigs and increased need to use antibiotics.[9]

The 1960s' awakening of environmental awareness with Rachel Carson's *Silent Spring* (1962) as an eye opener, and Ruth Harrison's *Animal Machines* (1964), focusing on the treatment of animals in modern food production, directed attention to the extensive use of drugs and lack of animal welfare (Kahn, 2016; Begemann et al., 2018; Kirchhelle, 2018). The environmental debate, along with resistance reports from both animal and human medicine, triggered the UK government to initiate the 'Swann Committee', focusing on antibiotics and resistance in human medicine and animal production. When the Swann report was delivered in 1969, it was stated that antibiotics for growth promotion drives resistance, thus threatening human and animal health (Wise, 2007; Kahn, 2016; Kirchelle, 2018; Begemann et al., 2018). The suggested solution was a division of antibiotic use; some should be accepted for routine use in feed (such as macrolides) while others should be reserved for human medicine (such as penicillin and tetracycline) (Swann et al., 1969; Wise, 2007; Kahn, 2016; Kirchhelle, 2018). The recommendation of the Swann Committee was partly followed, both in the UK and in the European Economic Community, and penicillin and tetracycline as growth promoters were phased out. However, along with the immediate need to alter the prevailing practice of group treatment, the short-term economic benefits were also considered, or as Kahn, (2016: 20) said:

9 Purchaser, slaughterhouse, interview 2018-10-29.

> However, the committee recognised the important role that low concentrations of antibiotics played in the growth rates of young pigs and poultry; this practise had commercial advantages.

In practice, the Swann recommendation created a back door. Antibiotics that should not be used for growth promotion could still be used for routine group treatment of animals, as long as they were provided by veterinarians for 'disease prevention' (Kahn, 2016; Begemann et al., 2018; Kirchhelle, 2018). Although the Swann recommendations influenced the policies of several European countries, the outcome became, as Kirchhelle (2018, p. 10) said, 'a patchwork' of different policy approaches that did not stop the embedding of antibiotics in animal-based food production. Instead of decreasing the regular use of antibiotics, increased and antibiotic dependent production systems continued to spread globally (Kirchhelle, 2018; Nathan and Cars, 2014).

> With the exception of Sweden, the ongoing availability of nontherapeutic AGPs, coccidiostats, and higher-dosed prescriptions minimised early bans' impact on European agriculture until the 2000s. (Kirchhelle, 2018:10)

III. The Swedish case – four decades of systemic change

What motivated 'insiders', pioneering pig meat farmers and managers at the Swedish Farmers' Association to engage in getting rid of routine group treatment of animals with antibiotics already in the early 1980s? And what measures were behind the radical reduction of consumption of antibiotics over the decades decreasing from around 50 mg/PCU to the contemporary 12 mg/PCU[10] (ESVAC 2020), a level that is still the odd exception within Europe, besides Sweden reported only in Norway and Iceland. The average within the EU and European Economic Area (EEA) was estimated by ESVAC (2020) to be 103 mg/PCU, while over 200 mg/PCU was reported from Italy, Spain and Cyprus. The significant differences in terms of consumption patterns within the EU, along with the low consumption figures in a few Nordic countries, has also caused interest among both researchers and global policy bodies. As stated in a World Economic Forum (WEF) Risk Report, 2013:33:

> Research is needed to understand how Nordic countries have made significant progress.

Below we will take a closer look at the context in which this progress emerged, on the concrete changes behind it, and last but not least, how it was received and the costs and benefits were distributed.

---

10 Population Correction Unit, amount of antibiotics for every kg of bodyweight at time of treatment.

III-1. Modernised pig meat production and an environmental shift

When the engagement in getting rid of everyday feeding with antibiotics took off in Sweden in the early 1980s, pig meat production, like Swedish agriculture at large, had been modernised. Animal-based food production in general relied on regular group treatments with antibiotics through feed and water, above all for pigs and poultry, were 'practically all piglets' (Wierup, 2001: 85) were given antibiotic growth promoters.[11] Sweden had adopted the suggestions made by the Swann committee to not use drugs reserved for human therapy as growth promoters. Furthermore, the Swedish (and Norwegian) welfare state had nurtured an organised scepticism towards guidelines produced by commercial bodies,[12] not least in relation to pharmaceutical firms (Kveim Lie, 2014). Still, antibiotics continued to be an established ingredient in purchased feed for pigs and poultry. As expressed by a manager at the Swedish Farmers' Association:

> The Swann Commission had evoked the awareness, but in practice little happened. The awareness was mostly seen as a motivation for continuous change of antibiotics.[13]

Sweden was not yet part of the EU, and output prices were negotiated between the state and the representatives for the farmers (Begemann et al., 2018; Martiin, 2015; Morell, 2011). The cooperative Swedish Farmer's Association was resourceful and through majority ownership had a significant role in relation to supplying and processing companies: feed suppliers, slaughterhouses, processing firms and equipment suppliers among others. As characterised by Flygare and Isacson (2011: 40):

> It had its own dairies, slaughter houses, wholesalers, and contractors, and it backed the local farming associations in matters of marketing, information, accounting, company law, data analysis, energy, veterinary medicine and, not least, planning for the future.

Awareness of the dark sides of routine treatments of animals with antibiotics, and the ambition to create change had strong links to what Larsson Heidenblad (2021: 12) characterises as an 'environmental shift', taking place during the second half of the 1960s, in the aftermath of the 'Silent Spring' debate. Voices concerned with what modernisation of agriculture meant for humans and nature reached the general public through increasing media coverage. Prominent scientists along with engaged representatives for governmental and non-governmental

---

11 Piglets were given olaquindox or mecadox (50 ppm) from the start of their feeding period until age of 10-12 weeks. Thereafter, they were given (avoparcin or virginimycin) until slaughter at the age of about 7 months (Wierup, 2001).

12 That is, drugs and chemicals should be based on corporate independent data in order to be trustworthy (Kveim Lie, 2014).

13 Retired Manager, Swedish Farmers Association, engaged since early 1980s. Interview, 2018-11-20.

organisations, contributed to the media debate and increased public awareness (Larsson Heidenblad, 2021; Begemann et al., 2018). In *Dagens Nyheter,* one of Sweden's largest newspapers, journalist Barbro Soller; who was later recognised as Sweden's first environmental journalist, brought attention to the role of biocides in agriculture and the treatment of farm animals (Larsson Heidenblad, 2021).[14] Her articles were published in 1971 in a book with the translated title *The Animal Factory,*[15] which can be considered a Swedish follow-up of Ruth Harrison's *Animal Machines* (Larsson Heidenblad, 2021).

The public debate over increased use of chemicals and toxic substances in agriculture and modern food production practices was also considered by the farmers themselves, who were in an era of rapid modernisation (Vail, Hasund, Drake 1994). Until the early 1960s, Swedish agriculture had been dominated by family-owned small farms and smallholdings spread all over the country[16] (Martiin, 2015; Flygare and Isacson, 2011; Morell, 2011). The radical restructuring was debated within the agricultural setting; some farmers thought it was too far-reaching. Although environmental demands were not promptly embraced, many farmers were concerned about the long-term consequences of the use of chemicals that accompanied the modernisation. The consequences of being exposed to toxic substances raised concerns. For example, in the wake of the intense media debate on biocides in the mid-1960s, changes made by individual farmers caused a decrease of about 50% in the consumption of biocide-processed seed a decade before it was legally banned in 1978 (Swedish Board of Agriculture 2017; Larsson Heidenblad, 2021).

In 1981, the newspaper *Dagens Nyheter,* through a series by journalist Thomas Michelsen, once again drew attention to the role of chemicals in modern farming, also published in a book with the translated title *'Mass food, Swedish agriculture in the age of chemicals'.*[17] Specific attention was directed to animal welfare and the role of antibiotics. The fact that about 30 tons of antibiotics per year were consumed in animal-based food production caused attention in the media, among the general public, and among animal-based food producers, especially in the pig meat setting.

A few years later, in 1985, the media debate was further fuelled when internationally recognised author Astrid Lindgren supported animal welfare and spoke against the over-medication of production animals, through a debate article in the same newspaper, *Dagens Nyheter.*[18] The article triggered a veterinarian, who shared the worries and already had discussions with Michelsen, to contact

---

14 In 1971 Barbro Soller was engaged by Swedish Television to develop its environmental news journalism.

15 *Djurfabriken.*

16 Only about 10% of farmland belonged to large estates, mainly located in the south of Sweden (Morell, 2011).

17 *Massmaten, svenskt jordbruk i kemiåldern.*

18 *Dagens Nyheter, Debatt, 1985-05-03.*

Lindgren. This lead to a cooperative venture where the veterinarian provided Lindgren with research-based data about the treatment and medication of production animals, which the author transformed into articles written in a fairy tale style, published by a supporting editor of one of Sweden's largest evening papers, *Expressen.*[19] The common theme of the articles was the great loss of animal welfare caused by the modernisation; a transition from 'animal care' to 'animal production', including over medication. The articles created great attention in the media in general, among the general public, and politicians.[20]

While representatives for animal-based pharma reacted negatively on the over medication accusation, and veterinarians were divided, from supporting to directly opposing, and the general public was upset. The latter caused serious worries in the political setting, not least among ruling Social Democrats.

III.-2. Farmers and a farmers' association engaged in change

When the debate was at its most intense, farmers and the Swedish Farmers' Association's work with getting rid of regular input of antibiotics was already running. In the early 1980s some pioneering pig meat farmers began experimenting with abandoning the regular input of antibiotics (and since antibiotics were a regular ingredient in feed, they had to pay a higher price for getting it antibiotic-free). As one of the pioneers expressed the worries:

> It was mainly experiences of how the soil is affected by chemicals that triggered the work. What happens with the soil in the long run, when manure from pigs fed with antibiotics is spread out on it?[21]

Through contacts on individual levels, the farmers got advice from researchers at the Swedish Agricultural University, with the introduction of batch-wise production as a mean to increase hygiene and precautionary health as one of the most important measures.

In 1981 the Swedish Farmers' Association presented its first antibiotic policy, suggesting voluntary exclusion of antibiotic growth promoters. Behind the Swedish Farmers' Association's engagement was a serious worry that the intense media debate should end up in lost consumers' trust in animal-based food. The antibiotic policy was followed by an extensive internal discussion, with voices both for and against it. Eventually, one of the pioneering pig meat producers presented a bill to the Farmers' Association, suggesting that it should push for a legal ban of growth antibiotics. The main argument was that in order to succeed with phasing out regular group treatment of animals with antibiotics, there must

---

19 *Dagens Nyheter*, 1985.05.05, *Expressen*, 1985.22.09; 1985. 10.27; 1986.01.03; 1987.01.14.

20 Later on Astrid Lindgren and Kristina Forslund published a book based on the debate (*Min ko vill ha roligt, Rabén & Sjögren, 1990*).

21 Interview, pig meat farmer, retired, engaged since 1969. 2019-04-25.

be the same conditions for all producers, that is, a forceful legal arrangement. Or as expressed by the farmer:

> We had to get a ban in order to get all with us.[22]

Along with the Swedish Farmers' Association, the argument was supported by the cooperatively owned Slaughter Association and the closely related Centre Party, formerly Farmers' party. In 1984 the Swedish Farmers' Association decided to approach the government with suggestions of a legal ban on antibiotics as a growth promoter, that is forbidding regular treatment of groups of animals through feed or drinking water. A similar bill was also presented by the Centre Party (see Nordéus, chapter 11 this volume). Since Sweden was outside the EU, and agriculture was state protected, the general assumption was that producers should be compensated for increased production costs.

The suggestion to forbid regular treatment of groups of animals through feed or drinking water also met resistance. The strongest argument against it was presented by representatives of the pharmaceutical industry, rather aggressively addressed directly to the pioneering farmers, to managers driving the issue within the Swedish Farmers' Association and engaged politicians within the Centre Party. In the political setting, the Right Wing party argued that the law was not necessary since the output, the food products, did not include any antibiotic residuals. Among those supporting a ban, a common experience was the initial doubtfulness and split opinions and among veterinarians. Some supported a ban, while others opposed it based on fear of both increasing animal infections and running production costs. As expressed by a manager at the Swedish Farmers' Association:

> Many veterinarians considered regular group treatment with antibiotics as very efficient for animal health and for production economy.[23]

Despite the controversies, as the first country in the world to do so, Sweden gained a new law in 1986 banning the use of antibiotic growth promoters (AGP), that is routine treatment with antibiotics in feed and water without a veterinary prescription (SFS, Swedish Code of Statutes 1985:295).[24]

> It was the farmers that asked for this, individually, through their organisations and politicians.[25]

From then forward antibiotics were only allowed by veterinary prescription for treatment of medical diseases. The Centre Party also suggested that the same ban

---

22 Interview, pig meat farmer, retired, engaged since 1969. 2019-04-25.

23 Retired Manager, Swedish Farmers' Association, engaged early 1980s. Interview, 2018-11-20.

24 https://www.riksdagen.se/sv/dokument-lagar/dokument/svensk-forfattningssamling/lag-1985295-om-foder_sfs-1985-295 25/04/2023.

25 Retired Manager, Swedish Farmers' Association, engaged early 1980s. Interview, 2018-11-20.

should be applied to imported meat, a requirement that was turned down by the Parliament.

A year later, the ruling Social Democrats presented a related bill on a new animal welfare law.[26] The law, which was passed in 1988, through its focus on precautionary health, became important for realising a decreased consumption of antibiotics. For pigs, the law among others implied that sows should be free ranged and have access to straw or similar bedding material. Besides the ability to access natural behaviour, including requirements of certain space for each animal, precautionary health was a key issue:

> The animals shall in the future be protected not only from suffering but also from disease. Animal husbandry must therefore be focused on keeping animals healthy.[27]

The law caused attention internationally, perhaps more as a curiosity than as an understanding of the role of precautionary health, and was summarised as follows in *The New York Times*, under the headline 'Swedish farm animals get a bill of rights':

> The Swedish welfare state, long renowned for its generosity and scope, is extending its reach into a new frontier: the barnyard.[28]

III-3. A ban followed by decades of continuous change

With the legal ban of regular treatment of antibiotics in place, it became evident what every day group treatment had covered up. When antibiotics were excluded, at most farms without preparations, infectious diseases and mortality among piglets increased, both in the suckling and weaning periods, leading up to an increased prescription of antibiotics (Bengtsson and Wierup, 2006).

However, after a few years the total consumption of antibiotics for animals began to decrease; from about 50 mg/PCU two years before it, to 31 mg/PCU in 1990. From 1995 forward group treatment began to decrease, representing about 10% of the total consumption of antibiotics, which during the last years has stabilised to around 12 mg/PCU (Swedres-Svarm 2018). With the ban in place, a systematic development process took off, and a number of changes were undertaken and later labelled 'biosecurity' and 'precautionary health' (Magnusson et al., 2019; Backhans et al., 2015).

Batch-wise, or so called 'all-in-all out' production routines, became practise along with sanitary periods between batches. This, in turn, implied changes in

---

26 Proposition 1987/88:93. https://www.riksdagen.se/sv/dokument-lagar/dokument/proposition/om-djurskyddslag-mm-_GB0393/html.

27 https://www.riksdagen.se/sv/dokument-lagar/dokument/proposition/om-djurskyddslag-mm-_GB0393/html. Author's translation.

28 *The New York Times* 1988.10.25.

stable design with specific sections for each stage of the breeding process. Animal purchasing routines were developed including quarantine of new animals to the herds. Visitors began to be controlled and had to adapt to basic hygiene routines. Vaccination programmes were introduced for both sows and piglets. As one farmer expressed it:

> It was a great feeling to get rid of routine antibiotics. The most important change allowing it was the batch-wise production, which made it possible to introduce totally new hygiene routines.[29]

Another farmer underlines the importance of changing the whole production system:

> The whole packet is different, with all-in-all-out, about 50% larger space per animal compared to Europe on average, and loose animals on straw.[30]

Although no farm is exactly like another, some common denominators can be outlined:

- Sows are free ranged in groups until about a week before litter.[31] Then the sow is moved to her own box, where she can still move freely and pursue her building instinct.[32] After littering, the sow and the piglets stay in the box 4-6 weeks.
- For the piglets, the weaning period is the most critical in terms of health, and the earlier the weaning, the higher the risk for infections such as diarrhoea, and for the need for treatment with antibiotics. Weaning, which in general is 4-6 weeks, is not allowed to take place until 90% of the piglets are more than 26 days old. After the weaning period, first the sow is moved out of the box while the piglets stay for some weeks more, in order to not be exposed to a new environment during this sensitive period. Then the piglets are moved to larger boxes for growing and fattening periods. When they are 6-7 months and weigh about 100 kilos they are slaughtered. Tail docking is not practiced and tail biting is rarely observed.[33]
- After each movement of animals, boxes are totally cleaned, and in between the boxes are mucked regularly. Before any new animal is brought into the herd

29 Interview, pig meat farmer, engaged since the late 1970s. 2018-10-28.

30 Interview, pig meat farmer, engaged since the early 1990s. 2020-02-10.

31 The productive life of a sow begins around the age of 1 year and includes a continuous cycle of insemination, gestation, farrowing after 114 days, and suckling for an average of 28 days.

32 All boxes must include a complete floor and all pigs must have access to bedding material such as straw. For pigs of all ages, 65-75% of the box floor should be complete, covered with bedding material and free from manure drain. The boxes must have a separate manure corner. On average, the sow with piglets has about 50% more space than EU legislation requires, the fattening pigs about 30% more, and slaughter pigs about 40% more. Furthermore, the boxes must have daylight and a certain air quality.

33 EU legislation forbids tail docking, but due to generous exceptions it is practiced in many production regimes.

it must spend some weeks in quarantine. No visitors are allowed to enter the stable without permission and then only with protective clothes and shoes. The transporters moving animals from stable to slaughter are not allowed entrance into the stable.

- All treatment with antibiotics or other drugs requires veterinary prescriptions, and each animal must have its own medical journal. The Swedish Board of Agriculture is the owner and organiser of the 'District Veterinaries'. This implies that the veterinaries are state employed. The District Veterinaries guarantee access to veterinaries over the whole country around the clock, and are also a state operative resource in case of outbreaks of infectious diseases.
- Besides the District Veterinaries, the consultancy company 'Farm and Animal Health'[34] owned by the Swedish Meat Producers' Association, Swedish Pig Farmers' Association, Swedish Beef Producers' Association, and Swedish Sheep Breeders' Association provide advice and knowledge dissemination on animal health and production economy.[35]

Along with the legal ban, creating a 'one-for-all' situation, pig farmers as well as veterinaries underlined the importance of a corresponding animal welfare law, as a mean to hinder 'production diseases' driving antibiotic consumption.

> If ambitions to reduce antibiotic consumption are not connected to animal welfare, either the input of antibiotics continues or the focus becomes directed to sterile production units.[36]

III-4. After three decades – supporting user reactions

When Sweden became an EU member in 1995 it became evident that the expectation made within the Swedish Farmers' Association that the abandoning of regular input of antibiotics; should constitute a 'market advantage', proved to be a rough overestimation. The expected positive reaction from the user side, the wholesale and retail industry as well as private and public consumers did not appear.

The EU entrance meant that the borders opened up for an influx of pig meat based products stemming from systems relying on significant higher consumption of antibiotics, and as estimated by the pig meat farmers and their cooperative organisation, at that time about 20% lower production costs. The Swedish production fell from about 4 million pigs and stabilised at about 2.6 million animals

34 Gårds och Djurhälsan.

35 https://distriktsveterinarerna.se/vart-uppdrag/om-distriktsveterinarerna; https://www.gardochdjurhalsan.se/radgivning-tjanster/gris/; http://www.sds-web.se/en/

36 Interview, Purchasing Manager, Slaughter house, engaged as purchaser since the early 1990s. 2018-11-16.

per year.[37] Behind this radical drop in production figures there was a dramatic restructuring of pig meat production units. Basically all small and medium size farmers closed down these activities, while the remaining increased their herd size and production volume. A purchaser at a slaughter house expresses the change as follows:

> It took a year after the EU entrance, then small producers began to disappear. Swedish pig meat farmers' focus on animal health, hygiene and restricted consumption of antibiotics could not match the cost level of imported meat.[38]

The membership also implied an EU questioning of the Swedish ban, something that became the outset of an extensive Swedish engagement related to European animal welfare and regular group treatment of animals with antibiotics. (For a detailed discussion, see Nordéus, chapter 11 this volume). As Nordéus shows, Sweden initially managed to get a temporary exemption of four years and, through an extensive involvement of politicians, researchers and civil servants, eventually got a permanent acceptance. Furthermore, the Swedish involvement contributed, as Nordéus puts it, remarkably to a similar EU ban in 2006.[39]

In the Swedish pig meat producing setting, the 2006 EU ban of growth antibiotics, motivated by food safety and public health, created recurrent expectations of gaining attention for what had stabilised as the 'Swedish Concept'; animal-based food production based on precautionary health and biosecurity without regular group treatment with antibiotics. However, the EU ban did not lead up to any significant change of the antibiotic consumption patterns in European animal-based food producing regimes, besides the already altering Nordic settings. Or as summarised in World Economic Forum's first risk report drawing attention to antibiotic resistance:

> [...] the amounts of antibiotics used to raise animals for food-production is still high, even in highly regulated markets like Europe. (WEF 2013: 33)

Hence, it was not the 2006 EU ban, but another media debate, taking off in 2014, that eventually created a reaction within the Swedish user setting that forced the retail industry to pay attention to the changes made within Swedish animal-based food production in general, pig meat production included.

The recurrent Swedish debate on animal welfare, or rather lack of it in the EU setting, started in spring 2014, in relation to the EU election. The overuse of antibiotics once again became a heated issue, but this time as something taking place outside the border, which both Swedish producers and consumers needed

---

37 https://jordbruksverket.se/download/18.28f4d91b172cdd65219b3acb/1592760271492/201801.pdf.

38 Interview, Purchaser, engaged since early 1990s. 2018-10-29.

39 Regulation 1831/2003/EC on additives for use in animal nutrition, replacing Directive 70/524/EEC on additives in feeding-stuffs. https://eur-lex.europa.eu/legal-content/EN/TXT/PDF/?uri=CELEX:32003R1831&rid=10.

protection from. The debate took off when Swedish Television, followed by basically all larger newspapers, directed attention to the Danish pig meat setting. The content of the program was directed by the fact that almost 90% of the pigs, but also many people exposed to pigs in their work, were carrying the multi resistant LA-MRSA bacteria, something that had caused a number of human deaths.[40]

The first reaction in the retail industry came from an individual ICA Hypermarket manager in southwest Sweden, who in April 2014 declared that Danish pig meat would be excluded from their assortment. The boycott was a protest against the lack of animal welfare including tail docking and against production regimes that triggered overuse of antibiotics. As the manager explained to the media, it was a way to put pressure on the industry at large to not accept anything.[41] The ICA Hypermarket announced that it would only market pig meat produced in settings where antibiotics were used to treat sick animals by veterinary prescription, that is, without regular group treatment.

The action was reported in the Swedish evening paper, *Expressen*,[42] and created attention both in other media and in the user setting. Three dominating wholesalers; Axfood, Coop and ICA, reacted rapidly. New, rather similar policies on animal welfare and antibiotics were developed and communicated publicly by private ICA and Axfood and cooperative Coop, and was later followed by the Swedish Wholesaler Association. As stated by ICA:

> Antibiotics may not be routinely used as a precautionary measure, but only after veterinary prescription. Antibiotics and hormones may not be used as growth promoters.[43]

The policies were followed by a significant change: the introduction of the branding 'Swedish meat', including consumer information about the restricted Swedish consumption of antibiotics used only as medical treatment by veterinary prescription.

Eventually, the increasing attention to antibiotic resistance and the systemic change made in the Swedish producing setting caused a significant change in terms of consumer reactions. Despite that the consumer price for Swedish pig meat was up to 20% higher per kilo compared to the EU average, the consumption pattern started to change. After 30 years of struggling with measures compensating for regular group treatment with antibiotics, the pig meat producers were eventually rewarded economically.

As one pig meat farmer, engaged since the early 1980s, stated:

---

40 https://www.svt.se/nyheter/utrikes/manga-grisbonder-drabbade-av-mrsa.
41 https://www.expressen.se/gt/ica-handlaren-vagrar-salja-danskt-flaskkott/.
42 https://www.expressen.se/gt/ica-handlaren-vagrar-salja-danskt-flaskkott/.
43 https://www.ica.se/XformFiles/ICA-djuromsorg-2017.pdf.

> The growing antibiotic resistance awareness is the best thing that happened to us. At last we are compensated for all the work we have done to get rid of antibiotics.[44]

However, the fact that the retail industry policies concerned the 'central assortment', implied space for purchase of meat from other types of producing regimes. The latter, together with the retail industry's long neglect of the efforts made, continued to cause displeasure among farmers. (That public procurement of animal-based food still is based on price, favouring import meat from other types of production regimes for hospitals, schools, etc., is another, still ongoing story.)

> It was not the retail industry but the media that evoked the consumer awareness. The retail industry is not interested in the production of animal-based food. It is focused on campaign products, not production.[45]

> On one hand they have their antibiotic policies, on the other hand they are marketing cheap import meat from antibiotic-dependent production. Although many consumers are aware, many are also prioritising low price.[46]

IV. Concluding discussion

It is time to return to the questions raised in the beginning of the chapter. How come that not only 'outsiders' called for a reduction of antibiotic consumption in animal-based food production, but that 'insiders', pig meat farmers and the Swedish Farmers' Association, mobilised to achieve change despite the understanding that this, at least in the short run, would imply higher production costs? What policy lessons can be learned concerning how to combat overuse of antibiotics in animal-food based production, and furthermore, how to support innovation journeys where the main costs appear in the production settings and the main benefits in global health systems, in the environment, and society at large?

If we start with considering the first question, the early1980s' increasing awareness of antibiotics and animal welfare, considered by both 'outsiders' and 'insiders', took place in a context with some specific characteristics. Rural Sweden was just a generation away. It was not until the 1960s that the traditional producing settings, dominated by family owned small farms and small holdings spread all over the country,[47] were transformed to larger units, based on labour efficient production practises (Martiin, 2015; Flygare and Isacson, 2011; Morell, 2011). That is, in the early 1980s, many Swedes had relatives with personal experiences

44 Interview, pig meat farmer, engaged since late 1970s. 2018-10-26.
45 Interview, pig meat farmer, engaged since the early 1990s.
46 Interview, manager pig meat farm, engaged since early 2010. 2020-08-25.
47 Only about 10% of the farmland belonged to large estates, mainly located in the South of Sweden (Morell, 2011).

with traditional small scale animal care. People in general were, as Begemann et al. (2018) underline, concerned about modern production techniques and their impact on nature and animals.

One of the most important forces behind the fact that the increasing awareness of antibiotic resistance was trickling down from an image level to materialise in terms of changes that could compensate for regular input of antibiotics, was that the main actors of a dense network of producers and suppliers related to pig meat production drew up and mobilised for change. Similar patterns, where businesses were able to provide innovative solutions with societal benefits puts pressure on governments, were observed in previous empirical studies on normative based change (Håkansson and Waluszewski, 2002). The Swedish Farmers' Association, as well as pioneering pig meat producers, required a 'one for all' production regime, something that could be achieved by a national legal ban. The expectations that the producers would be compensated for higher production costs was based on the circumstance that Sweden was not yet part of the EU, agriculture was state protected and adapted to domestic use (Begemann et al., 2018; Martiin 2015; Morell 2011). It was not until 1990 that the Swedish Parliament decided food production should be deregulated and adapted to EU policy, realised when Sweden joined EU in 1995 (Flygare and Isacson, 2011).

The materialisation of a production regime that could compensate for regular group treatment of pigs with antibiotics was facilitated by the fact that the Swedish Farmers' Association not only was a strong voice in relation to the legal ban of antibiotics, but also the owner of quite a few companies that together covered a number of resources related to pig meat production.

Its monopoly-like role was disputed, claimed to be forcing small farmers to adapt to what suited larger producers and the businesses owned by the cooperative (Morell, 2011). However, when it comes to carrying out the large amount of systemic changes leading up to a production regime that could compensate for regular input of antibiotics in pig meat production, the strong links between the cooperative owner and the owned companies became pivotal; a mobilising and coordination force behind a number of measures contributing to precautionary health and biosecurity.

The Swedish EU membership in 1995 and the influx of animal-based food products based on antibiotic-dependent production regimes illustrates that there are no simple economic mechanisms guaranteeing that the costs of normative based change are shared by those benefitting from it. It was simply not enough with consumers concerned about modern animal-based food production. Along with the debate on animal welfare and over medication there was also a debate on the high prices of food. The wholesale and retail industries focused on the price of the end-products, and did not promote what type of production regimes animal-based food was stemming from. It was not until 2014's media debate and the boycott of Danish pig meat made by the individual ICA retailer in 2014, that the retail industry engaged in making consumers aware of the difference between the production regimes behind the end-products.

The second question, concerning the policy lessons, also points to the importance of reaching and mobilising the main producers and users to achieve normative change. As both the Swedish 1986 ban and the EU 2006 ban reveal, a legal change in itself is not enough, it has to be accepted and adapted to in the producing and using setting, and not related to as something to bypass as smooth as possible. To gain influence over private actors' businesses, however, is not easily compatible with the contemporary EU policy regime, relying on independent market actors to transform knowledge to the innovations society needs (Eklund, 2013; Waluszewski, 2011).

The Swedish experience of combating overuse of antibiotics seriously questions the contemporary great trust in commercialisation of knowledge. Scientific knowledge concerning different types of risks caused by overuse and resistance has been available since the late 1940s through scientific publications and media reports. However, the reactions in terms of significant changes the producing and using settings related to animal-based food, leading up to radical reduction antibiotic consumption and of resistant infections, have mainly been restricted to Nordic countries (Kahn, 2016; Kirchhelle, 2018, van Boeckel et al., 2015) with Sweden, Norway and Iceland in the foreground. Although two other Nordic countries, Finland and Denmark, have reduced their antibiotic consumption, resistant infections such as LA-MRSA is still common in both these settings (Waluszewski et al., 2020).

Instead of a simple knowledge transfer, mobilising and organising across organisational and legal borders appears as a key ingredient involving public and private actors. The main content is not one major innovation, but rather numbers of systematic changes of social and material resources related to animal-based food production. Hence, a first policy lesson is the need for mobilisation of resourceful actors. A second, related lesson concerns the need to consider how to distribute the cost of change. In the Swedish setting, both the health care system in general and the people working in the pig meat setting have benefitted from the absence of, for example, LA-MRSA. However, for decades the cost of change was mainly an issue for the pig farmers.

World Economic Forum (2013, p. 33) conclude the discussion of overuse of antibiotics in the animal based food industry with underlining that 'research is needed' which can contribute to our understanding of the forces leading up to a significant progress. This chapter has revealed that the changes compensating for a regular input of antibiotics in Swedish pig production is the result of a 'never-ending-story' involving principally all main actors related to the producing network. Unfortunately, it took three decades to attract support from the retail industry. It is not given that systemic innovations with societal benefits are rewarding the actors carrying the main costs of change with economic benefits.

Bibliography

Backhans, A. (2015) 'Biosecurity level and health management practices in 60 Swedish farrow-to-finish herds', *Acta Veterinaria Scandinavica*, Vol. 57, Issue 14, pp. 3-11

Baraldi, E. (2016) 'Exploring the obstacles to implementing economic mechanisms to stimulate antibiotic research and development: A multi-actor and system-level analysis', *American journal of law & medicine*, Vol. 42, Issue 2-3, pp. 451-86

Begemann, S. (2018) 'How political cultures produce different antibiotic policies in agriculture: a historical comparative case study between the United Kingdom and Sweden'. *Sociologia Ruralis*, Vol. 58, Issue 4, pp. 765-85

Bengtsson, B., Wierup, M. (2006) 'Antimicrobial resistance in Scandinavia after a ban of antimicrobial growth promoters', *Animal Biotechnology*, Vol. 17, Issue 2, pp. 147-56

Van Boeckel et al. (2017) 'Reducing antimicrobial use in food animals – Consider user fees and regulatory caps on veterinary use', *Science Insights Policy Forum*, Vol. 357, Issue 6358, pp. 1350-1352

Van Boeckel et al. (2015) 'Global trends in antimicrobial use in food animals'. *PNAS*, Vol. 122, Issue 18, pp. 5649-5654

Cars, O., Hedin, A., Heddini, A. (2011) 'The global need for effective antibiotics – Moving towards concerted action', *Drug Resistance Updates*, Vol. 14, Issue 2, pp. 68-69

Carson, R., (1962) *Silent Spring*, Boston: Houghton Mifflin

Chandler, C.I.R. (2020) 'Current accounts of antimicrobial resistance: stabilisation, individualisation and antibiotics as infrastructure', *Palgrave Communications*, Online Vol. 5, Issue 53. https://doi.org/10.1057/s41599-019-0263-4

Czarniawska, B., Joerges, B. (1996) *Travel of Ideas: Translating Organizational Change*, Berlin/New York: Walter de Gruyter

Eklund, M. (2013) 'Science policy in a socially embedded economy', in Rider, S, Hasselberg, Y., Waluszewski, A, eds. 2013, *Transformations in Research, Higher Education and the Academic Market: The Breakdown of Scientific Thought*, Dordrecht: Springer Science & Business Media

Elzen, B. (2011) 'Normative contestation in transitions "in the making" – Animal welfare concerns and system innovation in pig husbandry", *Research Policy*, Vol. 40, Issue 2, pp. 263-75

Elzinga, A. (2004) 'The new production of reductionism in models relating to research policy', in *The Science-Industry Nexus. History, Policy, Implications*, Grandin, K., Wormbs, N., Widmalm, S. eds., MA: Science History Publications, pp. 2773-04

ESVAC (2020) *Sales of veterinary antimicrobial agents in 31 European countries in 2018: Tenth ESVAC report*, European Medicines Agency, European Surveillance of Veterinary Antimicrobial Consumption, (EMA/24309/2020)

Finlay M.R. (2004) 'Hogs, antibiotics, and the industrial environments of postwar agriculture', in Scranton, P, Schrepfer, SR eds, *Industrializing organisms. Introducing evolutionary history*. London: Routledge, pp. 237-60

Flygare, I.A., Isacson, M. (2011) 'The tension between modernity and reality, 1945-2010', in Myrdal, J., Morell. M. (eds.), *The Agrarian History of Sweden. From 4000 BC to AD 2000*, Lund: Nordic Academic Press, pp. 214-56

Fourcade, M., Ollion, E., Algan, Y. (2015) 'The Superiority of Economists', *Journal of Economic Perspectives*. Vol. 29, Issue 1, pp. 89-114

Geels, W.F., (2004) 'Understanding system innovations: a critical literature review and a conceptual synthesis', in *System Innovation and the Transition to Sustainability: Theory, Evidence and Policy*, Elzen, B., Geels, W.F., Green, K. eds, Cheltenham: Edward Elgar, pp. 19-47

Gradmann, C. (2016) 'Re-inventing infectious disease: Antibiotic resistance and drug development at the Bayer company 1945-80', *Medical History* Vol. 60, Issue 2, pp. 155-80

Grandin, K., Wormbs, N., Widmalm, S., eds (2004) *The Science-Industry Nexus*, Sagarmore Beach, MA: Science History Publications

Gould I.M. (2009) 'Antibiotic resistance: the perfect storm', *International Journal of Antimicrobial Agents*. Vol. 34, Suppl. 3, pp. 2-5

Håkansson, H. (2009) *Business in Networks*, Chichester: Wiley

———, Waluszewski, A. (2002) *Managing Technological Development. IKEA, the Environment and Technology*, London: Routledge

———, ——— 2007. 'Interaction: the only mean to create use'. In *Knowledge and Innovation in Business and Industry. The importance of using others*. Håkansson, H., Waluszewski, A, (eds), London: Routledge, pp. 147-67

Harrison, R. (1964) *Animal Machines: The New Factory Farming Industry*, London: Vincent Stuart Publishers

Hasselberg, Y., Rider, S, Waluszewski, A, (2013) 'Introduction', in Rider, S., Hasselberg, Y., and Waluszewski, A. eds, *Transformations in Research, Higher Education and the Academic Market. The Breakdown of Scientific Thought*, Dordrecht: Springer Science & Business Media, pp. 1-18

Hoffman, J., Outterson, K. (2015) 'What will it take to address the global threat of antibiotic resistance?', *The Journal of Law, Medicine and Ethics*, Special supplement to Vol. 43, Issue 2, pp. 6-11

Kahn, L.H., (2016) *One Health and the Politics of Antimicrobial Resistance*. Baltimore: John Hopkins University Press

Kirchhelle, C., (2018) 'Pharming animals: a global history of antibiotics in food production (1935-2017)'. *Palgrave Communications*, Vol. 4, Issue 1, https://doi.org/10.1057/s41599-018-0152-2

———, (2020) *Pyrric Progress. The History of Antibiotics in Anglo-American Food Production*, New Brunswick: Rutgers University Press

Kinch, M.S., (2014) 'An analysis of FDA-approved drugs for infectious disease: antibacterial agents', *Drug Discovery Today*, Vol. 9, Issue 9, pp. 1283-1287

Kveim Lie, A. (2014) 'Producing standards, producing the Nordic region: Antibiotic susceptibility testing, from 1950-1970', *Science in Context*, Cambridge University Press, Vol. 27, Issue 2, pp. 215-48

Larsson Heidenblad, D. (2021) *Den gröna vändningen: En ny kunskapshistoria om miljöfrågornas genombrott under efterkrigstiden*. Göteborg: Kriterium

Lawson, T. (2006) 'The nature of heterodox economics', Cambridge Journal of Economics, Vol. 30, Issue 4, pp. 483-505

Lhermie, G., Gröhn, Y.T. Raboisson, D., (2017) 'Addressing antimicrobial resistance: An overview of priority actions to prevent suboptimal antimicrobial use in food-animal production', *Frontiers in Microbiology*, Vol. 7, Online January 6, https://doi.org/10.3389/fmicb.2016.02114

Li, J., (2017) 'Current status and prospects for in-feed antibiotics in the different stages of pork production – A review'. *AJAS, Asian-Australian Journal of Animal Sciences.* Vol. 30, Issue 12, pp. 1667-73

Magnusson, U., Sternberg, S., Eklund, G., Rozstalnyy, A. (2019). *Prudent and efficient use of antimicrobials in pigs and poultry.* FAO Animal Production and Health Manual 23. Rome: FAO

Marglin, S.A. (2008) *The Dismal Science: How Thinking Like an Economist Undermines Community*, Cambridge, MA: Harvard University

Martiin, C. (2015) 'Modernized farming but stagnated production: Swedish farming in the 1950s emerging welfare state'. *Agricultural History Society*, Vol. 89, Issue 4, pp. 559-83

Michelsen, T. (1981) *Massmaten: svenskt jordbruk i kemiåldern*, Stockholm: Rabén & Sjögren

Mirowski, P. (2011) *Science-Mart. Privatizing American Science*, Cambridge, Mass.: Harvard University Press

Morell, M., (2011) 'Agriculture in industrial society, 1870-1945'. In Myrdal, J., Morell, M. eds, *The Agrarian History of Sweden. From 4000 BC to AD 2000*, Lund: Nordic Academic Press, pp. 165-213

Nathan, C., Cars, O. 2014. 'Antibiotic resistance. Problems, progress, and prospects', *New England Journal of Medicine*, 371, pp. 1761-1763

Outterson, K. (2015) 'Repairing the broken market for antibiotic innovation', *Health Affairs*, Vol. 34, Issue 2, pp. 277-85

Perna, A., Waluszewski, A. 2018. 'The interplay between innovation and accounting: The border-crossing Q-Linea project and its within-border accounting'. In *Accounting, Innovation and Inter-Organisational Relationships*, Carlsson-Wall et al. eds. Routledge: New York, pp. 12-28

Poel, Van De I.R (2000) 'On the role of outsiders in technical development'. *Technology Analysis & Strategic Management*, Vol. 12, Issue 3, pp. 383-97

Postma, M. (2015) 'Alternatives to the use of antimicrobial agents in pig production: A multi-country expert ranking of perceived effectiveness, feasibility and return on investment'. *Preventive Veterinary Medicine*, Vol. 118, pp. 457-66

Rider, S., Waluszewski, A. (2015) 'Crowding out Knowledge: Efficiency, Innovation and Higher Education', in *The Global Financial Crisis and Educational Restructuring (Global Studies in Education)* Peters, M., Paraskeva, J.M., Besley, T., eds, New York: Peter Lang, pp. 235-48

SFS, Swedish Code of Statutes (1985) Lag om foder, 1985:295, Stockholm: Svensk författningssamling

SFS, Swedish Code of Statutes (1988) Djurskyddslag, 1988:534, Stockholm: Svensk författningssamling

Slaugther, S., Rhoades, G. (2004) *Academic Capitalism and the New Economy: Markets, State, and Higher Education.* Baltimore: John Hopkins University Press

Soller, B., (1971) *Djurfabriken*. Stockholm: Rabén och Sjögren
Stokstad, E.L. (1949) 'The multiple nature of the animal protein factor'. *The Journal of Biological Chemistry*, Vol. 180, Issue 2, pp. 647-54
Swann, M. M. (1969) *Report of the joint committee on the use of antibiotics in animal husbandry and veterinary medicine*. London: Her Majesty's Stationery Office
Swedish Board of Agriculture/Jordbruksverket (2017) *Utsäde – skadegörare, analys och behandling*. Jönköping: Jordbruksverket
Swedres-Swarm (2018) *Consumption of antibiotics and occurrence of antibiotic resistance in Sweden*. Solna/Uppsala: Public Health Agency of Sweden and National Veterinary Institute
Thoms, U. (2012) Between promise and threat. *N.T.M. Journal of the History of Science, Technology and Medicine*, Vol. 20, pp. 181-214
Utterback, J.M., Abernathy, W.J., (1975) 'A Dynamic Model of Process and Product Innovation'. *Omega, The International Journal of Management Science*, Vol. 3, Issue 6, pp. 639-53
Vail, D.J., K.P. Hasund and L. Drake (1994) *The greening of agricultural policy in industrial societies: Swedish reforms in comparative perspective*. Ithaca, NY: Cornell University Press
Waluszewski, A. (2011) *The policy practitioners' dilemma. The national policy and the transnational networks*. Stockholm: Vinnova report 2011:07
———, Baraldi, C., Ciabuschi, F., (2018) 'Increase development and decrease use! Innovation controversies caused by antimicrobial resistance', in *Controversies in Healthcare Innovation Service, Technology and Organization*, Hoholm, T., La Rocca, A., Aanestad, M. eds, Basinstoke: Palgrave Macmillan, pp. 309-40
———, Harrison, D., Halinen, A. and Munksgaard, K. (2020) 'The "what's" and "how's" of some Nordic system innovations: compensating for routine group input of antibiotics in animal-based food production'. Proceedings of the 36th Annual IMP Conference, Örebro, Sweden, September 3-4
Ven, Van de, A., et al. (1999). *The Innovation Journey*. New York, Oxford University Press
WEF, World Economic Forum (2013) *Global Risks 2013. An Initiative of the Risk Response Network*, Cologny/Geneva: World Economic Forum
Wierup, M., 2001. 'The Swedish experience of the 1986 year ban of antimicrobial growth promoters, with special reference to animal health, disease prevention, productivity, and usage of antimicrobials'. *Microbial Drug Resistance*, Vol. 7: 2, pp. 183-90
Wise, R., (2007) 'An overview of the specialist advisory committee on antimicrobial resistance (SACAR)' *Journal of Antimicrobial Chemotherapy*, Vol. 60, Suppl. 1, pp. 5-7

Appendix

I. Respondents 2018-2021

1 Researcher, veterinarian, Sweden's National Veterinary Institute (SVA). Interview 2018-10-11

2 Researchers, veterinarian, Swedish University of Agricultural Science (SLU). Interview, 2018-11-19
3 Researchers, veterinarian, Swedish University of Agricultural Science (SLU). Interview 2020-08-20, 2020-08-27
4 Researcher, historian, Uppsala University. Interview, 2020-10-06
5 Pig farmer, retired. Active pig farmer since 1978, still working part time on the farm. Interview 2018-10-26
6 Pig farmer retired. Active pig farmer since 1978, still working part time on the farm. Interview 2018-10-26
7 Pig farmer, retired. Active farmer since 1969, still working part time on the farm. Interview, 2019-04-25, phone interview 2020-04-15
8 Pig farmer, Active farmer since 1990, former slaughter worker. 2018-06-14
9 Pig farmer, Active farmer since 1990. 2018-06-14
10 Pig farmer, Active farmer since early 1990s. Interview 2019-10-22
11 Pig farm manager, responsible for three units. Active since 2005. Interview 2019-11-20
12 Pig farmer, active since 1992, also engaged in Swedish Pig Farmers' Association Interview 2020-02-11
13 Pig farm manager. Active farmer since 2010. Interview 2020-08-25
14 Pig farmer and politician, retired. Active farmer since late 1970s, still part time working on farm. Phone interview, 2019-05-08
15 Purchaser at middle sized slaughterhouse since 1995. Former pig farmer. Interview 2018-10-29
16 Purchasing manager, Swedish unit of international slaughterhouse, since 1985, former pig farmer.
17 Manager, retired, Swedish Farmers' Association 1982-1993. Still engaged in the antibiotic/animal debate. Interview 2018-11-20, 2019-05-08
18 Manager Swedish Farmers' Association since 2003, former pig farmer. 2018-12-06.

Participating observation, 2019-05-08, main speakers

1 Researcher, anthropology
2 Researchers, sociology
3 Researcher, business studies
4 Researchers, agrarian history
5 Researcher, veterinary medicine
6 Researcher, consumer organisation

In the audience were researchers/experts from the following organisations participated: Uppsala University (Medicine, Technology, and Social Sciences), Swedish University of Agricultural Science (SLU), Sweden's National Veterinary Institute (SVA), ReAct, BI Norwegian Business School, University of Southern Denmark, Turku School of Economics.

NICOLAS FORTANÉ

14. Rise and fall of ecopathology

The thwarted development of a preventive approach to animal health (France, 1970s-1980s)

I. Introduction

From the 1950s onwards, the French agricultural landscape underwent profound changes. These included political and institutional changes related to the introduction of the so-called 'co-management' system, which favoured the empowerment of agricultural organisations and their progressive control over agricultural development policies (Muller, 1984). Social, technical and economic changes related to the intensification of agriculture also brought major transformations in the structuring of commodity chains and farm labour (Nicourt, 2013). In the livestock sector, these changes were mainly rooted in two laws, namely the 1954 law on the prophylaxis of bovine tuberculosis and the 1966 law on animal husbandry. The first formed the base of a 'sanitary tripod' entrusting the management of animal diseases to three central actors: the veterinary services of the Ministry of Agriculture, the agricultural profession and independent veterinarians (Berdah, 2008). The second overhauled the technical and economic organisation of livestock farming, particularly in terms of animal genetics, by introducing the pyramid model known as 'selection-multiplication-production', which has been particularly successful in the industrial sector (Vissac, 2002).

In the 1970s, some veterinarians began to take a critical look at the consequences of this productivist model. Animal health had worsened considerably due to these new farming conditions (Burgat & Dantzler, 2001), and the socio-economic situation of many livestock farmers had deteriorated markedly through individual and collective forms of impoverishment (Salmona, 1994). At the same time, these veterinarians criticized the veterinary profession for being both powerless and complicit in this 'alienation' of livestock farmers. They therefore sought to invent a veterinary medicine that would be better adapted to modern livestock farming, allowing better management of animal health (notably through prevention rather than care), and rethinking the role of veterinarians on farms.

Nicolas Fortané • IRISSO, CNRS, INRAE, Paris-Dauphine University, PSL, Paris, France: nicolas.fortane@inrae.fr

From Breeding & Feeding to Medicalization, Rural History in Europe, 17 (Turnhout: Brepols, 2023), pp. 337–360
BREPOLS PUBLISHERS 10.1484/M.RURHE-EB.5.135032

With this perspective in mind, a project to renovate veterinary knowledge and practices took form: ecopathology.[1]

However, the contradictions inherent in this critical stance towards productivist agriculture were apparent from the start. Although ecopathology contributed to challenging industrial livestock farming and resonated with the discourse of certain left-wing agricultural unions, it also carried with it the conditions for its reappropriation by dominant actors. By proposing to adapt veterinary medicine to productivist agriculture (even if this was initially to limit its harmful effects), ecopathologists provided their opponents with the means to make their raison d'être disappear. Thus, some twenty years after it first emerged, this form of veterinary medicine died out, at least partially.[2] Although ecopathalogy's critical view of the productivist model no longer made sense, some ecopathological concepts and techniques were reappropriated by the proponents of intensive livestock farming.

I.1. Tensions rooted in a long history

To understand the thwarted development of ecopathology, it is necessary to look at the trajectory of this form of alternative veterinary medicine and how it was shaped by the controversies regarding the relationships between veterinary medicine and animal health during the second half of the 20th century, and even beyond. Indeed, the criticism of industrial and productivist agriculture inherent in ecopathological knowledge and practices was related to the struggles of veterinary medicine to preserve its jurisdiction against competition from other professional groups such as farmers and farm advisors.

Two main trends have characterised the history of animal medicine since the end of the 19th century. First, it was built on a model close to human medicine, comprising an independent type of practice and a clinical approach (pathological, curative and individualised) to animal diseases. It then established itself as a specialist in 'communicable' diseases by appropriating the knowledge and tools of Pasteurian microbiology (an infectious conception of pathology), basing its legitimacy on 'public health expertise' rather than an 'agricultural expertise' (Berdah, 2012). Jurisdiction over animal health, which veterinarians

1 Ecopathology was an approach to animal health carried by a specific segment of the veterinary profession, from the 1970s to the 1990s. It focused on developing preventive tools and services to animal diseases, and to provide new forms of relationships between farmers and veterinarians. This chapter analyses the thwarted trajectory of this singular set of practices and knowledge in French veterinary medicine.

2 Indeed, even if this is beyond the (chronological) scope of this article, there is today a certain form of resurgence of ecopathological knowledge and practices, particularly in swine veterinary medicine. Furthermore, certain aspects of ecopathology have clearly persisted through an extremely marginal mode of practice in the veterinary profession (less than 3% of practitioners listed by the Veterinary Council), which is called "contractual medicine". For some details on this parallel history, see: Robinet, 2011; Combettes et al., 2012.

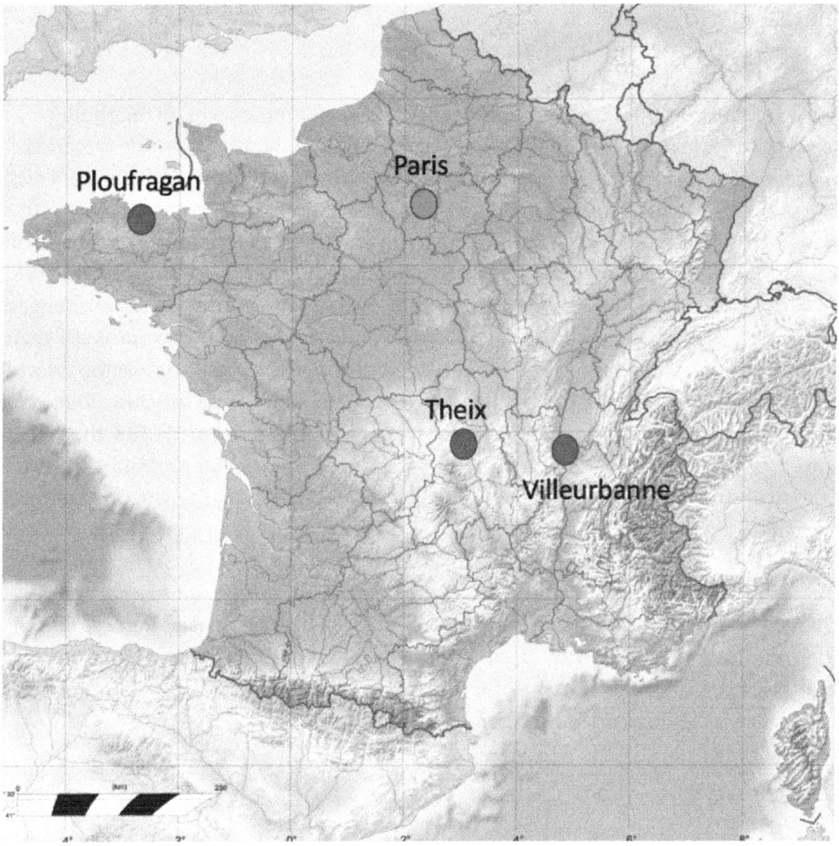

Figure 14.1. Location of research centers (in blue)

managed to monopolise after a long process of professionalisation, therefore has been oriented above all towards the objective of protecting human health rather than towards a logic of agricultural development and the supervision of animal husbandry practices (Hubscher, 1999). These issues were left to a discipline that veterinarians have gradually abandoned to agronomists, and which ultimately focuses much more on the performance of livestock farming than on animal health: "zootechnics"[3] (Hubscher, 1999; Jussiau, 1999). However, the industrialisation of agriculture in the second half of the 20th century led to the emergence of new farming conditions (batch farming, confinement of animals, increased herd size, etc.) which helped to reconfigure the relationship between animal health and zootechnics. Indeed, the jurisdiction of animal health gradually lost its autonomy

3 While this term does not have a direct equivalent in English, it means the science of animal husbandry.

in relation to other aspects of farming such as genetics, food and housing, allowing zootechnics to encroach on the historical territory of the veterinary profession.

It is thus impossible to understand the development of ecopathology as a critical look at the productivist model within a segment of the veterinary profession in the 1970s without analysing the competition between the different categories of actors involved on livestock farms. We will see that when this handful of veterinarians started to campaign for the development of a new form of animal medicine, they also aimed to reaffirm their prerogatives in the field of animal health based on an 'agricultural' type of expertise which challenged the expertise of farmers and farm advisors. Thus, following the example of their British counterparts a few years earlier, ecopathologists sought to develop a form of 'preventive veterinary medicine' by converting their 'fire brigade' role (i.e., intervening in farms only when pathological symptoms occur within the herd) into a position of 'advisors' capable of integrating animal health issues into the technical and economic management of farms (Woods, 2013). In this respect, ecopathology can be considered as an alternative form of veterinary medicine, not only because it broke with the clinical and infectious approach to animal diseases, but also because it underpinned a transformation of the veterinarian-farmer relationship and livestock management systems. All in all, the thwarted development of ecopathology as a critical discipline in the field of animal farming can only be understood within the context of the issues structuring the relationship between veterinary medicine and animal health.

I.2. Fieldwork

Ecopathology developed on three sites between the mid-1970s and the mid-1980s, then rapidly declined, before disappearing completely (at least under this name) around 1995. These three centres of ecopathology, whose trajectory this article aims to retrace, were an INRAE[4] unit at Theix (in Auvergne), an Anses[5] station at Ploufragan (in Brittany) and an association at Villeurbanne (near the Alps). Methodologically, this survey is based on fourteen semi-structured interviews conducted in 2012, five interviews from INRAE's Oral Archives and a corpus of some fifty documents, mainly consisting of grey and scientific literature produced by ecopathologists. This chapter is organised in three parts. I first go back to the origins of ecopathology before exploring in detail what this form of alternative veterinary medicine covers and explaining the conditions of its paradoxical disappearance.

---

4 National Research Institute for Agriculture, Food and the Environment.
5 French Food Safety Agency.

II. The origins of ecopathology, a political and medical project

Although ecopathology did not really start to develop until the second half of the 1970s, we have to go back to the end of the previous decade to see its beginnings. It was then that a small fraction of the veterinary profession began to take a critical look at intensive livestock farming and imagine new ways of practising veterinary medicine. The pioneers of ecopathology, i.e., the founders of the centres at Theix, Villeurbanne and Ploufragan, thus laid the foundations for a preventive approach to animal health.

II.1. Overhauling veterinary medicine

The seeds of ecopathology began to emerge in the post-1968 context through an attempt to renew veterinary education. This was when cultural and social criticism of veterinary medicine as an elitist profession, locked up in its past certainties and blind to the excesses of the productivist model, was formed (Fontaine, 1974). Michel Fontaine, a professor at the Lyon Veterinary School, was part of this movement. He criticized the veterinary school system for remaining distant from the realities of the field and for failing to adapt to the new livestock farming conditions. Fontaine was linked to left-wing agricultural trade unions (notably the Paysans Travailleurs – PT) which denounced the alienation of farmers by the industrialisation of agricultural practices, in particular through the development of integration contracts (Robinet, 2011). He developed an alternative conception of veterinary medicine, i.e., one critical of the productivist model that strives to provide farmers with tools to resist the various forms of domination exerted over them. In this respect, this attempt to redefine veterinary medicine not only had an important political significance, but also a technical dimension insofar as this critique was calling for a modification of veterinary knowledge and practices. It is therefore in this double sense (i.e., political and technical) that ecopathology can be considered as a form of alternative veterinary medicine.

> It was a time when there was real exploitation of the agricultural world by the technical elites who were the veterinarians, etc. And when I say 'exploitation', I weigh my words, it is exploitation in the sense that ignorance and lack of knowledge were exploited, in my opinion, for really financial purposes. (…)
> We realised that there was an incredible gap between the intensive livestock farming that was beginning to rise everywhere and the peasants who couldn't even master their working tools because they didn't understand them anymore.
>
> *An ecopathologist from Villeurbanne*

Eager to develop a space where this concept could be forged and circulated, Fontaine created Cefale in 1973.[6] It aimed to bring together researchers from different disciplines (zootechnics, pharmacology, veterinary sciences, etc.), veterinary practitioners and livestock farmers in a collective dedicated to training veterinary students in the field rather than locked up in schools as was the case at the time. Cefale was thus a place where a new approach to animal health was built and tested, not only in terms of medical methods and techniques (see below), but also by taking into account the economic and social issues of the agricultural world.

> Cefale says it wants to help livestock farmers understand their real problems, in all their complexity and in the socio-economic context in which they arise. This is why the approach is essentially based on surveys, technical investigations and direct contracts. (…)
> Cefale students are particularly concerned about those among the livestock farmers who are in difficulty, the most vulnerable and most threatened in their condition as farmers. Considering farmers' economic and social conditions on the basis of technical data, Cefale's members are led to a wider reflection on the current evolution of agriculture and rural development, environmental issues and consumer protection
>
> *Excerpts from Cefale's presentation brochure*[7]

This dynamic resonated with the mobilisations that took place in Larzac throughout the 1970s,[8] and a connection was made at the end of the decade with the creation of the 'vet group' of Larzac University, which a few years later became AVEM (Association of Veterinarians and Farmers of Millavois which recruited a large part of its members from Fontaine's students). The ambition of these diverse veterinary movements was to rethink the socio-economic organisation of veterinary medicine to free farmers from the various forms of exploitation to which they were subjected. Two alternative modes of practice were developing at this time, both in opposition to independent practice which was (and still is) the most prominent economic form of veterinary activity. The first alternative mode, which was to become the trademark of Cefale and the veterinarians of the Massif Central (among which AVEM was a 'historic' organisation) was the "Groupements Vétérinaires Conventionnés" (GVC).[9] The idea was that groups of veterinarians would be linked to groups of farmers through an annual agreement that set a fixed fee for the care of the animals. This system, which avoids payment on a fee-for-service basis, was designed as a form of risk insurance between

6 Centre d'études et de formation appliquées à l'élevage – Centre for Studies and Training Applied to Animal Husbandry.

7 Cited by Robinet, 2011.

8 Antimilitarist movements that took place in the Larzac mountains, in the Massif Central (centre of France).

9 Veterinary groups under contract.

veterinarians and livestock farmers. The second alternative mode, which was more strongly defended by the PT union, was to hire veterinarians as salaried employees of agricultural cooperatives. This system pushed the break with the classic type of veterinary medicine (i.e., independent practice) even further.

This conception of the socio-economic relations between livestock farmers and veterinarians was coupled with a reflection on their socio-technical relationships. These alternate veterinary movements were indeed reflecting on how to change livestock farming practices to become emancipated from the industrialisation process of agriculture. The political critique of the productivist model was in effect inseparable from a project aiming to renew the social relationships that structured the animal health jurisdiction (Fontaine, 1974). This implied re-establishing the relationship between veterinarians and farmers through new kinds of practices that would not contribute to the former exerting domination over the latter. In other words, since risks were mutualised by the contract linking vets and farmers, the knowledge and techniques (for managing animal health) that they mobilised also had to be adapted to this unique contractual relationship. Breaking away from the independent mode of practice implied a different approach to animal care than that of traditional veterinary medicine (Combettes et al., 2014).

As veterinary acts could no longer be punctual, individual and strictly curative, they had to be thought of as a long-term follow-up of herd health, especially as the conditions of modern animal husbandry tended to homogenise the characteristics of all of the animals on a farm. This approach, which later formed the basis of ecopathological knowledge and techniques as an alternative veterinary practice, was 'longitudinal monitoring'. It enabled vets to consider animal pathology from a population-based point of view and a preventive perspective. However, as we shall see below, this approach was not only a methodological innovation or a simple adaptation of veterinary knowledge and techniques to intensive livestock farming. It also was a way of involving farmers in veterinary practices. Farmers were particularly active in the longitudinal monitoring of their herds as they had to collect a lot of data and contribute to their analysis with the vets. In this sense, these alternative approaches to animal health were feeding both a political critique and a practical renovation of veterinary medicine.

> Overall, the idea that Fontaine and his students put in place is to say: we follow groups of animals and not individuals. That's a bit like that. And we follow over time, which is quite the only way to know about a problem (…). So, at the end of the day, for me, [ecopathology] is the discovery of longitudinal monitoring. It's the discovery of longitudinal monitoring of health, of an animal population. Whereas up until now we've been in one-off interventions.
>
> *An ecopathologist from Theix*

II.2. The 'pioneers' of ecopathology

Although the 'pioneers' of ecopathology did not strictly speaking come directly from the Cefale or GVC movement, they had links with them. The careers of the founders of the Theix (Michel Brochart) and Villeurbanne (Guy Rosner) ecopathology centres were indeed directly influenced by this alternative conception of veterinary medicine, which combined a preventive approach to animal diseases with a critique of the productivist model. Brochart was an INRA researcher who was very close to Fontaine, due to their geographical, generational and political proximity (Communist party in particular). Rosner was for a time close to left-wing political movements and livestock farmers who were members of the PT union, both when he was the veterinary advisor of the "Groupement de Défense Sanitaire (GDS) des Vosges"[10] and when he worked as a salaried vet for a farmers' union called "Syndicat d'action sanitaire des éleveurs du Don (SASED)" in the early 1970s. Moreover, most of the veterinarians with whom he worked after he created the Villeurbanne Ecopathology Centre were former Cefale students and/or GVCs members.

In terms of knowledge and practices, the pioneers of ecopathology quickly grasped the concept of herd health longitudinal monitoring. This enabled them to develop further what they meant by the idea of preventive veterinary medicine, namely their approach was not only population-based but also multifactorial (i.e., not specifically infectious) and interdisciplinary (i.e., not specifically veterinary). Brochart and Rosner indeed were involved and specialised for many years in two pioneering fields of veterinary science: nutrition and reproduction. Before the creation of the ecopathology laboratory in Theix in 1978, Brochart carried out research in these two fields, which enabled him to assess the limits of physio-pathological and microbiological approaches to animal health. Nutrition and reproduction were the fields of veterinary science where Pasteurian theories dominated the least and which had continued to maintain relationships with disciplines that historically had been neglected by veterinarians (zootechnics, rural economy). Rosner, although younger, was also very much involved in these particular aspects of animal health before creating the Villeurbanne ecopathology centre in 1984. Following his experience as a salaried veterinarian at SASED, he was notably a trainer in bovine insemination for several years and then conducted surveys on the valorisation of agricultural by-products in animal feed.

> Reproduction monitoring is perhaps the first [ecopathological] way of working, with which contractual veterinary practitioners with good ideas have worked. They did reproductive monitoring, so that means that we are back to a system… It is a bit the ancestor of what these contractual groups did, it was simply strictly related to reproduction monitoring, but it is the same idea (…)

---

10 In France, the GDS are the farmers' organizations in charge of implementing animal health policies at a local level, such as vaccination campaigns or disease surveillance and monitoring.

> So that was a prefiguration. All this longitudinal follow-up is a prefiguration of ecopathology. That's the thing for me.
>
> *An ecopathologist from Villeurbanne*

This peculiar technique, which prefigured the multifactorial approach at the heart of ecopathological expertise, contributed to decompartmentalizing animal health from strictly veterinary knowledge and practices. Interdisciplinarity, in particular the reconciliation of zootechnics and veterinary medicine, was indeed at the heart of the ecopathological project. It was thought to be a means of combining it with Fontaine's political critique of intensive agriculture, meaning that this new way of understanding animal health also had to help emancipate the peasant class by aiming to preserve farmers' income and agricultural jobs (Calavas, 1998). Ecopathology was an alternative form of veterinary medicine because it was trying to rebuild veterinary legitimacy over the animal health jurisdiction through agricultural expertise rather than public health expertise.

Articulating these multiple ambitions was not necessarily easy, but most of the early ecopathologists could find the keys to do so in a quite singular socialising experience: tropical veterinary medicine. Ecopathologists often have told me that they nourished their desire to break away from the classical patterns of veterinary medicine during some professional experience abroad early in their career. For some, this was during their military service, while for others it was during a stint at the Tropical Veterinary Medicine Institute (what veterinarians of the time commonly called 'going exotic'). Positions there were generally held for two or three years in African or Caribbean countries. They were expected to contribute to the development of local farming systems, to reconcile the logics of productivity and health protection (for both humans and animals), and were faced with the need to mobilise a wide range of knowledge and techniques (economics, agricultural and public health development, etc.).

Historically, tropical veterinary medicine has always constituted a system of expertise in which the jurisdiction of animal health was not monopolised by the veterinary profession (Brown & Gilfoyle, 2010). The conception of animal diseases developed by (future) ecopathologists and the role they were to play in this field were thus able to be expressed through the political and medical project of ecopathology at the turn of the 1970s, when the industrialisation of agriculture was rapidly growing and profoundly modified the social, economic and technical conditions of animal husbandry.

> So I applied for a position in 'cooperation'[11] and I was sent to Madagascar. As I had studied chemistry at university, I was appointed director of a chemistry lab, and at the same time I had a good knowledge of nutrition, animal husbandry and zootechnics. And there (…) I had a cross-cutting role, I did

11 This term means in French doing military service abroad.

> a bit of everything at the same time since I was a chemist, a nutritionist, a breeder, a pathologist. It was a great time.
>
> *An ecopathologist from Ploufragan*

All in all, understanding animal health from an ecopathological perspective is a twofold task. On the one hand, it means moving away from the clinical and infectious straitjacket of veterinary medicine towards a preventive perspective, defined as a population-based, multi-factorial and interdisciplinary approach to animal diseases. On the other hand, it means rethinking the place of the veterinary profession in the social organisation of animal farming, i.e., both the socioeconomic and sociotechnical relationships between the farmer and the veterinarian, and the relationship of veterinary medicine with other forms of expertise which tend to encroach on animal health in a context of the industrialisation of livestock production (zootechnics, rural economics, etc.). This is why ecopathology was not just a medical and a professional project but also a socio-political one that resonated with the positions of (or some of) its founders. In the beginning, ecopathology was highly connected to a left-wing political criticism of veterinary medicine and agriculture. Fontaine, Brochart and Rosner, for example, were linked to unions, parties and even social movements that were trying to free livestock farming from forms of domination to which humans, and in some cases animals, were exposed, and to empower farmers to resist the expropriation to which they were subjected and to take back ownership of their knowledge, tools and practices. However, for such an alternative form of animal health management to develop, the spaces of circulation and appropriation of new approaches and techniques needed to be restructured. This is precisely what was put in place with the creation of the three ecopathology centres, although in some places it meant making compromises with this initial political project.

III. Ecopathological surveys, a space for the circulation of veterinary knowledge and practices

Investigating the history of ecopathology involves not only an attempt to understand transformations in veterinary knowledge and practices, but also the way in which they are linked to the sociotechnical networks of animal health management. This is what we intend to show by understanding ecopathology as an alternative form of veterinary expertise. Although the situation of each centre presented some singularities, two cross-cutting characteristics emerged from the major ecopathological surveys that have been carried out. The first lay in the redefinition of the animal health jurisdiction, which ecopathologists intended to open up to actors other than just veterinarians. The second consisted of the invention, by the ecopathologists of Ploufragan, of a particular medical technique (i.e., a statistical approach to animal diseases) which allowed the reintegration of health-related problems into the global technical and economic management of livestock farming.

III.1. Sharing veterinary expertise with agricultural actors

One of the main ambitions of ecopathology was to imagine a way of managing animal health that would make it possible to break away from the forms of alienation that veterinary medicine tended to exert over livestock farmers within the framework of industrial and productivist agriculture. To this end, ecopathologists intended to share veterinary expertise with agricultural actors. However, this open approach also ensured that veterinary knowledge and practices would not be too marginalised by the other areas of expertise also contributing to the field of animal health. Ecopathology as it has developed in Theix and Villeurbanne is a perfect illustration of this.

In the early 1970s, the pioneers of a preventive approach to animal health sought to publicise ecopathology among agricultural actors, who would not only be the beneficiaries of, but also participants in, such a project. The "Fédération Nationale des Groupements de Défense Sanitaire Bovins (FNGDSB)"[12] was then interested in collaborating with the ecopathologists. GDS are animal health management organisations where vets and farmers work together, and that have played a historic role in the implementation of collective prophylaxis (Berdah, 2008). Brochart and Fontaine, and later Tillon (the founder of the Ploufragan ecopathology centre), took part in the FNGDSB congresses of 1973, 1975 and 1977, and clearly aroused enthusiasm. They argued that GDS should not be restricted to a public health logic with the control of regulated contagious diseases, but could also promote, in association with the veterinary profession, a form of agricultural expertise that would be more aware and better adapted to the new health-related problems emerging with the industrialisation of livestock farming (in particular, mastitis and lameness in cattle farming).

Following these congresses, the first "Continuous Eco-Pathological Survey" (EEPC) was set up by the INRA unit led by Brochart, which had just recruited a young veterinary researcher who played a leading role in the development of ecopathological surveys, Jacques Barnouin.[13] The latter coordinated the EEPC, whose main partner was the FNGDSB but with which other actors cooperated, such as milk control and artificial insemination bodies. Data collection involved farmers (who were freely recording events occurring in their herd) and GDS

12 This is the national federation of the GDS (who are in charge of the local implementation of animal health policies).

13 Although more concerned with the scientific aspect of ecopathology, rather than its practical or militant dimension, Barnouin shares some of the social properties of 'historical' ecopathologists, notably his training in interdisciplinarity and the multifactorial approach during an applied research experiment abroad (setting up longitudinal monitoring of cattle herds at the Institut Technique Agricole de Mostaganem in Algeria between 1973 and 1975).

technicians (who were making six visits per year to check the data and pass them on to INRA researchers). In line with a preventive approach to animal health, the data collected varied greatly in nature: climatic data, soil analysis, animal feed rationing, serological and biochemical analyses, etc. The EEPC thus constituted a system in which veterinary knowledge and practices broke away from traditional veterinary clinics (herd health longitudinal monitoring was the central technique of these surveys) and in which jurisdiction over animal health was ultimately shared between all of the actors around the farm.

> Initially, everyone was very cautious. But the idea is to use [ecopathology] as a basis, to make [ecopathology] penetrate through GDS (…). That was the original idea. So they were given a part of the animal health management and Brochart tried to convince them that they had to stop doing just tuberculosis and brucellosis, that we had to look elsewhere and that there was a possibility of considerably increasing the profitability of livestock farms, to study livestock diseases, to understand their risk factors and to reduce their economic impact. That's how we set up this continuous ecopathological survey, the observatory farm. The first is 78-82.
>
> *An ecopathologist from Theix*

A few years later, Rosner set up a slightly different system at the Ecopathology Centre of Villeurbanne, but with the same objective of opening up veterinary knowledge and practices to agricultural actors. Thanks to his past professional experience (GDS des Vosges, SASED), he endeavoured to take animal health away from the exclusive control of the veterinary profession and lead it towards the territory of technical advice, which was then in the hands of the agricultural actors (Chambers of Agriculture, technical institutes, etc.) (Rémy, 2006; Compagnone et al., 2009). Taking up the principles of ecopathology, which aimed to decompartmentalise animal health and legitimise veterinary expertise on farms, Rosner developed his own survey networks to ensure the construction, dissemination and appropriation of alternative veterinary knowledge and practices.

The particularity of Rosner (and, therefore, of the Villeurbanne Ecopathology Centre) lay in an approach based on continuing education that made it possible to develop a pedagogical technique for co-constructing social demand (inspired by the 'group pedagogy' of post-war agricultural reformers – Muller, 1984). Meeting regularly in Villeurbanne, all of the participants contributed to the entire survey process, from choosing the pathologies studied to defining survey protocols and analysing the results. The main challenge was indeed the dissemination of knowledge and know-how resulting from the ecopathological surveys. Rosner supported the idea of circulating and sharing veterinary expertise to prevent it from becoming "an instrument of political domination and financial enrichment" (Calavas, 1998). Two specific tools were put in place to this end: the 'Ecopatho Guidelines' and the 'Ecopatho Transfer programmes'. Thought to be "disease prevention programmes, enabling development and technical and economic support organisations for livestock farming to use and apply the results of surveys" (Calavas & Rosner, 1997), they once again embodied this conception of ecopathology as a form of alternative veterinary medicine, which considered animal health as a territory fundamentally shared between the veterinary profession and agricultural actors (livestock farmers, technicians and other advisors). However, at the same time, these spaces for the circulation and appropriation of veterinary expertise made it possible to renew the foundations and legitimacy of veterinary expertise.

III.2. The invention of a new veterinary technique

The creation of the Porcine Pathology Station (SPP) in Ploufragan in 1972 happened in a context that was quite different from the other two ecopathology centres. The history of agriculture in Brittany is marked by Christian trade unionism. Although the latter shared with the left-wing agricultural movements the objectives of emancipating the peasant class by taking back control over the means of production, it differed from them in its relationship to the industrialisation of livestock farming, which it considered to be a means of modernising and empowering the agricultural world (Nicourt, 2013). SPP aimed to support the Brittany pig industry in a move towards greater industrialisation of production (which was then in deficit) by managing the new animal health problems which accompanied this intensive organisation of livestock farming (in particular reproductive disorders in sows and respiratory and digestive syndromes in piglets) (CVG, 2006). The ecopathology put in place at Ploufragan was very much due to the particularities of this historical and geographical context, as well as to the characteristics of the three main facilitators: a veterinarian (Jean-Pierre Tillon), an agronomist (François Madec) and a statistician (Joël Josse). They are the ones who set up the technique that would constitute the heart of ecopathological expertise.

In 1975, a pathology that was not described in the veterinary literature affected a large part of the pig population in Brittany: the thin sow syndrome. Madec and Josse decided to undertake a statistical survey, using geometric data analysis methods (in particular factorial correspondence analysis) to compare husbandry management practices between affected and unaffected herds. Having identified a certain number of variables between these farms, they set up prevention measures with the farmers based on a modification of certain factors associated with the presence of the syndrome. The result was apparently spectacular and has often been mentioned in heroic tales portraying ecopathologists as 'health detectives', much in the same way as the great myths that have forged the history of epidemiology. This episode was an important moment for the development of ecopathology as it concretely contributed to formalising the ecopathological approach. This was to mainly focus on so-called 'livestock diseases' and apprehend them from a population-based, multi-factorial and interdisciplinary perspective (Madec & Josse, 1980). Very quickly, the ecopathologists of Ploufragan not only argued for an abandonment of the Pasteurian conception of animal diseases, but also for a link between veterinary medicine and zootechnical and economic approaches (Tillon, 1981).

However, compared to the two other Ecopathology Centres, Brittany ecopathologists went much further in their attempt to break away from classical veterinary clinics by inventing a statistical method that made it possible to redefine what could be considered a disease. It not only consisted of using statistical modelling techniques to identify a cluster of causes with a pathological event (which would have been first detected by the classic categories of veterinary clinics), but above all of using these statistical techniques to determine the biological, zootechnical and economic events that could be qualified as pathological. Consequently, the ecopathologists in Brittany were no longer considering animal health as an autonomous veterinary jurisdiction. They were linking it intimately to the realm of the technical and economic performances of farms. Alongside genetics, animal feed, housing and so on, animal health had become one of the factors influencing farm productivity, and the identification of health risk factors was from then on meant to prevent animal diseases as much as to optimise farmers' yields.[14] Once again, it was by opening up to other forms of knowledge and practices that veterinary expertise sought to ensure its legitimacy in the (now heteronomised) jurisdiction of animal health.

Although this statistical technique directly renewed the knowledge and practices of veterinary medicine, it was only possible, as in Theix with the EEPCs and in Villeurbanne with the Ecopatho Transfert programmes, to do so by setting up a system organising the production of data, the supervision of farms and, *in fine*, the circulation of veterinary expertise. This system was gradually constructed with the establishment of the SPP's 'survey networks', which numbered more than 600 farms in the mid-1980s (compared with around thirty when the Station was set up ten years before) (Calavas, 1998).

All in all, although this statistical method is undeniably the trademark of ecopathology (it subsequently was used in all three centres), we can see the paradox it could represent with regard to the initial political project of an alternative form of veterinary medicine. It is true that ecopathology favoured the emergence of a preventive approach to animal diseases, revived the agricultural expertise long neglected by the veterinary profession, and ultimately empowered farmers to better

14 The annual program of the Journées de la Recherche Porcine (JRP) in the 1970s, 1980s and 1990s is a good indicator of the integration of health-related issues in the definition of ('good') animal husbandry practices (https://www.journees-recherche-porcine.com/). JRP are organised by the Pork Institute, created in 1961 and whose mission is to support operators in the pig sector (farms, cooperatives, slaughterhouses, etc.). In the early 1970s animal health issues were minor topics in JRP, but they developed considerably from the end of the decade (SPP members were almost systematically the authors of these communications and publications), and from 1985 onwards they became the second most numerous topic in JRPs (after animal feed but before genetics and housing). It should also be noted, as Jocelyne Porcher points out, that the work presented at JRPs is strictly aimed at increasing the productivity of pig farms and, with very few exceptions, does not include any critique of the industrial model: Porcher, 2010.

cope with the industrialization of agriculture. However, in a way, ecopathology accompanied rather than challenged the development of productivist agriculture. This was particularly visible at the SPP in Ploufragan, which was explicitly at the service of this conception of agriculture, but it also occurred in the other two centres through the reappropriation of ecopathological knowledge and techniques by agricultural actors. Thus, despite a fairly rapid rise in power, which can be explained in particular by the challenges posed to veterinary medicine by the recomposition of the agricultural world, ecopathology was quickly faced with its own contradictions, which curbed the development (and hopes) of an alternative form of veterinary medicine.

IV. The paradoxical disappearance of ecopathology

Beyond the specific situations of each centre, two trends explain the fall of ecopathology. On the one hand, it was challenged by the emergence and rapid growth of another veterinary discipline, one also based on a preventive approach to animal diseases but oriented towards public health expertise, namely veterinary epidemiology. On the other hand, a recomposition of the French agricultural landscape from the 1990s onwards promoted the decompartmentalization of animal health and prevented veterinary medicine from acquiring a perfectly legitimate (i.e., monopolistic and uncontested) agricultural expertise. The whole paradox is this: if and when ecopathology reaches its objectives (namely the appropriation of its unique herd health management knowledge and techniques by the farmers and agricultural organisations), then it loses its raison d'être.

IV.1. Ecopathology overtaken by epidemiology

The first cause of the fall of ecopathology was the parallel development of animal epidemiology, which began in the mid-1980s to operate in a competitive mode. For a decade or so, these two forms of veterinary expertise remained undifferentiated, in the sense that they were not distinguished by their representatives. Ecopathology and epidemiology thus were able to feed off each other (particularly in terms of statistical techniques – Bénet, 1982) and 'work' towards their institutionalisation in a joint manner. However, when public authorities began to consider the need to encourage a preventive approach to animal diseases, ecopathologists were unable to defend their conception of veterinary medicine, either in the eyes of INRA or the French Ministry of Agriculture. In addition, some Chairs of the Transmissible Diseases department at the Maisons-Alfort

Veterinary School succeeded in imposing Pasteurian epidemiology based on the 'contagion' model on a national scale.[15]

At the end of the 1970s, almost twenty years after the creation of the Veterinary Research Department at INRA, animal health was a quite important field within the Institute, even though veterinary scientists did not get the same academic prestige as the geneticists and microbiologists. Two research centres competed for the department's leadership: Theix and Nouzilly. Although some of their respective teams were working on the same topics (in particular, nutrition and reproduction), their directors did not share the same conception of animal pathology: Brochart, as we already have seen, opposed the dominant Pasteurian approach, which considered microbes to be the primary causes of diseases (Cranney, 1996); his opponent, Michel Plommet, a bacteriologist and well-known brucellosis specialist, was, on the contrary, a supporter.

This opposition within INRA over the conception of animal pathology, associated with the development of Pasteurian epidemiology at the Alfort School, contributed to the marginalisation of the Theix ecopathology laboratory. Indeed, Plommet was linked with Alforians whose epidemiological approaches constituted for him the continuity of the collective prophylaxis in which he was directly involved during the fight against brucellosis in the early 1970s. This marginalisation of the ecopathologists within animal health research was accentuated when, in the following decade, discussions began that would lead to the creation of epidemiological surveillance networks for animal diseases. Ecopathologists, who were few in number and even absent from the areas where exchanges between public authorities, agricultural organisations and veterinary research and teaching institutes were taking place (INRA and Ministry working groups, Veterinary schools, etc.), gradually lost the little support that they had had, first and foremost that of the FNGDSB, which ceased its participation in the Theix ecopathological surveys in 1986 (Plommet, 2004).

The creation of the French Food Safety Agency in 1988 then marked the beginning of a federation of the various networks that were established in the 1970s and 1980s. These included prophylactic systems co-managed by the GDS and veterinary services, various surveillance networks that were still in their infancy, and the networks set up by ecopathologists to study livestock diseases (Dufour, 1993). All of these systems have gradually been reoriented towards epidemiological surveillance. In other words, the focus of veterinary public health expertise shifted from directly improving farm technical-economic performance to ensuring that the national livestock was free of disease to allow the trade of animal products and to prevent disease transmission (including to humans). Faced with such a transformation of public policies in the field of animal health, which began at the end of the 1980s and took shape over the next two decades

15 This article is obviously not intended to trace the history of animal epidemiology. Nevertheless, a few elements are important for understanding the failure of sustainable institutionalisation of ecopathology.

with the reform of the Food Safety Agency and of the Ministry of Agriculture, veterinary expertise was strengthened in the field of public health (Alam, 2009), but it did not really manage to develop the agricultural skills that ecopathology was seeking to promote.

The fall of ecopathology was not, however, just about the reshaping of the field of animal health at an institutional level. The development of epidemiology also put the legitimacy of ecopathology under pressure in the academic field. Although factorial correspondence analysis proved its worth in terms of (zoo)technical advice for farmers, it began to be scientifically criticised starting in the late 1980s. This was related to the rise of epidemiology in veterinary research and teaching that promoted forms of statistical techniques and knowledge other than the ones derived from geometric data analysis used by ecopathologists, in particular logistic regression methods imported from Anglo-Saxon veterinary epidemiology.[16]

> They [the Alforians] used to tell us: "Listen, little ones, you'll have to stop this if you want to publish". And we said, "Who cares, it's a good method". And it's almost a conspiracy, well not a conspiracy but we said to ourselves: "they don't accept my publications because, because people are against the French school of data analysis". (…) They've put us under pressure from an academic point of view, that is to say that we didn't understand.
>
> *An ecopathologist from Villeurbanne*

Faced with this situation, ecopathology had little choice. It had to transform and adapt, even if this meant losing its original ambitions to develop a form of agricultural expertise within veterinary medicine. The three ecopathology centres then experienced very contrasting situations. Theix began to convert to epidemiology: the laboratory changed its name in 1998, and in the following decade contributed to the institutionalisation of the discipline within INRA alongside a Nantes research unit that had developed in the 1990s. Ploufragan, because of its function of technical support to the pig industry, has yet to totally abandon the ecopathological approach insofar as it had a certain effectiveness in the prevention of livestock diseases of multifactorial origins. However, as the SPP never was the bearer of the political project of the 'pioneers' Fontaine and Brochart, it did

16 This study only tackles the case of France, so it is not possible to say whether or not similar processes took place in other countries. However, it seems that ecopathology also developed in the former French colonial empire through veterinarians and agronomists who were using this approach to develop local animal farming in a context of industrialization of agriculture in Southern countries. Experiences of colonial veterinary medicine were an important step in the trajectory of many of the pioneers of ecopathology (Tillon, Barnouin) and their tools and theories have been applied in other contexts by some of their students and colleagues who spent their careers in Southern countries (Faye et al., 1994; Faye, Lancelot, 2006), in particular through the French Research Institute for Tropical Agriculture (CIRAD). Ecopathology seems to have lasted longer in such contexts, and modern forms of tropical veterinary medicine based on participative approaches seem to have retained strong links with the initial project of ecopathology, but, again, it is not in the scope of this chapter to study this part of the story.

not allow ecopathology to survive as an alternative veterinary medicine (to the productivist model). On the contrary, it did contribute to taking ecopathological expertise out of the hands of veterinary medicine to the benefit of the technical and economic organisations of the pig industry. Lastly, Villeurbanne did not succeed in carrying out any of these forms of reconversion and disappeared completely.

Nevertheless, the decline of ecopathology did not correspond to a pure and simple disappearance, but more specifically referred to the end of a form of expertise that was *strictly* veterinary. In this sense, the aborted history of ecopathology is perfectly in line with the processes initiated since the end of the 19th century (development of public health expertise rather than agricultural expertise), which the heralds of this alternative conception of veterinary medicine were finally unable or unwilling to stop.

IV.2. Domestication of ecopathology

"We sawed off the branch we were sitting on". These words came up several times in my interviews with ecopathologists. They evoke the appropriation of ecopathological knowledge and practices by agricultural actors, in particular the statistical techniques of geometric data analysis that enabled them to integrate health issues into the technical and economic management of the farm, without, of course, being trained in veterinary medicine. It could certainly be argued that this was Fontaine's political project, namely to open up veterinary expertise to livestock farmers and to break away from the clinical and independent veterinary system. However, it must be said that this decompartmentalization of animal health on the one hand ultimately supported the industrialization process of agriculture (contrariwise to the pioneers of ecopathology, who were fighting it) and, on the other hand, failed to re-establish the legitimacy of an 'agricultural' veterinary expertise. Ecopathology certainly succeeded in renewing knowledge and practices on animal diseases, but it lost the ambitions of its pioneers in the 1990s. In a way, ecopathology was 'domesticated' by agricultural actors and ceased to be an alternative form of expertise. By opening up too much, ecopathologists lost their ability to distinguish veterinary medicine from other types of animal health approaches and, therefore, to remain the legitimate holders of this jurisdiction. The disappearance of the centre of Villeurbanne was quite emblematic of this trend.

Created in the mid-1980s to encourage the development of agricultural expertise within veterinary medicine and to share knowledge and techniques with all those involved in animal health, the Villeurbanne centre benefited from significant local political support. However, changes in the trade union landscape and agricultural development policies that occurred ten years later, in particular the transformation of technical advisory systems (Labarthe & Laurent, 2013), quickly made its presence unnecessary. From the moment that cooperatives, technical institutes and even Chambers of Agriculture were able to conduct their own statistical surveys and livestock management practices were standardised (by eliminating the least productive farms), animal health issues became less salient and the ecopathology centre lost its main *raison d'être*. As it was no longer able to provide a specific expertise, funding quickly dried up and the centre was forced to close its doors.

> When funding began to dwindle in 1990-95, we looked for external contracts and we carried out investigations for others. We may have continued, by reducing, improving and selling everything. (...) But some of the services were already provided by economic organizations such as cooperatives, etc., which either charged the farmers directly or integrated the price of these services into the feed, etc.
>
> *An ecopathologist from Villeurbanne*

At the end of this (brief) adventure, the individual trajectories of the ecopathologists from Villeurbanne perfectly illustrate the double vice in which ecopathology was clamped. While some converted to epidemiology, either at INRA or in national veterinary laboratories, others joined different agricultural organisations and started a career as technical advisors. The domestication of ecopathology thus marked the appropriation of preventive veterinary medicine by agricultural actors, but the inevitable corollary was not only the abandonment of the political criticism of productivist agriculture which was at the core of the ecopathological project, but also a failure to fully legitimise agricultural expertise within veterinary medicine. Thus, although the decompartmentalization of animal health was thought to be a way of curbing the social, economic and technical domination of livestock farmers by the agricultural elite, these trends were ultimately unable to resolve their contradictions and to promote the sustainable development of ecopathology. It is hard not to think of the words of Michel Fontaine who, within the framework of Cefale in 1974, even before ecopathology as such was formed, already saw this domestication possibly coming:

> The desire for a path towards the integration of technical and scientific activities into the productive environment, for a modification of current social relations and for an attempt to bring them closer to the conflicts over production processes, is poorly expressed. Moreover, this desire is not perceived by the vast majority of the farmers interviewed and is countered by the very functioning of the organisations that the farmers have given themselves and accepted. (...) From then on, a question arises for those seeking dialogue with the rural world: at what point and according to what type of study does the work carried out become an integrated service within the set of instituted services? In other words, who is going to take ownership of this work if it is done seriously? The independent professions? The organisations providing technical advice? The firms looking for advertising arguments? The administrations acting for a 'better agriculture' etc.?
>
> *Fontaine, 1974,* op.cit., *p. 447-48*

V. Conclusion

In sum, ecopathology was born between the mid-1970s (Theix and Ploufragan) and the mid-1980s (Villeurbanne), with each of these three sites having some specific features in terms of research, technical support and local agricultural development. These three experiences have made it possible to characterise what ecopathology has been. In terms of knowledge, it consisted of a population-based, multifactorial and interdisciplinary approach to animal pathology. In terms of practices, it referred to the technique of longitudinal monitoring and geometric data analysis (identification of risk factors), allowing a link between health and (zoo)technical aspects and aiming at the prevention (rather than the cure) of animal diseases. In terms of sociotechnical networks, it presupposed a change in the relationships between veterinarians and those involved in animal farming (farmers, agricultural technicians, advisors). Ecopathology thus constituted an alternative form of veterinary medicine, tending towards a decompartmentalization of the jurisdiction of animal health and backed by a political critique of intensive livestock farming.

However, ecopathology failed with regard to its initial ambitions. On the one hand, while it certainly managed to provide farmers with the means to resist the forms of social, economic and technical domination that the productivist model engendered, this ultimately was achieved by supporting the conditions of industrial livestock farming (and adapting animal health management to them) rather than by challenging them. On the other hand, the appropriation (and domestication) of ecopathological knowledge and techniques by agricultural actors has prevented veterinary medicine from breaking away from its clinical and infectious straitjacket and from developing an agricultural form of expertise. The recurrent debates on the need to enhance the value of (zoo)technical advice and preventive approaches in veterinary activity (for example in the case of antimicrobial resistance controversies – Fortané, 2021) can thus be seen as a contemporary manifestation of this thwarted development of ecopathology.

All in all, the history of ecopathology could remind us that the construction and legitimation of a jurisdiction (the one of veterinary medicine, with its knowledge, tools and practices) is a continuous and ecological process. Contrariwise to a teleological conception of professionalization considering that final achievement would be obtained through market closure and monopoly, veterinary medicine is always reshaping its borders and foundations (Bonnaud and Fortané, 2018). Rather than considering recurrent challenges facing a profession as crises, they should be perceived as regular manifestations of tensions that define the nature of a given activity (Woods, 2011). Animal health is indeed a domain where various actors and approaches compete and co-exist, in a way that is also framed by an evolutive environment (agricultural policies, industrialization and corporatization of livestock farming, etc.). Veterinary medicine has certainly established a solid legitimacy over the last two centuries on the management of animal health, but it must continue to re-establish it on renewed bases. Ecopathology was a moment where these tensions were expressed in a certain way, but it certainly does not mark the resolution of these tensions, which keep manifesting themselves through current issues and controversies. It is certainly the role of historians, sociologists and anthropologists of veterinary medicine to keep track of the ongoing and everlasting tensions that shape animal health.

List of acronyms and translation

Anses	Agence Nationale de Sécurité de l'Alimentation (French Food Safety Agency)
AVEM	Association Vétérinaires-Eleveurs du Millavois (Farmers-Vets association from the Millavois)
Cefale	Centre d'études et de formation appliquées à l'élevage (Centre for Studies and Training Applied to Animal Husbandry)
CIRAD	Centre de coopération internationale en recherche agronomique pour le développement (French research institute for agriculture and developement)
EEPC	Enquête EcoPathologique continue (Continuous ecopathological survey)
FNGDSB	Fédération Nationale des Groupement de Défense Sanitaire Bovins (National federation of bovine health groups)
GDS	Groupement de Défense Sanitaire (Animal health group)
GVC	Groupement Vétérinaire Conventionné (Contractual Veterinary Group)
INRA	Institut National de Recherche Agronomique (French institute for agricultural research)
JRP	Journée de la Recherche Porcine (Conference on Porcine Research)
PT	Paysans Travailleurs (Farmer union)
SASED	Syndicat d'action sanitaire des éleveurs du Don (Farmer union for health programs)
SPP	Station de Pathologie Porcine (Center for Porcine Pathology research)

Bibliography

Alam Thomas, « La vache folle et les vétérinaires. Récit d'une victoire inattendue et paradoxale sur le terrain de la sécurité sanitaire des aliments », *Revue d'études en agriculture et environnement*, 90 (4), 2009, p. 373-98

Bénet Jean-Jacques, « Compte-rendu du stage à la Station de Pathologie Porcine de Ploufragan (20-24 septembre 1982) », *Bulletin de l'AEEMA*, 1982, p. 82-92

Berdah Delphine, « Suivre la norme sanitaire ou 'périr' : la loi de 1954 sur la prophylaxie collective de la tuberculose bovine », *in* C. Bonneuil et al. (dir.), *Science, chercheurs et agriculture. Pour une histoire de la recherche agronomique*, Paris, Quae/L'Harmattan, 2008, p. 203-22

———, « Entre scientifisation et travail de frontières : les transformations des savoirs vétérinaires en France, XVIIIe-XIXe siècles », *Revue d'Histoire Moderne et Contemporaine*, 59 (4), 2012, p. 51-96

Bonnaud Laure, Fortané Nicolas, « L'Etat sanitaire de la profession vétérinaire. Action publique et régulation de l'activité professionnelle » *Sociologie*, vol 9 (3), 2018, p. 253-58

Brown Karen, Gilfoyle Daniel (eds.), *Healing the herds: disease, livestock economy and the globalization of veterinary medicine*, Athens, Ohio University Press, 2010

Burgat Florence, Dantzer Robert (dir.), *Les animaux d'élevage ont-ils droit au bien-être ?*, Paris, INRA éditions, 2001

Calavas Didier, *Ecopathologie des animaux d'élevage. Principes, démarches, implications*. Thèse d'université, Université Claude Bernard, Lyon 1, 1998

———, Rosner Guy, « Institutionnalisation d'une recherche-action en santé animale : l'expérience du Centre d'Ecopathologie Animale », *in* C. Albaladejo et F. Casabianca (dir.), *La recherche-action. Ambitions, pratiques, débats*, Versailles, INRA, 1997, p. 91-114

Combettes Patrice et al., « La santé animale sociale et solidaire », *Le sociographe*, 5, 2012, p. 261-84

Compagnone Claude (dir.), *Conseil et développement en agriculture. Quelles nouvelles pratiques ?* Versailles, Quae, 2009

Cranney Jean, *INRA, 50 ans d'un organisme de recherche*. Versailles, INRA, 1996

Dufour Barbara, « Naissance et développement de l'épidémiosurveillance animale en France », *Epidémiologie et santé animale*, 23, 1993, p. 83-100

Fontaine Michel, « Des étudiants en quête de paysans », *Les Temps Modernes*, 340, 1974, p. 440-49

Faye Bernard et al., *Ecopathologie animale : méthode et application en milieu tropical*, Montpellier, CIRAD, 1994

———, Lancelot Renaud, 'Ecopathological Approach in Tropical Countries. A Challenge in Intensified Production Systems', *Annals of the New York Academy of Science*, 1081 (1), 2006, p. 137-46

Fontaine Michel, « Des étudiants en quête de paysans », *Les Temps Modernes*, 340, 1974, p. 440-49

Fortané Nicolas, 'Antimicrobial resistance: preventive approaches to the rescue? Professional expertise and business model of French "industrial" veterinarians', *Review of Agricultural, Food, and Environmental Studies*, 102, 2021, p. 213-38

Hubscher Ronald, *Les maîtres des bêtes. Les vétérinaires dans la société française (XVIIIe-XXe siècle)*, Paris, Odile Jacob, 1999

Jussiau Roland et al., *L'élevage en France. 10 000 ans d'histoire*, Paris, Educagri, 1999

Labarthe Pierre, Laurent Catherine, 'Privatization of agricultural extension services in the EU: Towards a lack of adequate knowledge for small-scale farms?', *Food Policy*, 38, 2013, p. 240-52

Madec François, Josse Joël, *Application d'une méthode d'étude sanitaire globale à la prévention des troubles digestifs du sevrage du porcelet*, Rapport, SPP de Ploufragan, 1980, 72 p

Muller Pierre, *Le technocrate et le paysan. Essai sur la politique française de modernisation de l'agriculture de 1945 à nos jours*, Paris, Editions Ouvrière, 1984

Nicourt Christian, *Être agriculteur aujourd'hui. L'individualisation du travail des agriculteurs*, Versailles, Quae, 2013

Plommet Michel , « Histoire de la commission scientifique de la FNGDSB », *Epidémiologie et santé animale*, 46, 2004, p. 137-48

Porcher Jocelyne, *Cochons d'or. L'industrie porcine en question*, Versailles, Quae, 2010

Rémy Jacques (dir.), *Conseiller en agriculture*, Paris, INRA, 2006

Robinet André, *Larzac, Millau, Grands Causses. Elevage et partage des savoirs*, Paris, L'Harmattan, 2011

Salmona Michèle, *Souffrances et résistances des paysans français*, Paris, L'Harmattan, 1994

Tillon Jean-Pierre, *Étude sanitaire globale des élevages industriels de porc*, Rapport, SPP de Ploufragan, 1981

Vissac Bertrand, *Les vaches de la République. Saisons et raisons d'un chercheur citoyen*, Paris, INRA, 2002

Woods Abigail, 'The Lowe report and its echoes from history', *Veterinary Record*, 169 (17), 2011, p. 434-36

———, 'Is prevention better than cure? The rise and fall of preventive veterinary medicine, 1950-1980', *Social History of Medicine*, 26 (1), 2013, p. 113-31